ALPHA OLEFINS APPLICATIONS HANDBOOK

CHEMICAL INDUSTRIES

A Series of Reference Books and Textbooks

Consulting Editor
HEINZ HEINEMANN
Heinz Heinemann, Inc.,
Berkeley, California

Volume 1: Fluid Catalytic Cracking with Zeolite Catalysts, *Paul B. Venuto and E. Thomas Habib, Jr.*

Volume 2: Ethylene: Keystone to the Petrochemical Industry, *Ludwig Kniel, Olaf Winter, and Karl Stork*

Volume 3: The Chemistry and Technology of Petroleum, *James G. Speight*

Volume 4: The Desulfurization of Heavy Oils and Residua, *James G. Speight*

Volume 5: Catalysis of Organic Reactions, *edited by William R. Moser*

Volume 6: Acetylene-Based Chemicals from Coal and Other Natural Resources, *Robert J. Tedeschi*

Volume 7: Chemically Resistant Masonry, *Walter Lee Sheppard, Jr.*

Volume 8: Compressors and Expanders: Selection and Application for the Process Industry, *Heinz P. Bloch, Joseph A. Cameron, Frank M. Danowski, Jr., Ralph James, Jr., Judson S. Swearingen, and Marilyn E. Weightman*

Volume 9: Metering Pumps: Selection and Application, *James P. Poynton*

Volume 10: Hydrocarbons from Methanol, *Clarence D. Chang*

Volume 11: Foam Flotation: Theory and Applications, *Ann N. Clarke and David J. Wilson*

Volume 12: The Chemistry and Technology of Coal, *James G. Speight*

Volume 13: Pneumatic and Hydraulic Conveying of Solids, *O. A. Williams*

Volume 14: Catalyst Manufacture: Laboratory and Commercial Preparations, *Alvin B. Stiles*

Volume 15: Characterization of Heterogeneous Catalysts, *edited by Francis Delannay*

Volume 16: BASIC Programs for Chemical Engineering Design, *James H. Weber*

Volume 17: Catalyst Poisoning, *L. Louis Hegedus and Robert W. McCabe*

Volume 18: Catalysis of Organic Reactions, *edited by John R. Kosak*

Volume 19: Adsorption Technology: A Step-by-Step Approach to Process Evaluation and Application, *edited by Frank L. Slejko*

Volume 20: Deactivation and Poisoning of Catalysts, *edited by Jacques Oudar and Henry Wise*

Volume 21: Catalysis and Surface Science: Developments in Chemicals from Methanol, Hydrotreating of Hydrocarbons, Catalyst Preparation, Monomers and Polymers, Photocatalysis and Photovoltaics, *edited by Heinz Heinemann and Gabor A. Somorjai*

Volume 22: Catalysis of Organic Reactions, *edited by Robert L. Augustine*

Volume 23: Modern Control Techniques for the Processing Industries, *T. H. Tsai, J. W. Lane, and C. S. Lin*

Volume 24: Temperature-Programmed Reduction for Solid Materials Characterization, *Alan Jones and Brian McNicol*

Volume 25: Catalytic Cracking: Catalysts, Chemistry, and Kinetics, *Bohdan W. Wojciechowski and Avelino Corma*

Volume 26: Chemical Reaction and Reactor Engineering, *edited by J. J. Carberry and A. Varma*

Volume 27: Filtration: Principles and Practices, second edition, *edited by Michael J. Matteson and Clyde Orr*

Volume 28: Corrosion Mechanisms, *edited by Florian Mansfeld*

Volume 29: Catalysis and Surface Properties of Liquid Metals and Alloys, *Yoshisada Ogino*

Volume 30: Catalyst Deactivation, *edited by Eugene E. Petersen and Alexis T. Bell*

Volume 31: Hydrogen Effects in Catalysis: Fundamentals and Practical Applications, *edited by Zoltán Paál and P. G. Menon*

Volume 32: Flow Management for Engineers and Scientists, *Nicholas P. Cheremisinoff and Paul N. Cheremisinoff*

Volume 33: Catalysis of Organic Reactions, *edited by Paul N. Rylander, Harold Greenfield, and Robert L. Augustine*

Volume 34: Powder and Bulk Solids Handling Processes: Instrumentation and Control, *Koichi Iinoya, Hiroaki Masuda, and Kinnosuke Watanabe*

Volume 35: Reverse Osmosis Technology: Applications for High-Purity-Water Production, *edited by Bipin S. Parekh*

Volume 36: Shape Selective Catalysis in Industrial Applications, *N. Y. William E. Garwood, and Frank G. Dwyer*

Volume 37: Alpha Olefins Applications Handbook, *edited by George R. Lappin and Joe D. Sauer*

Additional Volumes in Preparation

ALPHA OLEFINS APPLICATIONS HANDBOOK

edited by

George R. Lappin

Joe D. Sauer

Ethyl Corporation
Baton Rouge, Louisiana

CRC Press is an imprint of the
Taylor & Francis Group, an **informa** business

First published 1989 by Marcel Dekker, Inc.

Published 2019 by CRC Press
Taylor & Francis Group
6000 Broken Sound Parkway NW, Suite 300
Boca Raton, FL 33487-2742

First issued in paperback 2019

ISBN 13: 978-0-367-45111-0 (pbk)
ISBN 13: 978-0-8247-7895-8 (hbk)

Visit the Taylor & Francis Web site at
http://www.taylorandfrancis.com

and the CRC Press Web site at
http://www.crcpress.com

Library of Congress Cataloging-in-Publication Data

Alpha olefins applications handbook.

(Chemical industries ; v. 37)
Includes index.
1. Olefins. I. Lappin, George R.,
II. Sauer, Joe D. III. Series.
TP248.89A52 1989 661'.814 89-1502
ISBN 0-8247-7895-2 (alk. paper)

Preface

Alpha olefins are extremely versatile chemical intermediates to a wide variety of industrial and consumer products. Through the practice of chemistry and chemical engineering, derivatives of alpha olefins have been developed that touch our lives in many ways. In some applications, only a low concentration of one of the higher olefins means greatly improved properties. In other applications, the alpha olefins represent most of the product. Whether present in small concentrations or large, they are used because they are most effective in a particular application. Sometimes products based on alpha olefins are the only choice, but more often the derivatives are chosen for their particular cost effectiveness as compared to other products.

Over the 25 years that alpha olefins have been commercially available, one new use after another has been developed. For example, the development of biodegradable surfactants for laundry detergents has been facilitated by the availability of alpha olefins. These cost-efficient derivatives mean effective cleaning in their end-use applications with minimal environmental impact on waterways in highly populated areas. Tougher linear low-density polyethylene plastic films containing alpha olefin comonomers mean less damage to goods in shipment and efficient disposal of refuse. Plasticizer esters with improved properties mean polyvinyl chloride plastics that are usable over a wider temperature range in automobiles, so they are less likely to crack in winter than were the older plastics. Improved crack resistance in high-density polyethylene bleach bottles means less breakage, with the accompanying losses and messes for the grocer and the consumer. Synthetic lubricants made from alpha olefins mean better automobile fuel economy and longer times between oil changes. Alpha olefin derivatives are used to produce longer-lasting paper and rubber seals in the space shuttles; they are also used to help maintain sanitary conditions in hospitals and to assist in extracting more crude oil from the ground.

We forecast that many new uses will be developed because of expansion of the literature and because of industrial and academic scientists' work with samples in all parts of the world. We hope that the reader will be challenged by the successes and ideas presented here to discover and develop still more uses.

The purposes of this book are to discuss the applications of higher linear alpha olefins containing 4 to 30 carbon atoms, to describe current commercial uses of alpha olefins, to indicate potential new uses, to document methods of production, and to provide physical property and general property data on the olefins, but not necessarily on the derivatives. Within this book, we have concentrated on generally recognized commercial applications for alpha olefins. We exclude the lighter alpha olefins ethylene and propylene. Butene is discussed since it is a key polyethylene comonomer and is produced in traditional alpha olefin plants. Some of the applications discussed can use other olefins such as the nonlinear olefins or the non-alpha olefins, and the properties of derivatives of these non-alpha olefins are compared to the properties of the derivatives of alpha olefins where appropriate. These properties are included for the benefit of those who work with and understand the physical and toxicological properties of olefins.

Chemistry is more interesting when related to applications, so we have endeavored to illustrate the relative value of the different applications by presenting some market information. This market information sets the stage for the applications, with emphasis placed on those applications that have the most commercial value and on those that are growing rapidly. Because the book is organized around applications, one reaction may be discussed in more than one chapter. For example, the oxo reaction will be discussed in the chapters on basic chemistry, plasticizers, detergents, and synthetic acids.

By 1930, many alpha olefin reactions had been discovered. These reactions could not be utilized fully, however, until the alpha olefins were readily available at economic prices. The higher linear alpha olefins first became available in commercial quantities in 1962 when Chevron began production based on cracking C20 and higher waxes. Before 1962, they were produced by dehydration of primary alcohols. Production from alcohols ceased in about 1966. Chevron had expanded their capacity in 1964. Gulf was the first to produce alpha olefins based on ethylene when they started operation in 1966. Widespread availability at a reasonable price resulted in many large volume uses for alpha olefins. Since 1965, alpha olefins has grown from a small, expensive specialty with under 10 million pounds utilized annually to a 2 billion plus pound per year basic petrochemical building block.

Alpha olefin production is possible by many routes. Oligomerization or chain growth of ethylene is the primary source of linear alpha olefins today and is likely to be the basis for such production for many decades. The trend away from production from waxes or paraffins has occurred for reasons of economics and quality. The entry of each additional supplier of linear alpha olefins has been positive for the further growth of alpha olefin markets. Both Ethyl in the United States and Mitsubishi in Japan entered the market in 1970; Shell U.S., in 1978; and Shell U.K., in 1983. Each entry added to the development of new markets: Chevron investigated many applications, Gulf initially supplied olefins for synthesis of tertiary amines, and Ethyl became a large supplier to plasticizer and detergent producers. Both Ethyl and Shell helped alpha olefin sulfonates grow, while Shell and Chevron pioneered efforts in the use of surfactants for enhanced oil recovery.

This book represents the efforts of some of the most experienced people in their industries. Each contributor has been selected either by the editors or by others knowledgeable about their particular segment of the chemical world. Whereas all the contributors are experts in their fields, each in a way represents the efforts of many other workers. The editors can count at least 200 people in Ethyl who have contributed in production or applications development over the last 25 years. Much of the technology goes back to Professor Karl Ziegler's group at the Max Planck Institute for Coal Research, Mulheim, Germany. Some of the key people at Ethyl include M. F. Gautreaux, A. O. Wikman, R. A. Moser, A. E. Harkins, C. W. Lanier, J. R. Zietz, M. E. Tuvell, W. T. Davis, W. D. Taylor, and T. H. Bramfitt.

We hope this book will have widespread appeal for all in the chemical industry, including people in education, research, design, purchasing, production, sales, marketing, and management. Finally, we acknowledge the assistance of family members, friends, and co-workers for their patience with us in this endeavor.

George R. Lappin
Joe D. Sauer

Contents

Contributors

James E. Borland	Ethyl Corporation, Baton Rouge, Louisiana
Glenn T. Carroll	Pennwalt Corporation, King of Prussia, Pennsylvania
Gaylon L. Dighton	Dow Chemical U.S.A., Plaquemine, Louisiana
Patrick C. Hu	Ethyl Corporation, Baton Rouge, Louisiana
Christine L. Koski	Celanese Chemical Company, Inc., Dallas, Texas
William Y. Lam	Petroleum Additives, Ethyl Corporation, St. Louis, Missouri
George R. Lappin	Ethyl Corporation, Baton Rouge, Louisiana
Osamu Okumura	Lion Corporation, Tokyo, Japan
David Paul	Monsanto Chemical Company, St. Louis, Missouri

Edmund F. Perozzi Petroleum Additives, Ethyl Corporation, St. Louis, Missouri

Ann M. Pettigrew Ethyl Corporation, Baton Rouge, Louisiana

Michael N. Pinkerton Pinkerton International, Inc., Baton Rouge, Louisiana

Joe D. Sauer Ethyl Corporation, Baton Rouge, Louisiana

Ronald L. Shubkin Ethyl Corporation, Baton Rouge, Louisiana

R. Vijay Srinivas Pennwalt Corporation, King of Prussia, Pennsylvania

James R. Strong Celanese Chemical Company, Inc., Corpus Christi, Texas

Jerry D. Unruh Celanese Chemical Company, Inc., Corpus Christi, Texas

John D. Wagner Ethyl Corporation, Baton Rouge, Louisiana

Clive Wilne Imperial Chemical Industries Plc, Middlebrough, Cleveland, England

Izumi Yamane Lion Corporation, Tokyo, Japan

ALPHA OLEFINS APPLICATIONS HANDBOOK

CHAPTER 1

Commercial Aspects

GEORGE R. LAPPIN Ethyl Corporation, Baton Rouge, Louisiana

1 INTRODUCTION

Linear even-carbon-number alpha olefins having four or more carbons are a basic petrochemical building block of over 2 billion pounds, including 790 million pounds per year of captive requirements. In addition, an estimated 400 million pounds per year of butene-1 are produced by non-higher alpha olefin processes. First produced in 1964 in significant commercial quantities at attractive prices, alpha olefins have grown rapidly and are expected to continue to grow as the literature on uses of alpha olefins continues to expand (see Figs. 1 to 3). The general markets are frequently categorized as (1) PVC plasticizers, (2) household detergents, (3) linear low- and high-density polyethylenes (as comonomers), (4) lubricants, and (5) other uses. Significant markets include detergents, sanitizers, plastics, lubricants, paper, metalworking, oil recovery, agriculture, and various specialty applications (see Fig. 4).

This chapter is an overview of the commercial aspects of alpha olefins, including routes for production and major markets. As these areas are explored in more detail in later chapters, the reference chapter is shown after each main heading in this chapter.

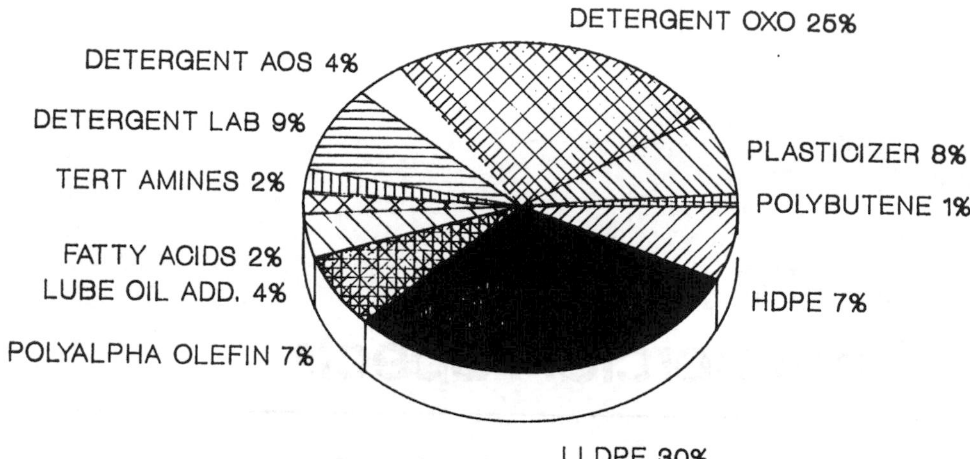

Figure 1 Alpha olefin markets, projected for 1990.

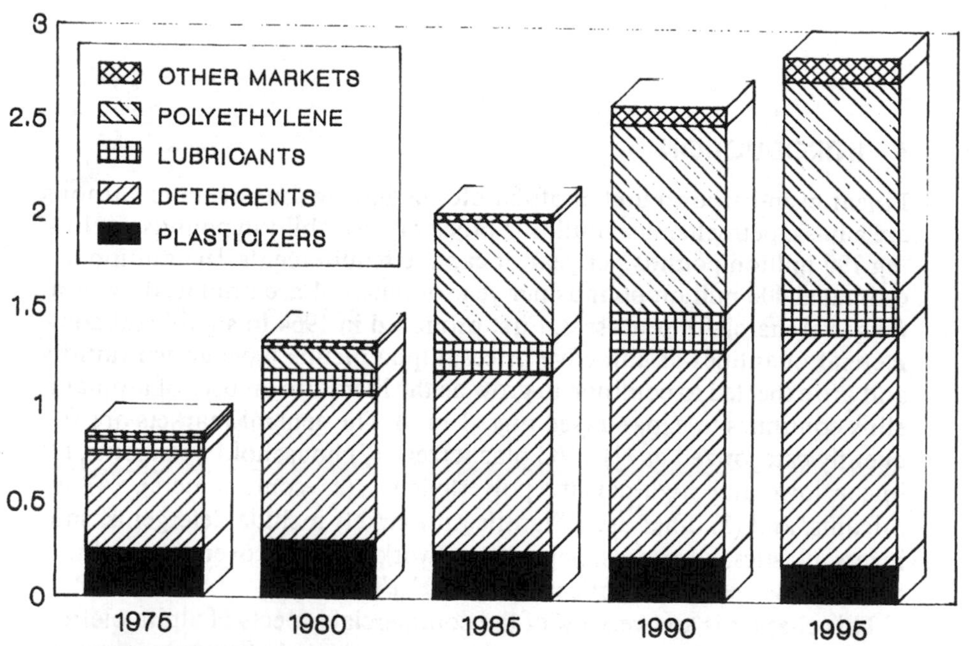

Figure 2 Alpha olefin market growth, excluding the USSR and its allied countries (billions of pounds per year).

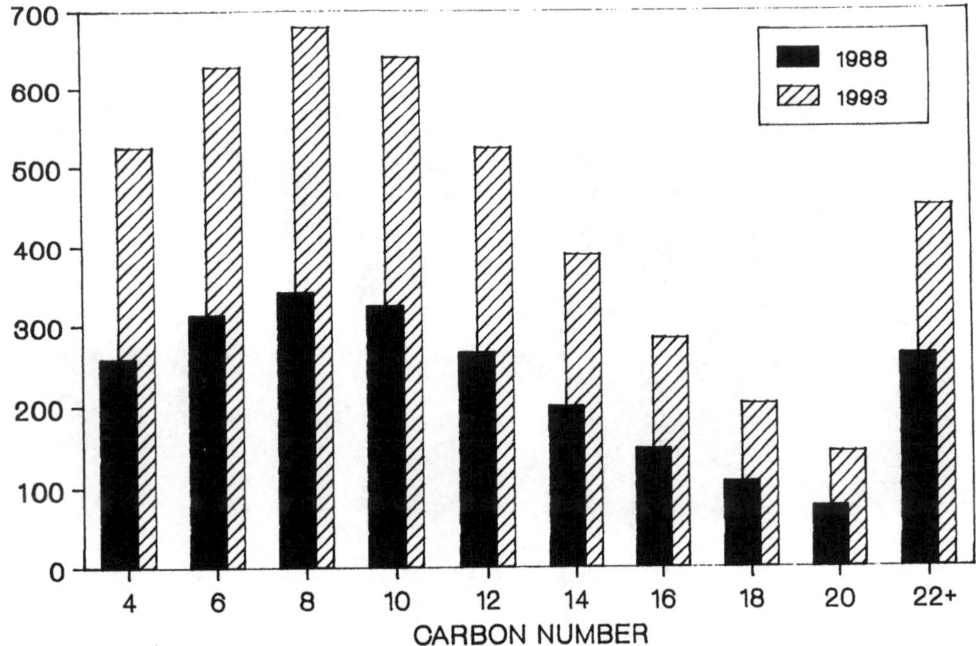

Figure 3 Alpha olefin capacities estimated for 1988 and 1993 (millions of pounds per year).

2 ROUTES TO LINEAR ALPHA OLEFINS (CHAPTER 3)

Alpha olefins are produced in ethylene-based plants operated by Ethyl, Shell, Chevron/Gulf, and Mitsubishi (see Table 1). Enichem Augusta (formerly Liquichimica) in Italy produces odd and even linear internal olefins by extraction from partially dehydrogenated kerosene fractions.

Petrochemical operations during the past 20 years show a trend away from wax- or paraffin-based plants and toward ethylene-based plants. For economic reasons, Chevron's wax-cracking plant in California was shut down in September 1984. For several reasons, including economic ones, Shell has curtailed its wax-cracking operation in Europe. Wax-cracking plants yield lower-quality olefins and some undesirable fractions. For example, it is estimated that Shell fed 3 pounds of wax and produced 2 pounds of olefins per pound they sold or used as a primary alpha olefin before they started up their more efficient Shell Higher Olefin Process (SHOP) plant based on ethylene in 1983. Shell restarted their paraffin chlorination-dehydrochlorination (CDC) plant in the United States in mid-1985, apparently because it was a low-cost investment and a readily

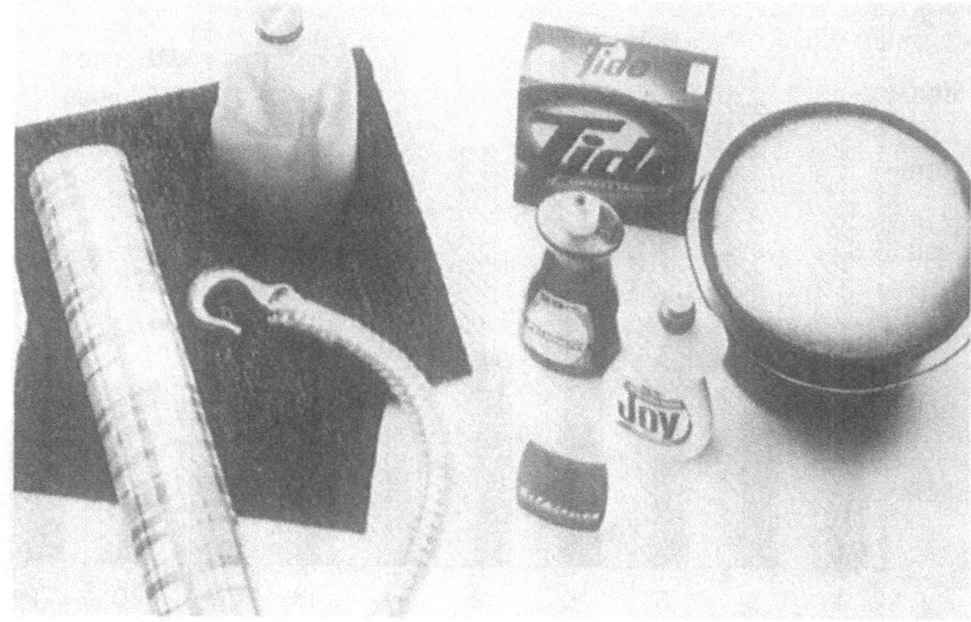

Figure 4 A few of the products that utilize alpha olefin derivatives.

available source of capacity. This CDC unit, which had been started up in 1967, was shut down in 1981 when Shell's U.S. SHOP unit was adequate for demands.

Chevron and Gulf pioneered large-volume commercial production of alpha olefins, with plants initially operating in 1962 and 1966, respectively. In a friendly merger aimed at defeating an unfriendly takeover bid by an investment group, Chevron bought Gulf Oil in 1984. This move resulted in merging the initial suppliers of alpha olefins.* Chevron's wax-cracked olefins plant at Richmond, California, had a capacity of about 90 million pounds per year and was operated from 1964 through September 1984. Chevron had a smaller unit in operation in 1962. Gulf's Cedar Bayou, Texas, plant began operation in 1966 with a capacity of 120 million pounds per year, expanding to 200 million pounds per year in 1981 and then to 250 million pounds per year in 1986. Chevron is building a new

* The original Gulf ethylene-based alpha olefin plant will be referred to as the Chevron plant or the Chevron/Gulf plant, but the Ziegler process first developed by Gulf will be called either the Gulf or Chevron/Gulf process.

Table 1 Estimated 1987 Alpha Olefin Capacity

Total capacity (million pounds per year)

Producer	Base capacity	Captive demand	Net merchant
Ethyl U.S.	950	100	850
Chevron/Gulf U.S.	250	30	220
Shell U.S.	650	350	300
Shell U.K.	450	300	150
Mitsubishi Japan	66	10	56
Total	2366	790	1576

Percent of capacity in C_6–C_{18} range

Producer	Base	Captive
Ethyl U.S.	84	93
Chevron/Gulf U.S.	71	90
Shell U.S.	65	65
Shell U.K.	65	65
Mitsubishi Japan	71	100

C_6–C_{18} capacity (million pounds per year)

Producer	Base capacity	Captive demand	Net merchant
Ethyl U.S.	798	93	705
Chevron/Gulf U.S.	178	27	151
Shell U.S.	423	228	195
Shell U.K.	293	195	98
Mitsubishi Japan	47	10	37
Total	1737	553	1185

plant at Cedar Bayou, Texas, at a cost of $87.5 million. It will raise production to 450 million pounds a year by early 1990 and is designed to be easily expandable to 750 million pounds a year, depending on market conditions. Gulf licensed the original technology from Ziegler in Germany. The process uses catalytic amounts of triethylaluminum to chain-grow ethylene and simultaneously displace the olefins in situ. The product contains 1.4% paraffins because the aluminum alkyl is hydrolyzed at the end of the reaction. The alpha olefins are of good quality, containing some vinylidene olefins and essentially no internal olefins. Gulf's process has limited flexibility with regard to carbon number distribution, and Gulf has used inventory control to handle excess olefins. Approximately 71% of the population is in the C_6–C_{18} main alpha olefin product range with about 15% each of C_4 and C_{20}–C_{40} constituting the remainder. Chevron/Gulf is strong in ethylene supply, which is a key to their alpha olefin position. Captive requirements include polyalpha olefins and comonomers for high-density polyethylene.

Mitsubishi is a Gulf licensee with a plant in Japan producing 66 million pounds per year and a market limited to Japan. Approximately half the olefins used in Japan were imported from the United States in the years 1978–1987. Idemitsu has announced plans to build an olefin plant in Japan with a capacity of 110 million pounds per year. In the last 10 years, ethylene has been more expensive in Japan than in the United States, which is one reason the United States has expanded alpha olefin production faster than Japan; other reasons are the faster growth in U.S. requirements and the fact that U.S. plants have been larger and therefore more economical.

Ethyl Corporation began production of alpha olefins in December 1970 by adding to its primary alcohols complex. Like Gulf, Ethyl licensed the original technology from Ziegler in Germany. As was necessary with the alcohol process technology, Ethyl extensively modified the Ziegler olefin technology. The Ethyl process is similar to that of Gulf, as both use triethylaluminum. However, the Ethyl process is reported to use separate growth and olefin elimination steps. Ethyl reports ability to peak carbon numbers of both the alcohols and the olefins, selectively creating the most desired products. Approximately 82% of the production is in the C_6–C_{18} main alpha olefin product area, with about 15% as C_4 and under 2% as C_{20}–C_{40}. The exact distribution varies almost daily as market demands for olefins and alcohols change. Ethyl's capacity in million pounds per year grew from 200 in 1971 to 400 in 1974, 450 in 1976, 500 in 1980, over 800 in 1981, and over 950 in 1987. Captive requirements include alkyldimethylamines, lubricants, aluminum alkyls, primary alcohols, and alkenylsuccinic anhydride.

Shell's captive demands for higher olefins include oxo alcohols in the

United States, United Kingdom, and France and linear alkylbenzene (LAB) in Europe, Australia, and South Africa. Feedstocks for these have included SHOP olefins, wax-cracked olefins, and paraffin dehydrochlorination olefins. The two SHOP plants produce over 1 billion pounds of alpha olefins annually, of which an estimated 650 million pounds per year are used internally, leaving an estimated 400 million or more pounds for sales to the alpha olefin merchant market. Shell has 160 million pounds per year of linear odd and even internal olefin capacity in the United States based on their CDC unit. Shell had perhaps 600 million pounds per year of wax-cracked olefin capacity in Europe, most or all of which, as noted earlier, is currently unused. Shell's position in ethylene oxide is a key to their strength in detergents in the United States. Shell is a large-volume producer of ethylene. Since 1980, various portions of Shell have announced plans to build SHOP plants in the United States, Canada, and other parts of the world. Shell is building a new SHOP plant at Geismar, Louisiana, which is scheduled to begin operation in 1989, with 535 million pounds per year of capacity.

Enichem Augusta (formerly Liquichimica) in Italy produces odd and even linear internal olefins by dehydrogenation of normal paraffins extracted from kerosene. These are satisfactory as feedstock to linear alkylbenzene plants and to oxo alcohol plants. They produce linear alkylbenzene and alcohols in their complex.

Butene-1 from alpha olefin production is sold or used by Gulf, Mitsubishi, and Ethyl. Shell can recycle their butene-1 to their disproportionation reactors and use it in polybutylene production. Exxon, Texas Petrochemicals, DuPont of Canada, Huels in Germany, Shell in Europe, and other companies produce butene-1 via separation from various butylene streams. A recently developed system involves feeding ethylene plant butane-butene (B-B) streams into methyl *t*-butyl ether (MTBE) plants, with the butene-1 being a coproduct of the unit. Adding butene-1 as a coproduct may double the cost of an MTBE plant. Economics appear to favor nonethylene-based routes, so alpha olefin producers sell the butene-1 at generally lower prices than other alpha olefins. Butene-1 must be produced along with all other alpha olefins, and it would appear that the alpha olefin producers will always price their butene-1 to sell in competition with butene-1 from refinery or ethylene plant streams.

Many indirect types of competition exist for the alpha olefin products, including coconuts, leather, glass, and paper. The oil from coconuts can be made into primary alcohols. Leather, glass, and paper are alternatives to plastics. Derivatives of alpha olefins have been replacing many of these in specific end-use areas.

3 POLYETHYLENE (CHAPTER 4)

Alpha olefins are used in both high-density and linear low-density polyethylene as comonomers to improve their properties. In high-density polyethylene (HDPE) the olefins are used at up to the 2% level to impart stress-crack resistance, particularly for household bleach bottles. There are many grades of polyethylene, and comonomers provide one of the key differences. Many HDPE grades use no comonomer, resulting in an average comonomer usage level under 1% of the total HDPE usage.

Linear low-density polyethylene (LLDPE) is the largest volume end use of C_6–C_{18} alpha olefins today. To a large degree, the near-term future of alpha olefins depends on the rate of LLDPE growth, the amount of comonomer used, and which of the three—butene, hexene, or octene—will be most useful. Various forecasters predict that LLDPE will penetrate to between 30 and 50% of the combined global market for LLDPE and the conventional high-pressure low-density-pressure polyethylene (LDPE). It has been estimated that LLDPE's share was about 18% in 1985. The analysis in Table 2 is based on LLDPE penetrating to 30% by 1990, which would require 890 million pounds of alpha olefins. If it reaches 50%, much new comonomer capacity will have to be added, as comonomer demand would reach 1.5 billion pounds.

Table 2 Global Demand for LLDPE, LDPE, and HDPE

	1985	1990
Penetration of LLDPE into LDPE (%)	18	30
Combined demand LDPE/LLDPE (billion pounds per year)	30	36
Estimated amount of LLDPE (billion pounds per year)	5.4	10.8
Total comonomer at 8% of LLDPE (million pounds per year)		
LLDPE butene-1	260	435
LLDPE hexene-1	65	260
LLDPE octene-1	100	155
LLDPE 4-methyl-1-pentene	5	40
Total	425	890
HDPE demand estimate (million pounds per year)		
HDPE butene-1	100	120
HDPE hexene-1	55	85
Total LLDPE/HDPE comonomer	580	1090

4 PLASTICIZERS (CHAPTER 5)

Plasticizers for poly(vinyl chloride) (PVC) may be produced starting with an oxo reaction of C_6, C_8, and C_{10} alpha olefins with carbon monoxide and hydrogen to produce a semilinear alcohol. The alcohol is then reacted with phthalic anhydride or other acid to produce a C_7-C_9-C_{11} phthalic ester or other ester. The ester is used as a plasticizing agent in flexible PVC, a market of several billion pounds per year for esters of this nature.

The C_6–C_{10} olefin requirements for plasticizers vary greatly with economic cycles and peaked in the United States at about 200 million pounds per year in 1979. U.S. demand correlates with housing starts and automobile sales although, on a direct basis, these represent only one-third of the market for the flexible PVC. The linear esters provide more permanence at high temperatures and make the PVC less likely to crack at low temperatures. These desirable properties may also be obtained by using more expensive, less available acids such as azaleic with less expensive 2-ethylhexanol. The semilinear ester values are set by the price of dioctyl phthalate (DOP). Rationalization of the ester supply has been occurring for 15 years and is likely to continue. Monsanto was the largest overall supplier until they sold their Texas City plant to the Sterling group in a leveraged buyout in mid-1986. BASF acquired the plasticizer business portion of the Texas City plant from Monsanto and has alpha olefins tolled by Sterling to plasticizer alcohols and esters. Monsanto remained in the butyl phthalate ester business. Other U.S. suppliers include Exxon, Tenneco, and USS Chemicals, which offer various linear, semilinear, and branched esters.

5 SURFACTANTS (CHAPTERS 6–9)

Detergents as a class are less subject to economic cycles than plastics are. But detergents have shown slower growth than plastics. Growth is occurring as alpha olefin derivatives replace other less biodegradable surfactants.

In the United States, Shell is the key producer of oxo alcohols for detergents, for which an estimated 350 million pounds per year of alpha olefins are consumed. Oxo alcohols in Europe represent a market of 400 million pounds per year for alpha olefins in the European detergent alcohol market. The resulting alcohols are sold as oxo alcohol ethoxylates. Shell is the largest producer, and this accounts for perhaps 100 million pounds per year of their internal demands for C_{10}–C_{14} olefins. ICI is the next largest producer in Europe. BASF and Exxon-CdF Chimie consume the remainder. The total market is growing at 2% per year, and ethoxylates are penetrating at another 1% per year, for a total growth rate of 3% per year.

Linear alkylbenzene (LAB) is increasing in importance as a market for alpha olefins as developing countries switch to biodegradable detergents. One large potential is in Latin and South America, where a switch from propylene tetramer to C_{10}–C_{14} alpha olefins is occurring for a portion of the requirements. For the longer term, companies in these regions may build paraffin-based plants, which tend to be more economical to operate but require significant investment to build. If all this region were to convert their entire alkylbenzene production to LAB, the total olefin demand might well be more than 200 million pounds per year.

Alpha olefin use in alpha olefin sulfonates (AOS) reached 74 million pounds per year in 1985 and is expected to grow to 105 million pounds per year in 1990. AOS provides superior performance in cold, hard water and works with alcohol ether sulfates in light-duty detergent applications. It is used in heavy-duty laundry powders in Japan. There are plans for using AOS in India, China, and other parts of the world. Concern about skin sensitivity has been resolved by the avoidance of bleaching of AOS with hypochlorite.

6 TERTIARY AMINES (CHAPTER 9)

The tertiary amine market represents a requirement for 50 million pounds per year of alpha olefins, with growth to 60 million pounds per year projected by 1990. Uses for the tertiary amines include light-duty liquid detergents, disinfectants, sanitizers, and oil recovery. These olefin derivatives, in the form of quaternary salts, play an important role in the battle against infectious diseases. They kill bacteria, fungi, and viruses involved in pneumonia, gonorrhea, diphtheria, influenza, polio, and many other contagious diseases.

7 ENHANCED OIL RECOVERY (CHAPTER 10)

Steam was used in California to recover heavy crude oil when crude oil was $3 per barrel. Thermal recovery of heavy oil continues during high and low crude oil prices. Enthusiasm for thermal recovery methods necessarily decreased as crude oil prices fell. C_{16}–C_{18} AOS, C_{16}–C_{18} dimer AOS, and C_{16}–C_{18} alkylaryl sulfonate have all been shown to increase oil recovery in individual tests of their use as foam-diverting agents by Shell, Texaco, Chevron, Stanford University, the University of Southern California, and the U.S. Department of Energy. The foams are used to divert the steam into oil-rich portions of the formation that the steam might otherwise bypass. Testing continues, aimed at identifying the overall effectiveness of the foam procedure and the best surfactant for this application. Although use of alpha olefins in foam-diverting agents is small today, the maximum potential for alpha olefins could be of the order of 100 million pounds per year. Development will probably be slow as significant investment in materials and personnel are required to implement the use of the foam. Heavy oils are also recovered by thermal methods in Canada, Venezuela, and Indonesia, and these could greatly increase the potential for alpha olefins.

Alpha olefin derivatives may eventually play a role in various other enhanced oil recovery (EOR) activities, including carbon dioxide and micellar-polymer. It is difficult to attempt to forecast demands for alpha olefins in these areas because these uses lack either or both the economic incentives and the technical indications of potential success.

8 FATTY ACIDS (CHAPTER 11)

Fatty acids (C_7 and C_9) were first produced commercially in the United States from alpha olefins (C_6 and C_8) in 1980 by Celanese (now Hoechst Celanese). Celanese's synthetic acid plant, the only one in the United States at present, produces 40 million pounds per year. These acids are used in the same areas as the other fatty acids in the C_7 and C_9 range, although the synthetics are preferred for some end uses because of their consistent quality and availability. They are used in plasticizers, synthetic lubricants, metal salts for grease thickeners and paint driers, surfactants and cosmetics, alkyd resins for coatings, high-water-based cutting fluids, and many other products. Total demand for the C_6 and C_8 alpha olefin raw material is estimated to be of the order of 20 million pounds per year, with growth of 3 to 4% expected in current areas and with additional growth likely but difficult to predict for developing areas.

9 LUBRICANTS (CHAPTER 12)

The current market for olefins in lube oil detergents is estimated to be 120 millions pounds per year. Shell, Ethyl, and Chevron have captive demands estimated to be 70% of the total. This market continues to grow faster than the total lube oil market as detergent concentrations increase and as "natural" supplies decrease. The natural lube oil sulfonates were coproducts from treatment of oils with sulfuric acid to remove unsaturates and produce white oils. For environmental reasons, hydrogenation is replacing sulfuric acid, and the result is fewer natural sulfonate supplies.

10 SYNTHETIC LUBRICANTS (CHAPTER 13)

Total demand for alpha olefins in synthetic lubricants grew from 65 million pounds per year in 1980 to 100 million pounds per year in 1985. The market is projected to grow at 10 to 15% per year to 200 million pounds per year in 1990. It could grow faster, based on the renewed emphasis in Detroit on 5W30 crankcase oils and their use in transmission fluids.

11 MERCAPTANS (CHAPTER 14)

Key mercaptans produced from the higher alpha olefins are based on butene, hexene, octene, decene, and dodecene. The butyl mercaptans are used to produce cotton defoliants, the octyl mercaptans are used to produce a water repellent, and the dodecyl mercaptans are used in rubber and detergents.

12 PAPER, FABRIC, AND LEATHER (CHAPTER 15)

Alkenylsuccinic anhydride (ASA) based on C_{16}–C_{20} alpha olefins is growing rapidly worldwide as fine-quality paper and gypsum board paper mills switch to alkaline sizing for economic reasons. The fine-quality area is divided into uncoated free sheet and coated free sheet. Growth occurred first in coated free sheet. Growth spread to the gypsum boards next, and now growth appears to be occurring in uncoated free sheet. Alpha olefin usage is forecast to increase from 12 million pounds per year in 1985 to 21 million pounds per year in 1990. The market potential for alpha olefins may be as large as 100 million pounds per year worldwide, but it is too early to predict the maximum. Small amounts of alpha olefins are used to produce other ASAs of various carbon numbers, which are used in petroleum additives, food additives, and other applications. Literature sources list over 20 uses for ASAs of various types.

Alpha olefins may be converted to leather tanning agents by reacting them with sulfonating agents. To be effective, only partial conversion on the order of 20% is achieved, as the unreacted portion acts as an oil and the converted portion acts as a detergent during the removal of the hair from the animal skin. U.S. demand for alpha olefins and olefin-paraffins is of the order of 3 million pounds per year. Many other materials have been used, such as methyl esters and fish and vegetable oils. Growth is unlikely in this area.

13 MISCELLANEOUS USES (CHAPTER 16)

Miscellaneous uses for higher alpha olefins include various small outlets, such as epoxides, chlorinated paraffins, waxes, fuel additives, and drag flow improvers. Total alpha olefin volume in this category is on the order of 30 million pounds per year, and this could grow significantly.

14 TOXICOLOGY (CHAPTER 17)

Alpha olefins are practically nontoxic by oral and dermal routes of exposure. They are expected to be only minimally irritating to the skin and eyes. The lower members, including ethylene, propylene, and butene, are characterized as simple asphyxiants or weak anesthetics.

15 STORAGE AND HANDLING (CHAPTER 18)

Proper storage and handling is essential to assure maintenance of the excellent quality of the olefins. Exposure to air will result in absorption of oxygen and water, with resulting problems when these interfere in various reactions. Peroxides will develop with time after exposure to air.

CHAPTER 2

Basic Chemistry

JOE D. SAUER Ethyl Corporation, Baton Rouge, Louisiana

1 INTRODUCTION

This chapter provides a general review of the basic chemistry of C_4–C_{30}+ linear alpha olefins. Synthesis of these olefin materials is covered in detail in subsequent chapters. This chapter focuses on reactions that have been demonstrated to provide access to various classes of chemical compounds from olefin feedstocks. Ethylene and propylene are generally excluded from the scope of this book, but they will be included where they help show a useful trend.

2 STRUCTURE OF OLEFINS

Bonding at the carbon–carbon double-bond system can be explained by using an sp^2 hybridized orbital model. Thus carbon atoms in the double bond have three equivalent bonding orbitals. Bond angles are therefore roughly 120° between any pair of sp^2 orbitals. Atoms are arranged about the double bond in a planar trigonal fashion. The fourth bonding electron of each carbon atom is in a p orbital, perpendicular to the plane of the three bonds. A simple representation of both bond geometry and orbital configuration is shown.

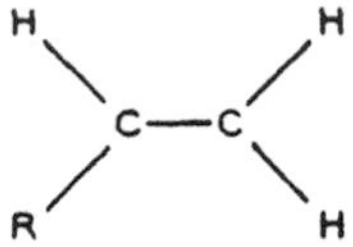

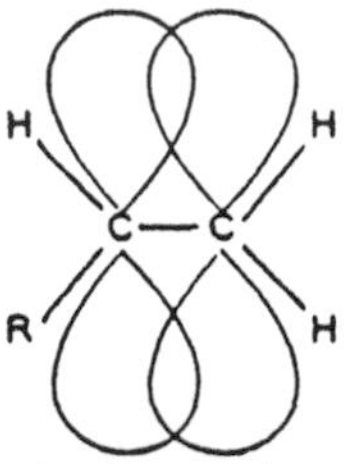

This bonding picture should also indicate that the sp^2-sp^2 sigma bond in the carbon–carbon double bond is slightly shorter and considerably stronger than a typical carbon–carbon single bond. In spite of this total strength, due to the overlapping π–π bond, the double bond behaves as a high-electron-density site in the molecule and can be extremely accessible to electron-seeking reagents.

In many olefinic systems, geometric isomerization at the double bond is possible; that is, one can have cis- and trans-isomer pairs (alternative nomenclature: Z- and E- isomer pairs). With alpha olefins, this is not possible, since the double bond is at the terminus of the molecule. It is possible, however, after double-bond isomerization or rearrangement of the molecular backbone, for products arising from alpha olefins to exhibit this property.

3 NOMENCLATURE

Common names are not usually used with olefins, except in the case of the simple compounds: ethylene, propylene, and butylene. Occasionally, one encounters olefins named as derivatives of ethylene: for example, monoalkylethylene (1-olefin) and 1,1-dialkylethylene. Typically, however, the rules of the IUPAC system are followed. The rules are available for reference purposes in most libraries. See Table 1 for examples of 1-olefin nomenclature.

4 PHYSICAL PROPERTIES

Specific physical properties of even-carbon-number linear 1-olefins are available in the appendix. Ethylene has a melting point of −169°C and a boiling point of −102°C. Boiling points increase as the carbon number increases in the homologous alkene series in a very regular fashion with approximately a 30°C increase per $-CH_2-$ group added. Melting-point variations are not as easily described, but an increase of 10° per $-CH_2-$ group can serve as a guideline for an approximate melting-point value for the higher, linear homologs. The liquid olefins typically exhibit densities of about 0.7 g/mL at room temperature. Dipole moments, although measurable, are typically small for the olefins. Obviously, addition of

Table 1 Olefin Nomenclature

Chemical formula	Carbon chain length	IUPAC name	Common usage name(s)[a]
$CH_2=CH_2$	2	Ethene	Ethylene
$CH_2=CH-CH_3$	3	1-Propene	Propylene; methylethylene
$CH_2=CH-(CH_2)-CH_3$	4	1-Butene	Butylene; ethylethylene
$CH_2=CH-(CH_2)_2-CH_3$	5	1-Pentene	Pentylene; α-*n*-amylene; propylethylene
$CH_2=CH-(CH_2)_3-CH_3$	6	1-Hexene	α-*n*-Hexylene
$CH_2=CH-(CH_2)_4-CH_3$	7	1-Heptene	α-*n*-Heptylene
$CH_2=CH-(CH_2)_5-CH_3$	8	1-Octene	α-*n*-Octylene; 1-Caprylene
$CH_2=CH-(CH_2)_6-CH_3$	9	1-Nonene	α-*n*-Nonylene
$CH_2=CH-(CH_2)_7-CH_3$	10	1-Decene	α-*n*-Decylene
$CH_2=CH-(CH_2)_8-CH_3$	11	1-Undecene	α-*n*-Undecylene
$CH_2=CH-(CH_2)_9-CH_3$	12	1-Dodecene	α-*n*-Dodecylene
$CH_2=CH-(CH_2)_{10}-CH_3$	13	1-Tridecene	α-*n*-Tridecylene
$CH_2=CH-(CH_2)_{11}-CH_3$	14	1-Tetradecene	α-*n*-Tetradecylene
$CH_2=CH-(CH_2)_{12}-CH_3$	15	1-Pentadecene	α-*n*-Pentadecene
$CH_2=CH-(CH_2)_{13}-CH_3$	16	1-Hexadecene	α-*n*-Hexadecylene; Cetene
$CH_2=CH-(CH_2)_{14}-CH_3$	17	1-Heptadecene	α-*n*-Heptadecene
$CH_2=CH-(CH_2)_{15}-CH_3$	18	1-Octadecene	α-*n*-Octadecene
$CH_2=CH-(CH_2)_{16}-CH_3$	19	1-Nonadecene	α-*n*-Nonadecene
$CH_2=CH-(CH_2)_{17}-CH_3$	20	1-Eicosene	
$CH_2=CH-(CH_2)_{19}-CH_3$	22	1-Docosene	
$CH_2=CH-(CH_2)_{21}-CH_3$	24	1-Tetracosene	
$CH_2=CH-(CH_2)_{23}-CH_3$	26	1-Hexacosene	
$CH_2=CH-(CH_2)_{25}-CH_3$	28	1-Octacosene	
$CH_2=CH-(CH_2)_{27}-CH_3$	30	1-Triacontene	

[a]Taken from *The Merck Index*, 9th ed., Merck & Co., Rahway, N.J., 1976); also, *The Condensed Chemical Dictionary*, 9th ed., Van Nostrand Reinhold, New York, 1977.

other groups around the double bond can drastically alter all of the properties mentioned.

5 CHEMICAL REACTION CLASSIFICATION

In general, alpha olefins will react in the expected manner for any substrate containing a carbon-carbon double bond. Furthermore, because the double bond in an alpha olefin is at the end of the molecular chain, it often behaves as though it is more accessible to the incoming reagents. In many systems, then, reactivity appears to be higher than typical alkenes, where the double bond is either at an internal position or is flanked by bulky substituents.

As an additional benefit to the industrial user, linear alpha olefins provide a molecular system with the reactive group on the end of a straight chain. Any functional group modifications, then, can result in end products that are also linear, straight-chain molecules. This situation can lead to certain benefits with respect to the general physical properties, especially to the overall biodegradability of long-chain functional products used as surfactants.

For the purposes of this chapter, reactions of a carbon–carbon double bond will be classified in the following major groups:

1. Addition reactions
   a. Electrophilic additions
   b. Free-radical additions
2. Substitution reactions

Within this framework, addition reactions are those which result in products no longer having a carbon–carbon double bond. The substitution reactions, however, afford products that have a double bond still intact, although rearrangements may have taken place.

6 MECHANISTIC CONSIDERATIONS

Keeping in mind the common types of reactions that olefins undergo, we find a brief review of general mechanisms in order. With addition reactions, the product obtained has two new single bonds (sigma) and no longer possesses the original carbon–carbon double bond. The two general processes most often encountered are electrophilic addition and free-radical addition. The type of product obtained in each case will be dictated by the stability of the intermediate species formed. Figures 1 to 3 illustrate simple pathways for each type of addition.

Over the span of many years, reactions such as these have been carried

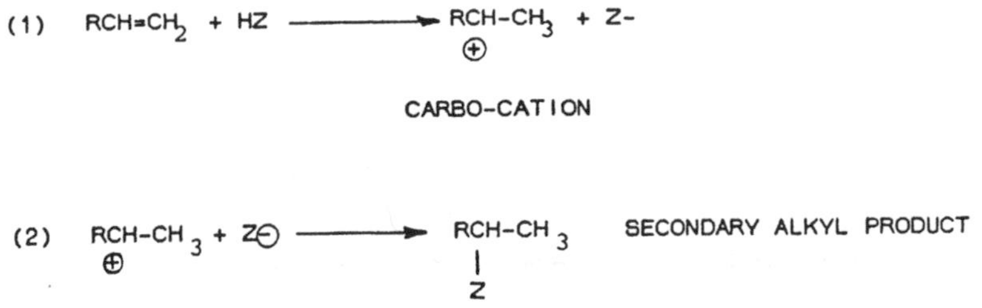

(1) $RCH{=}CH_2 + HZ \longrightarrow RCH^{\oplus}{-}CH_3 + Z^-$

CARBO-CATION

(2) $RCH^{\oplus}{-}CH_3 + Z^{\ominus} \longrightarrow RCH(Z){-}CH_3$ SECONDARY ALKYL PRODUCT

[STABILITY OF CARBO-CATIONS: TERTIARY > SECONDARY > PRIMARY > METHYL.]

Figure 1. Electrophilic addition. Example: Alpha olefin reacting with Lewis acid affords the most stable intermediate carbocation which, in turn, reacts with the conjugate base present to produce a secondary product (follows Markowni-kov's rule).

out with a wide variety of olefinic substrates. Some of the more common reactions, as indicated by larger numbers of literature citations, at least, yield final products that are important to many areas of modern chemical processes. A short list of common products arising from reactions of various olefins is shown below. Obviously, this list is not intended to be all-inclusive. It will be noted that many of these transformations have not become important commercially, for a variety of reasons, including cost of reagents, usefulness of final product, and physical and chemical properties.

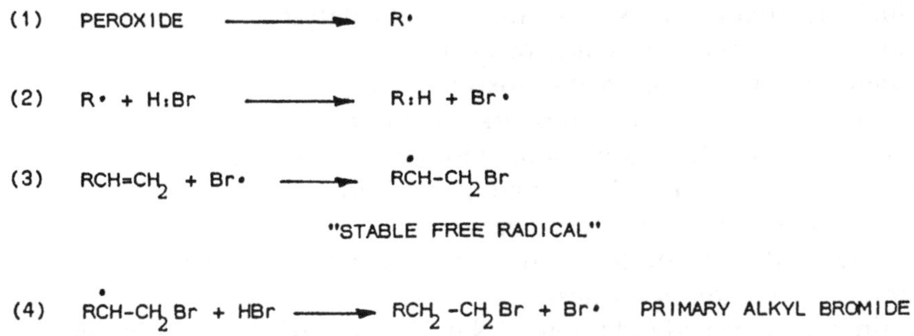

(1) PEROXIDE $\longrightarrow$ R•

(2) $R\bullet + H{:}Br \longrightarrow R{:}H + Br\bullet$

(3) $RCH{=}CH_2 + Br\bullet \longrightarrow R\dot{C}H{-}CH_2Br$

"STABLE FREE RADICAL"

(4) $R\dot{C}H{-}CH_2Br + HBr \longrightarrow RCH_2{-}CH_2Br + Br\bullet$ PRIMARY ALKYL BROMIDE

Figure 2. Free-radical addition. Example: Alpha olefin reacting with free radical to afford the most stable intermediate free radical, which, in turn, reacts with reagent to form the primary final product (anti-Markownikov mode of addition).

(1) FREE RADICAL $X\cdot$

(2) $R\text{-}CH_2\text{-}CH{=}CH_2 + X\cdot \longrightarrow R\text{-}\dot{C}H\text{-}CH{=}CH_2 + HX$

ALLYL FREE RADICAL

(3) $R\text{-}\dot{C}H\text{-}CH{=}CH_2 + X_2 \longrightarrow R\text{-}CH(X)\text{-}CH{=}CH_2 + X\cdot$

[STABILITY OF FREE RADICALS: ALLYL > TERTIARY > SECONDARY > PRIMARY > METHYL > VINYL]

Figure 3. Free-radical substitution. Example: Alpha olefin reacting with free radical to afford the most stable intermediate free radical, the allyl free radical, which, in turn, reacts with reagent X to form the final product and a new free radical, X.

Typical Products Directly Available from Olefinic Precursors

Alcohols (addition of water; hydration)
Alcohols (hydroalumination/oxidation)
Alcohols (hydroboration/oxidation)
Alcohols (oxidation in presence of formic acid, followed by saponification)
Alcohols (oxidation via selenium oxide catalyst)
Alcohols (*see* aldehydes, oxo reaction, subsequent reduction over catalyst)
Alcohols, secondary and tertiary (alcohols addition with free-radical source)
Aldehydes (oxidation using various metal catalysts)
Aldehydes (ozonolysis, reductive workup)
Aldehydes (via hydroformation/oxo reaction)
Alkanes, branched (addition of alkyl halides)
Alkanes, substituted (addition of complex alkyl halides)
Alkenes, branched (catalytic oligomerization)
Alkylation (see Chapter 8 for details)
Alkynes (halogenation followed by bis-dehydrohalogenation)
Amines (reduction of nitriles)
Amines, N-alkylated (addition of NaCN to sulfuric acid/olefin adducts)
Boranes, trialkyl (addition of boron hydrides)
Bromides, alkyl (addition of hydrogen bromide)
Carboxylic acids (acid-catalyzed addition of formic acid)

Carboxylic acids (ozonolysis, oxidative workup)
Chlorides, alkyl (addition of hydrogen chloride)
Chloroalkylethers (addition of alpha chloro ethers in the presence of metal halides)
Chlorohydrins (addition of HOCl in the presence of chromyl chloride)
Chloroketones (addition of acetyl chloride/acetic anhydride)
Chloronitriles (addition of cyanogen chloride)
Chlorosulfonyl chlorides (addition of sulfuryl chloride)
Dimers, olefin (acid-catalyzed dimerization of mono-olefin)
Dinitroalkanes (addition of dinitrogen tetroxide)
Disulfides, alkyl (addition of sulfur, disulfur dichloride)
Dithiocyanates (addition of dithiocyanogen)
Epoxides (reaction with peracids)
Esters, complex (addition of carboxylic acids)
Glycols (hydroxylation via permanganate or peracids)
Halohydrins (addition of HOX)
Hydroperoxides (addition of hydrogen peroxide)
Iodides, alkyl (addition of hydrogen iodide)
Isomerization to alternate olefins
Ketones (ozonolysis of internal olefins, reductive workup)
Ketones, unsaturated (*see* chloroketones, subsequent dehydrochlorination)
Mercaptans (addition of hydrogen sulfide)
Metal complexes, Miscellaneous (addition of metal salts, e.g., copper, silver, etc.)
Nitriles (addition of ammonia)
Nitriles, secondary (addition of hydrocyanic acid)
Nitriles, unsaturated (oxidation in presence of ammonia and catalysts)
Nitroalkyl nitrates (addition of dinitrogen tetroxide and subsequent oxidation)
Nitroalkyl nitrites (addition of dinitrogen tetroxide)
Nitroolefins (treatment with nitrating agents)
Nitrosochlorides (addition of nitrosyl chlorides)
Oligomers (acid or coordination catalysts)
Ozonides (addition of ozone)
Paraffins (addition of hydrogen)
Peroxides (addition of oxygen)
Phosphines, alkyl (addition of phosphine)
Phosphinic, alkyl, and phosphonic, alkyl, acids (addition of phosphorus)
Phosphonates, dialkyl alkyl (addition of dialkyl phosphates)
Polymerization (see Chapter 4 for details)
Silanes, alkyl (addition of silicon hydride)
Succinic anhydrides, alkenyl (Ene reaction with maleic anhydride)

Sulfates, alkyl (addition of sulfuric acid)
Sulfonates, alkane (addition of bisulfite reagents)
Sulfonates, alkene (reaction with SO_3)
Sulfonates, alkyl (addition of SO_3-type reagents followed by saponification)
Sulfonates, hydroxyalkane (reaction with SO_3 followed by hydrolysis)
Sulfones, poly (addition of sulfur dioxide)
Sulfonic acids, alkyl (addition of SO_3-type reagents followed by hydrolysis)
Sultones (addition of sulfur trioxide or complexed SO_3-type reagents)
Thiones (addition of sulfur)
Vicinal dibromides, alkyl (addition of bromine)
Vicinal dichlorides, alkyl (addition of chlorine)

7 MOST COMMON COMMERCIAL OLEFIN REACTIONS

The most common transformations involving olefins in industrial areas undoubtedly include the following:

1. Hydroformylation (oxo reaction)
2. Oligomerization/polymerization
3. Miscellaneous simple addition reactions
4. Alkylation reactions
5. Sulfations/sulfonations
6. Oxidations

It is useful from a practical standpoint to consider each of these broad classifications separately and to describe further, typical transformations carried out in each category.

The hydroformylation reaction, as the first example, can be carried out in several ways to generate various end products, including alcohols, aldehydes, carboxylic acids, and esters. These products are comprised of materials that have some branching present in their molecular framework, as a consequence of the actual site of addition of the incoming CO on the olefin backbone. Also, this chemistry, as practiced, may be carried out in single steps, or as additional modifications after the initial hydroformylation step has occurred. Some of the possibilities are summarized in Figure 4.

Another major class of common industrial syntheses dependent on olefinic feedstocks provides many useful and varied products. This class, oligomerization and polymerization, is displayed briefly in Figure 5. Subtle modifications and differences in actual practice of the related art result in this area being far too complex to discuss in detail in this chapter. Indeed, volumes of reference material are available for most of the general routes

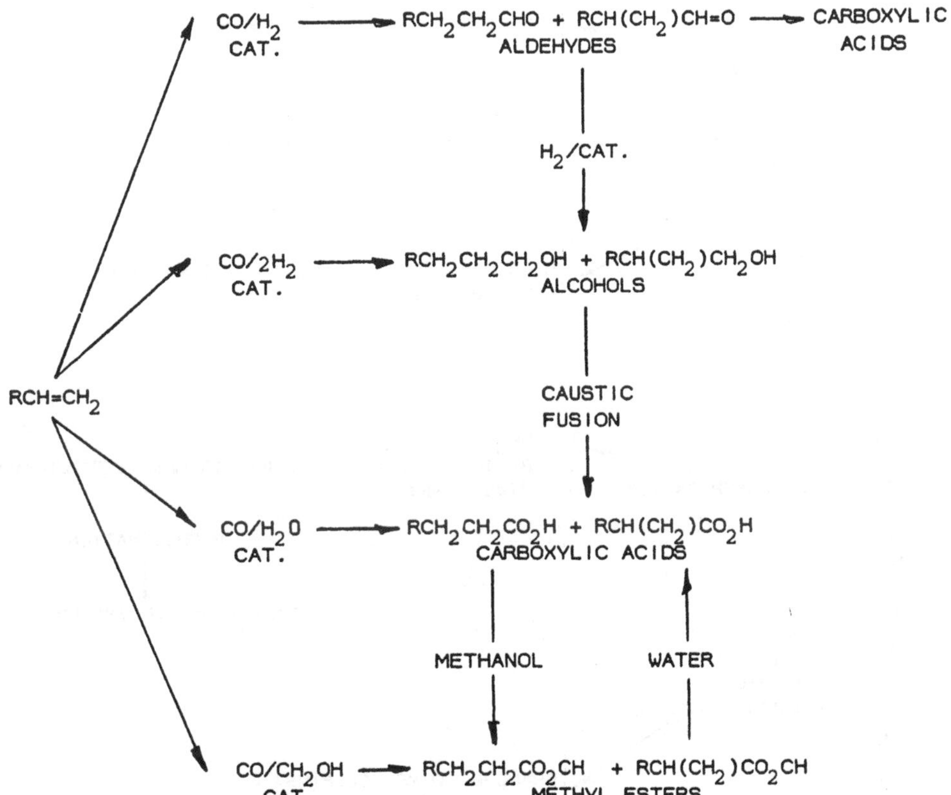

Figure 4. Hydroformylation reactions.

described as subsets within this category. This group comprises the largest single technological usage of alpha olefin intermediates today, and in all probability, the levels of use will continue to rise in the future. Additional information on these processes is given in other chapters.

A very diverse group of olefinic reactions can be described as "miscellaneous simple addition reactions." In this family, some reagent can be considered as being added across the carbon-carbon double bond of the olefin. In most instances, this addition will be controlled by the stability of the particular intermediate involved (i.e., cationic or free radical), giving a predominant product. Some examples of commercial interest are illustrated in Figure 6.

Although not strictly necessary for an understanding of "olefin chemistry and reactions," it seems appropriate at this point to develop a few of the additional modifications possible by using a multistep synthesis ap-

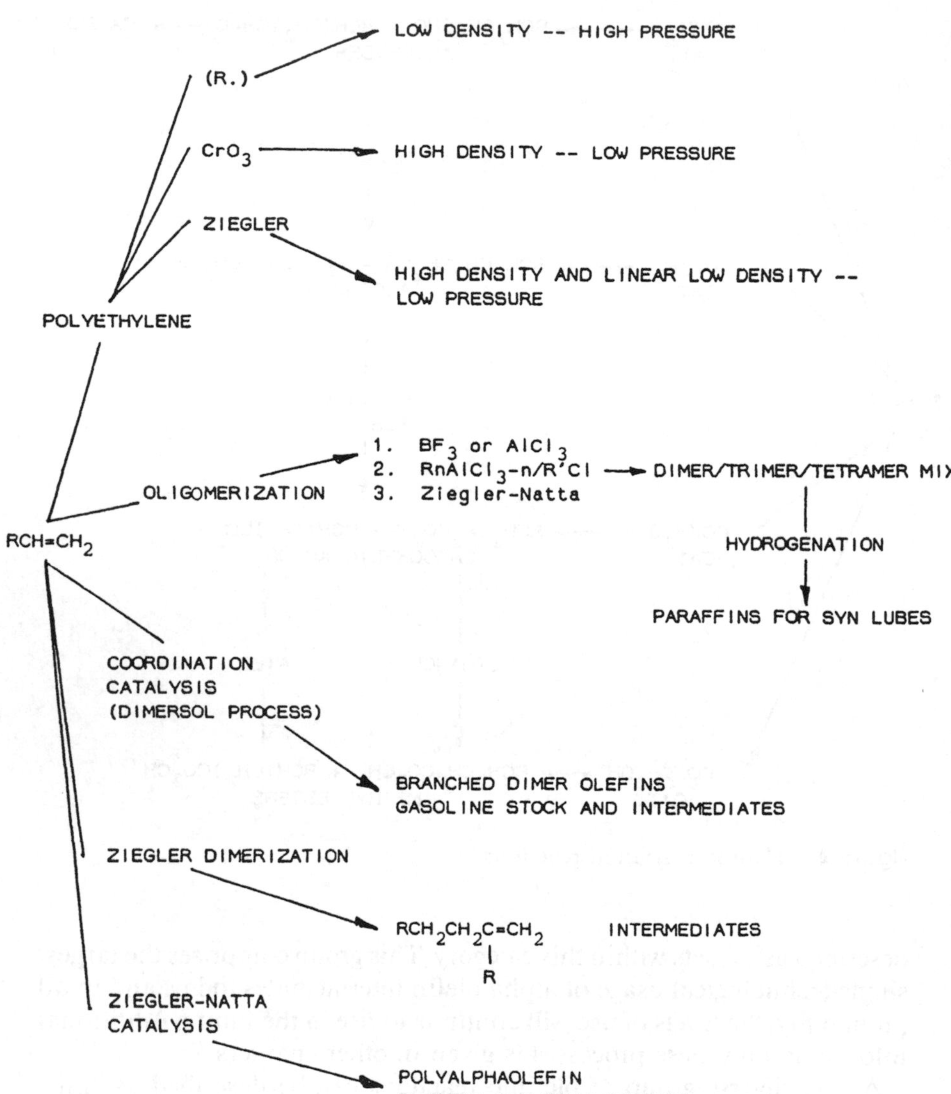

Figure 5. Oligomerization and polymerization reactions.

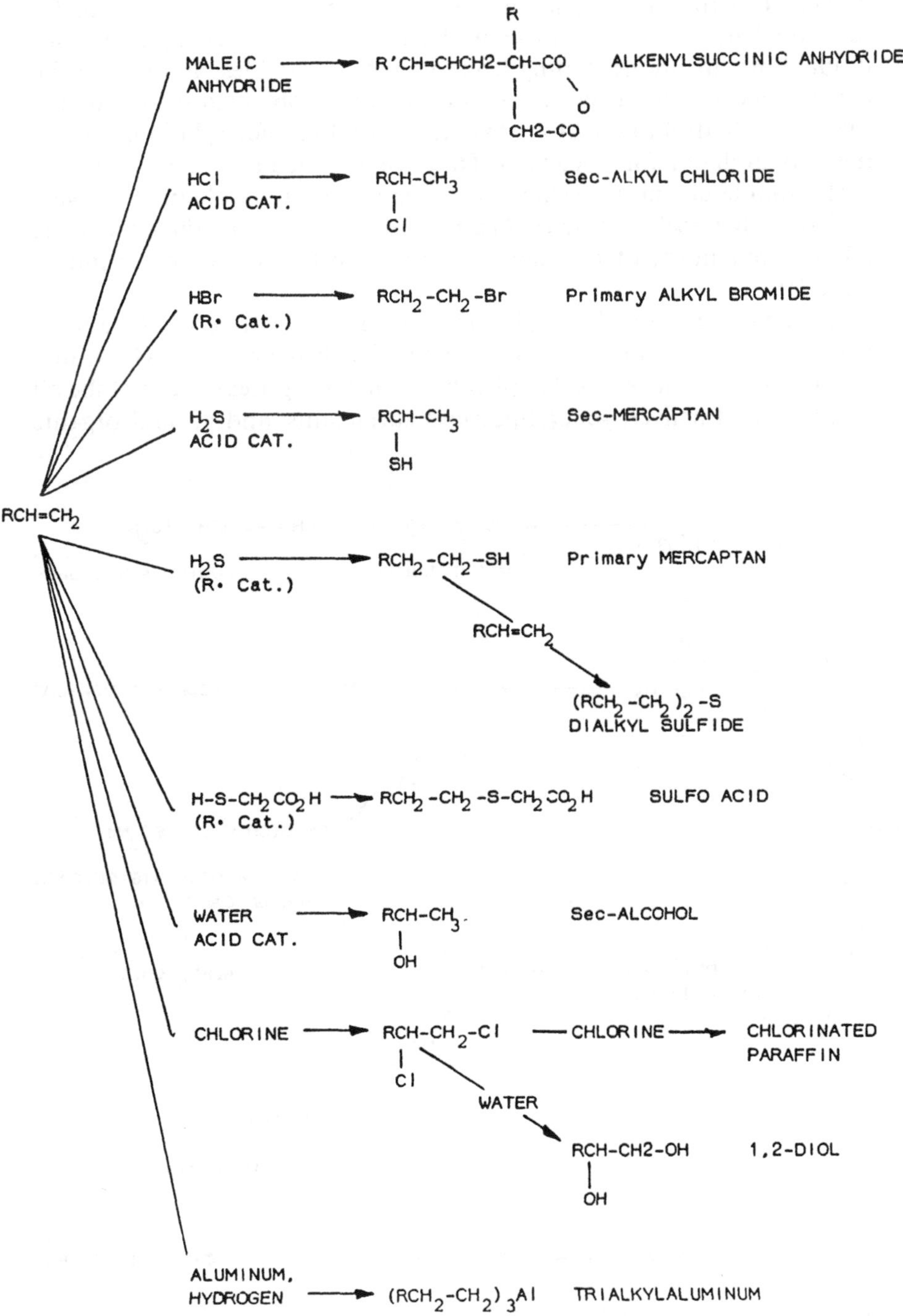

Figure 6. Miscellaneous simple addition reactions.

proach. To gain an appreciation for the extreme versatility of the olefin functional group, Figures 7 to 9 have been included to review possible additional modifications starting from alcohols, bromides, and carboxylic acids/esters. These classes of compounds can be obtained readily by the hydration, hydrobromination, oxidation, or hydroformylation, respectively, of alpha olefin precursors. They are included to demonstrate how easily simple chemical reactions can transform alpha olefins to a wide variety of chemical derivatives. Again, these figures are not intended to list all (or even most) of the modifications possible for these functional materials.

Alkylation reactions for olefins are considered separately in Chapter 8 because of the commercial value of the finished products. These materials, alkylates, have broad applications, including areas such as lube oil additives and actives, general-purpose surfactants, and general organic

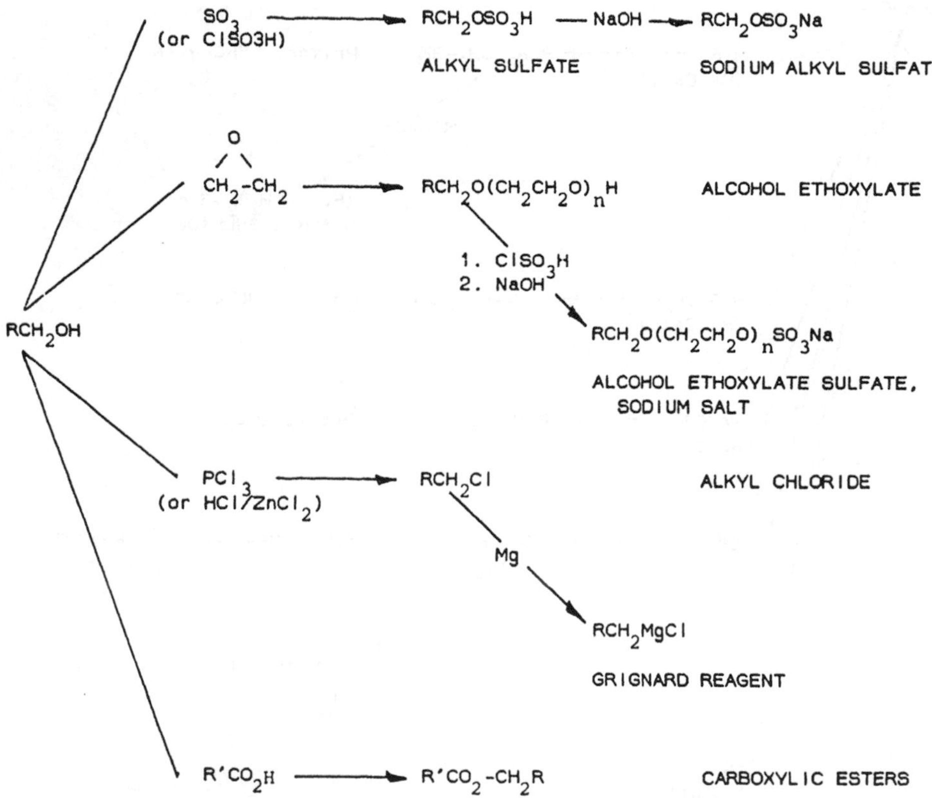

Figure 7. Simple multistep synthesis examples: alcohol reactions.

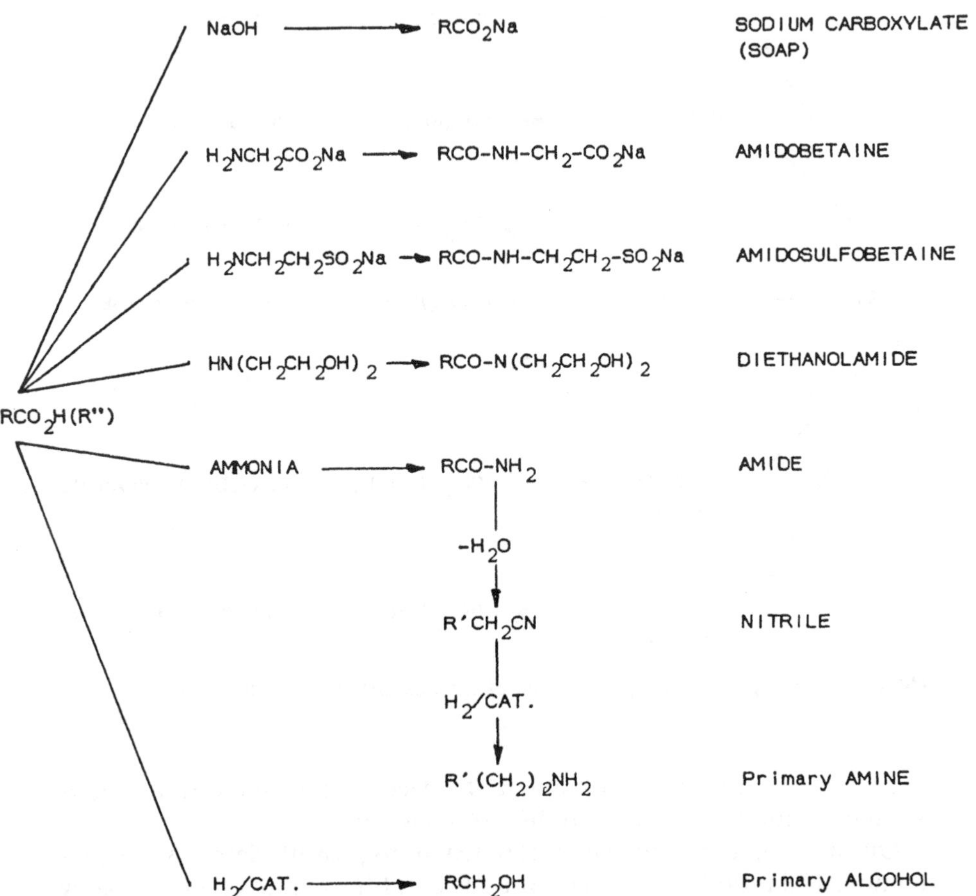

Figure 8. Simple multistep synthesis examples: carboxylic acid/ester reactions.

intermediates for synthetic purposes. In many instances, the choice of catalyst can make major differences in the physical properties of the final product obtained. More details on this subject can be found in Chapter 8. A general overview of these transformations is given in Figure 10.

A major subset of alpha olefin reactions that are commercially important involve the introduction of functional groups with sulfur atoms. Some of these are grouped, for convenience, in Figure 11. As with many of the derivatives of olefins, several of the end products have valuable applications in the area of surfactants and detergents. In many instances, the reaction paths are complicated and involve intermediates not normally isolated. In these cases, for simplicity, we are showing final products only, and often separate steps (hydrolysis, saponification, digestion, etc.) are

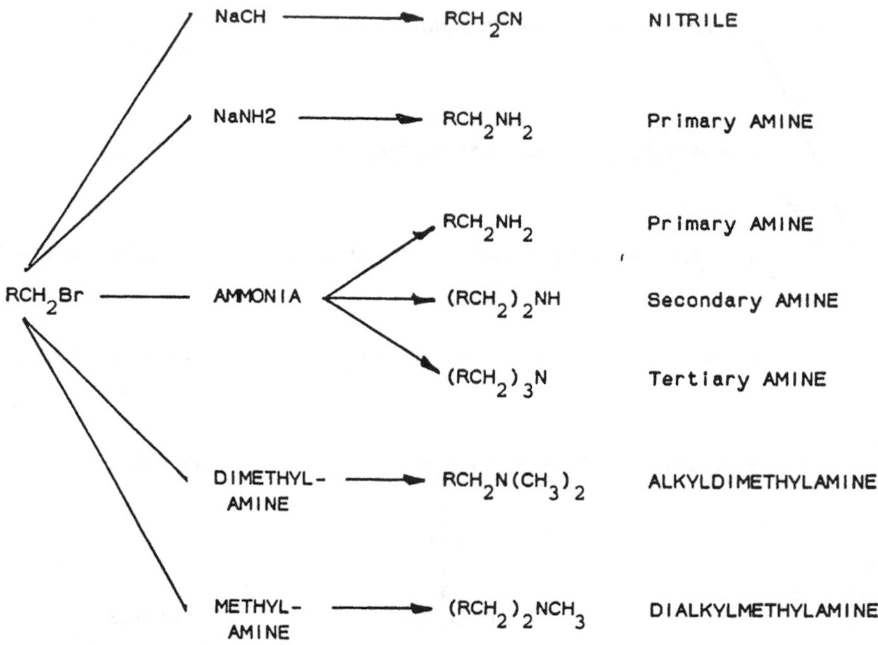

Figure 9. Simple multistep synthesis examples: alkyl bromide reactions.

necessary in processing to convert the feed olefin and various intermediates within the process to the products shown.

One major component of this set of reactions possibly deserves more attention. The introduction of continuous SO_3 film sulfonation technology affords a practical, commercial route to alpha olefin sulfonates (AOS). This process involves the treatment of liquid alpha olefins with an SO_3/air mixture. A slight excess of SO_3 and efficient heat removal affords a surfactant intermediate, which upon aging and neutralization is of suitable quality for use as a major active in a variety of formulated products.

Alpha olefin sulfonates have been shown to be less irritating to skin and the least toxic orally of the main anionic surfactants. Furthermore, AOS-type surfactants are stable and demonstrate good detergency, good solubility, and good biodegradability. More detail is available in Chapter 8.

Again, for convenience, several oxidation-type reactions have been collected as a series for consideration. These are displayed in Figure 12. Additional modifications for some of the products derived from initial oxidation of olefins are also shown in the related Figure 13. This array amplifies the potential versatility of the alpha olefin as a precursor to a

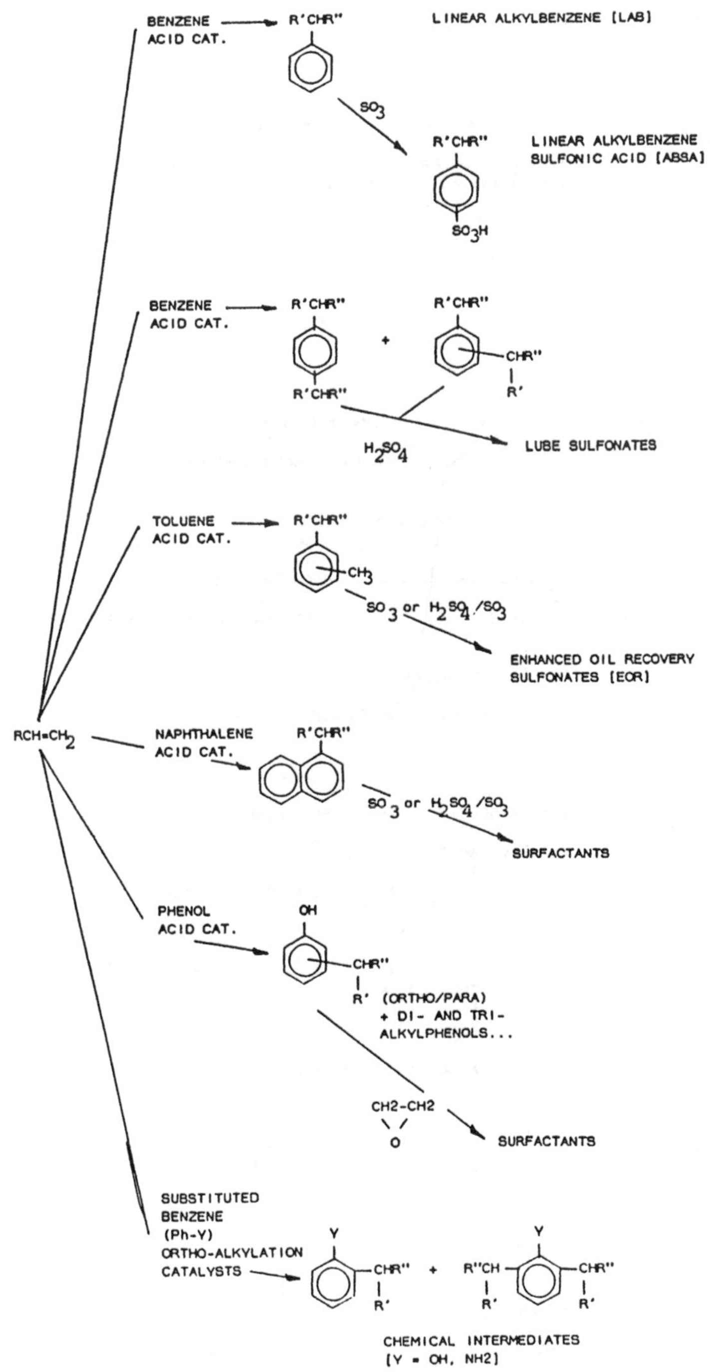

Figure 10. Alkylation reactions.

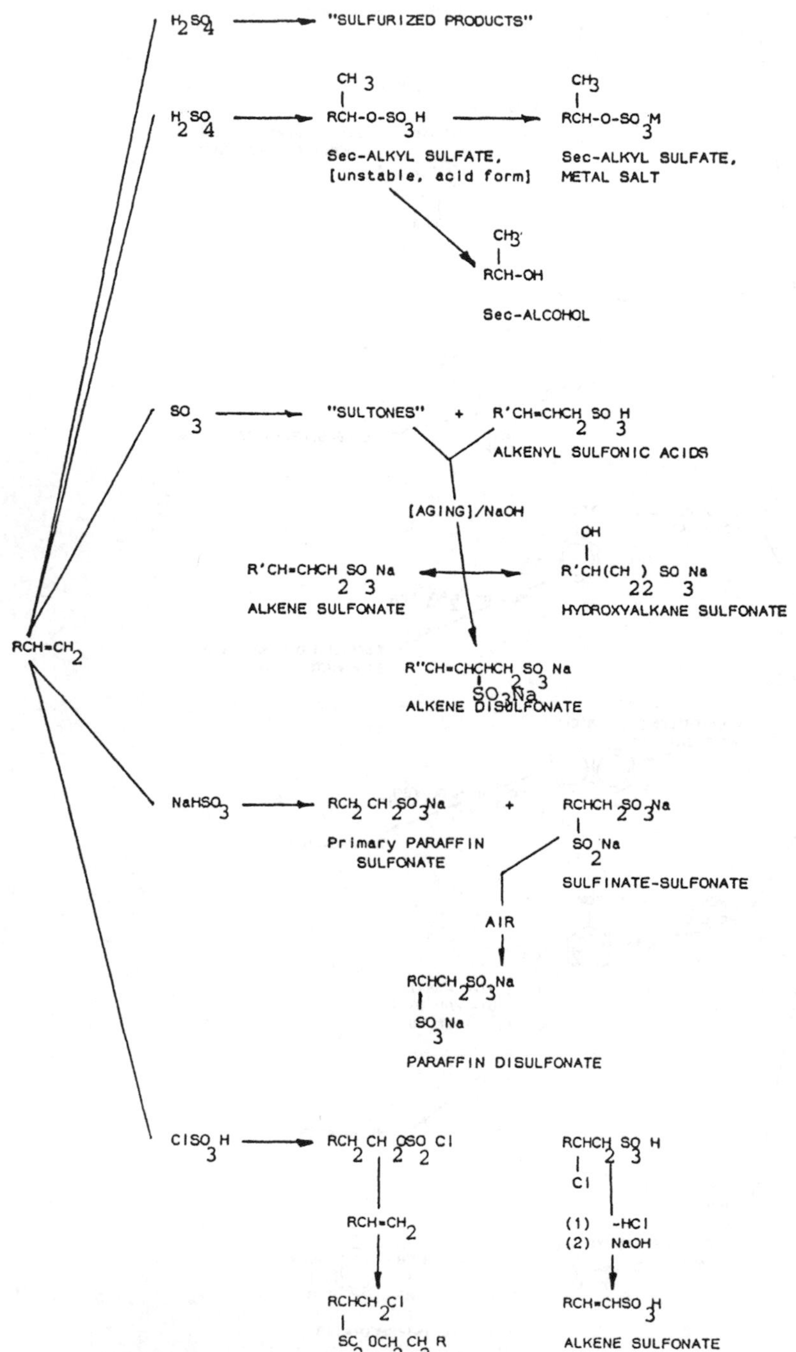

Figure 11. Sulfates and sulfonates.

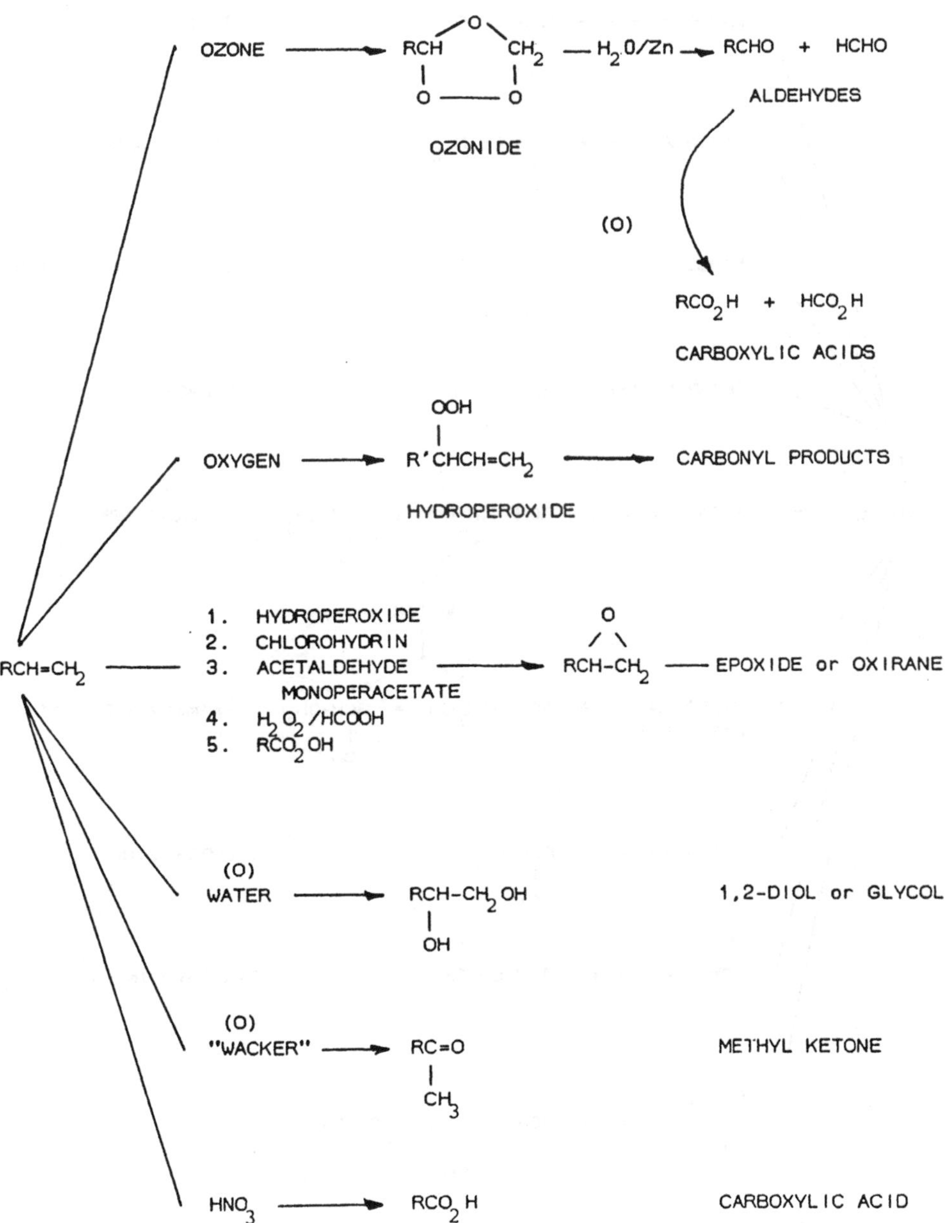

Figure 12. Oxidation reactions.

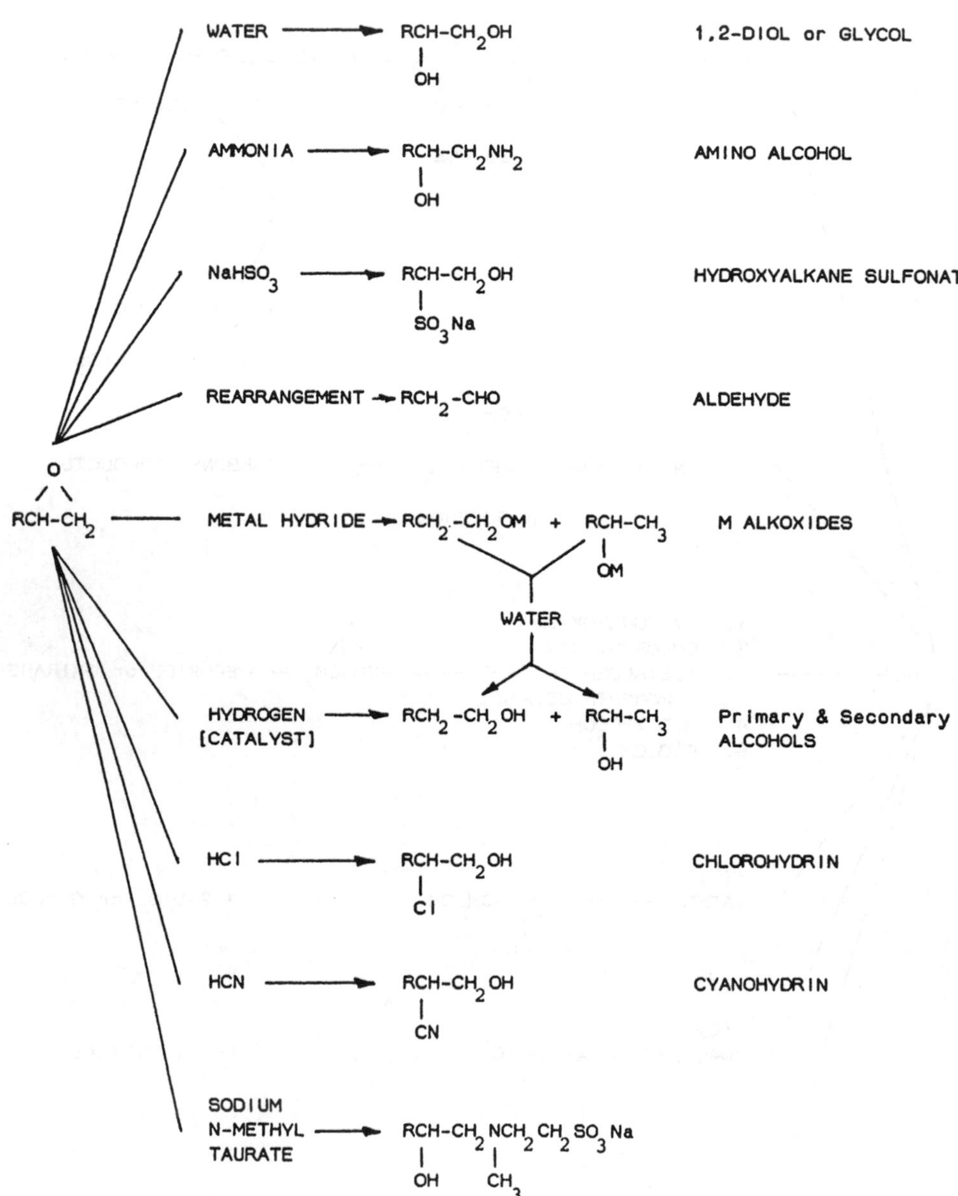

Figure 13. Epoxide reactions.

variety of end products. Obviously, applications for these chemicals are extremely varied and diverse.

8 CHARACTERIZATION OF OLEFINIC MATERIALS

Olefins in general, and alpha olefins in particular, can be characterized by a variety of both classical and instrumental methods. Basic techniques are described in the following examples, which outline typical characterization methods.

8.1 Spectroscopic Methods

Infrared Spectroscopy

Typical absorptions for alpha olefin (in 1/cm); (M, medium; S, strong) are as follows:

M ~3200
M ~3100
S 2936–2916: asymmetric CH_2 stretch
M 2863–2843: symmetric CH_2 stretch
M ~1820
M 1648–1638: vinyl-H
M 1455–1435: deformation CH_2
M ~1400
S 995–985: (with overtone) vinyl-H
S 915–905: vinyl-H
W 724–722: CH_2 rocking

Nuclear Magnetic Resonance Spectroscopy

Typical spectral chemical shifts for alpha olefin (ppm, internal TMS standard, neat olefin, coupling not specified) are as follows:

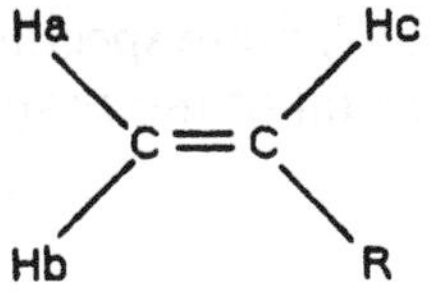

Ha Hc
C=C
Hb R

Ha: 4.85
Hb: 4.95
Hc: 5.30

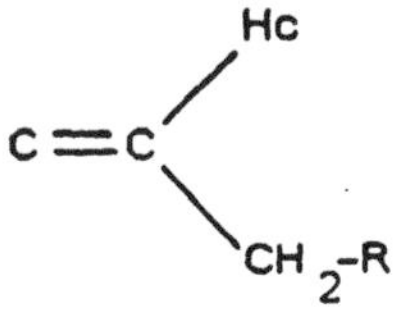

Hc
C=C
CH_2-R

CH_2: 1.98

```
     Hc
    /                  (CH2)n:   1.16
C=C
    \                  CH3 :     0.78
     C-(CH2)n-CH3
```

8.2 Classical Methods: Determination of Olefin Structure by Degradation Techniques

Ozonolysis

Treatment of an olefin of unknown structure with ozone followed by reductive workup (H_2O/Zn) results in mild cleavage of the carbon-carbon double bond and the subsequent generation of carbonyl-containing fragments (ketones and aldehydes).

Permanganate Oxidation

Treatment of an olefin of unknown structure with $KMnO_4$ results in a vigorous oxidation to afford carboxylic acids, ketones, and CO_2 (from terminal $=CH_2$ groups).

```
R1     R2                                 R1           R2
  \   /                                     \          /
   C=C        O3         H2O/Zn              C=O  +  O=C
  /   \     ------->   ---------->          /          \
R3     R4                                 R3           R4
```

9 LITERATURE REVIEW; SUMMARY AND CONCLUSIONS

The mass of published patents and technical literature dealing with alpha olefins is staggering and can be reviewed only briefly. This chapter is intended to highlight typical commercial uses and transformations involving olefins in general and alpha olefins in particular. Much more specific information on selected individual topics is given in several of the chapters within.

CHAPTER 3

Routes to Alpha Olefins

GEORGE R. LAPPIN Ethyl Corporation, Baton Rouge, Louisiana

1 INTRODUCTION

Linear alpha olefins with four or more carbons are produced from ethylene by Ethyl, Chevron, Shell, and Mitsubishi. The current suppliers continue to announce plans to increase capacity to meet developing needs. Shell is building a new plant of capacity 535 million pounds per year at Geismar, Louisiana, to be on-stream in 1989; Chevron is adding 200 million pounds per year of capacity at Cedar Bayou, Texas, to be on-stream in 1990; and Ethyl plans to reduce the bottlenecks at its current facility at Pasadena and/or to build a new facility. A plant based on Ethyl technology is being built in the USSR. A plant based on Gulf technology is planned to begin operation in Czechoslovakia in 1992. Idemitsu has announced plans to build a plant in Japan of capacity 110 million pounds per year. Various other potential suppliers of linear alpha olefin have considered or announced plans to build plants but have yet to build them.

As for wax-cracking facilities, Chevron shut down a wax-cracking olefin plant in 1984 after they acquired Gulf, with its ethylene-based alpha olefin plant. Shell had three wax-cracking olefin plants in Europe, at least two of which are no longer operating. The paraffin dehydrogenation routes to linear internal alpha olefins used for feed to linear alkylbenzene plants will receive little discussion in this book.

In considering the ethylene-based routes, it should be recognized there are just two approaches for producing alpha olefins, which in this chapter are called catalytic and stoichiometric. Others have called the catalytic approach a one-step process or the geometric distribution method, and the stoichiometric approach a two-step process or the statistical distribution method. In the following discussions, the chemistry is discussed and then the processes are described for the types of alpha olefins produced by the three distinct commercial processes as originated by Ethyl, Gulf, and Shell. The Mitsubishi process was licensed from Gulf, and the USSR process was licensed from Ethyl. The three distinct processes are referred to by the names of the companies that first commercialized them: Gulf, Ethyl, and Shell. Limited information suggests that the Idemitsu process uses the catalytic approach.

2 CHEMISTRY

The chemistry is similar for the commercial processes for production of alpha olefins from ethylene, but the products and product distributions are different. Economics vary among the processes and depend on the cost of ethylene, the size of the unit, the complexity of the process, and the ability to sell or use all of the products at good values. A world-scale plant is now at least 300 million pounds per year in capacity.

The Ethyl, Shell, and Chevron/Gulf ethylene oligomerization processes use different chemistries, solvents, reactor temperatures, and reactor pressures. The different conditions result in different reactor configurations. Since oligomerization is an exothermic reaction, heat must be removed, and either the reactors are heat exchangers (Ethyl and Chevron/Gulf) or heat exchangers are in the system (Shell).

Ethylene polymerization during oligomerization tends to foul or plug equipment and the commercial processes must provide for removal of this pluggage in some fashion. The ability to avoid pluggage or to handle it when it occurs represents one of the key proprietary areas for the producers. The basic flowsheets appear to be similar, as all must feed a catalyst and ethylene at pressures over 70 atm, all must distill and store the final products, all must have furnaces to supply heat and cooling towers to remove heat, and the excess ethylene at the end of the reaction must be separated and recycled using compressors. In addition, the processes may benefit from the use of a solvent.

It is because of the similarity of the basic equipment that economics appear to vary little per pound of total alpha olefins produced in the same-size plants. The Chevron/Gulf process has almost 30% of its products in the less valuable C_4 and $C_{20}+$ ranges. The Shell process, including the isomerization and disproportionation (I/D) unit, requires the highest investment, but it is best able to utilize alpha olefins that are produced but which do not have markets. Shell can utilize alpha olefins in the production of large quantities of odd and even linear internal olefins. In its total complex, Ethyl can exchange unneeded or less valued alpha olefins, such as butene, for those more in demand, such as decene. Thus the Ethyl process can produce a range most tailored to market demands.

The Ethyl and Chevron/Gulf processes are based on Ziegler aluminum alkyl chemistry. This chemistry is discussed in detail in *Comprehensive Organometallic Chemistry* [1] and in the March 1960 *Annalen der Chemie* [2] (please note that we have included only a few general references at the end of this chapter, and these are grouped by process). The *Annalen der Chemie* issue describes the early work in Ziegler's laboratory on ethylene chain growth and olefin elimination from aluminum alkyl compounds. Licensing Ziegler olefin and aluminum alkyl chemistry proved to be just the starting point for those who developed it into chemical processes. Gulf was the first with a commercial alpha olefin plant based on ethylene. Ethyl modified its original Ziegler approach through six generations of pilot plants and commercial plants, starting with their alcohol plant in 1965. Shell developed a unique nickel complex catalyst system and uses a solvent such as benzene or 1,4-butanediol.

Huang, in a 1986 presentation, explained the basic reactions, which are similar for both the Al- and Ni-based chemistries.

Chain Growth Reaction

$$M{-}(C_2H_4)_n{-}C_2H_5 + C_2H_4 \xrightarrow{k_g} M{-}(C_2H_4)_{n+1}{-}C_2H_5$$

where M denotes catalyst and n is an integer.

Displacement Reaction

$$M{-}(C_2H_4)_n{-}C_2H_5 \xrightarrow{k_d} M{-}H + CH_2{=}CH{-}(C_2H_4)_{n-1}{-}C_2H_5$$

$$M{-}H + C_2H_4 \xrightarrow{k_a} M{-}C_2H_5$$

In the chain growth reaction ethylene is inserted into the bond between the metal atom and the adjacent carbon. The displacement reaction consists of two steps—dehydrometalation to form the alpha olefin and ethylene hydroalumination to provide a new ethylene growth site. Mechanisms of the Al- and Ni-based systems differ and are discussed in the general references.

In the Ziegler processes practiced by Chevron, Mitsubishi, Ethyl, and soon the USSR and Czechoslovakia, M represents the aluminum atom in an aluminum alkyl. In the catalytic process practiced by Shell, M represents the nickel in Shell's catalyst. Other variations have been reported,

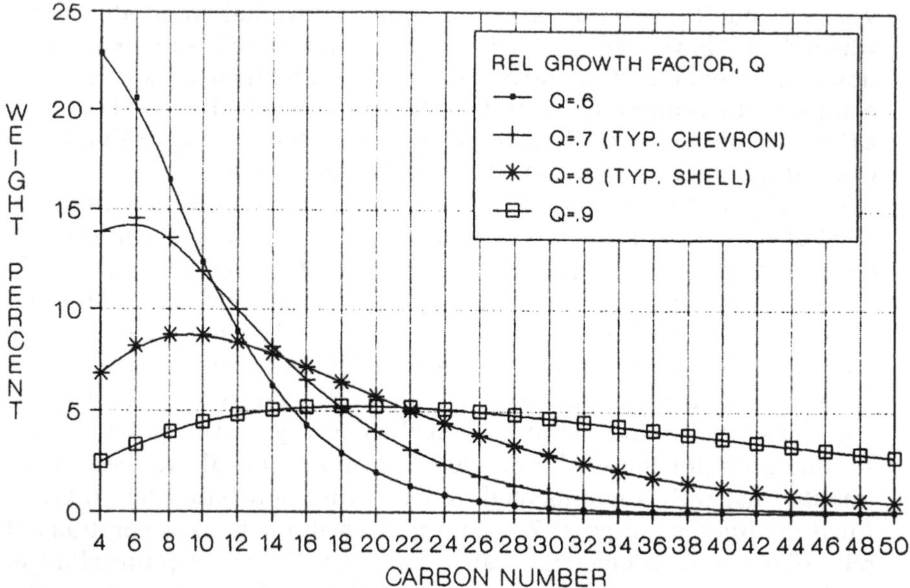

Figure 1. Theoretical catalytic oligomerization: Chevron/Gulf, Mitsubishi, and Shell.

but all ethylene oligomerizations appear to proceed through these basic steps (see Fig. 1).

2.1 Stoichiometric Chain Growth: Aluminum Alkyl State

In the aluminum alkyl-based reactions, the state of the aluminum alkyl varies, as the generally used triethylaluminum is a tetramer at low temperatures, a dimer at chain growth temperatures, and a monomer at higher temperatures. This association of molecules causes the triethylaluminum to behave and distill as if it were a much heavier material. Although predominantly associated, there is some dissociation to monomeric species, and only monomer participates in the rate-controlling steps of chain growth and displacement. Tributylaluminum is less associated than triethylaluminum, so that various physical as well as chemical changes occur as the ethylene oligomerization proceeds. Higher aluminum alkyls and branched aluminum alkyls are almost completely dissociated and they react and behave as monomers. Introduction of one ethylene molecule into a triethylaluminum monomer, followed by association with another triethylaluminum molecule, would form a dimer containing five ethyl groups and one butyl group. After further growth, the monomeric trialkyaluminum molecule might contain one butyl group (C_4), one decyl group (C_{10}), and one octadecyl group (C_{18}) attached to one aluminum atom. It can be noted that alkylaluminum groups are quite dynamic in a mixture of different trialkylaluminum compounds, and a constant exchange of alkyl groups takes place between aluminum atoms.

2.2 Stoichiometric Chain Growth: Poisson Distribution

The Ethyl process employs a stoichiometric chain growth step forming an aluminum alkyl mixture in which alkyl group composition is described by the Poisson equation.

$$x_p = \frac{n^p e^{-n}}{p!}$$

where x_p is the mole fraction of alkyl groups in which p ethylene units have been added and n is the average number of ethylene units added per equivalent of aluminum. Examples of product distributions from stoichiometric ethylene chain growth are shown in Figures 2 and 3.

2.3 Catalytic Chain Growth: Geometric Distribution

Kinetic reaction equations have been developed to show the theoretical product distributions in the catalytic or one-step chain growth processes which feature simultaneous chain growth and displacement reactions. The Chevron and Shell processes use this type of system (see Fig. 4).

Gulf operates at high temperatures and Shell relies on the displacement efficiency of the nickel catalyst to achieve desired rates of displacement relative to growth. These processes yield alpha olefins in a geometric distribution of carbon numbers as described by the equation

$$Q_N = \frac{X_N}{X_{N-2}}$$

where Q is defined as a geometric molar weight factor, X_N as the moles of an olefin having n carbons, and X_{N-2} as the moles of the preceding olefin having two fewer carbons. The Q as defined here is like K in the Freitas and Gum article [9]. Q varies with reaction conditions and will be set by producers to best meet demands. Based on modeling efforts, it appears that Q may vary from about 0.5 to about 0.75 for Chevron and from about 0.6 to about 0.85 for Shell. For example, on a molal basis and at a Q of 0.7, the amount of C_6 would be 0.7 times the amount of C_4 and the amount of C_8 would be 0.7 of the amount of C_6 and 0.49 (0.7 × 0.7) times the amount of C_4.

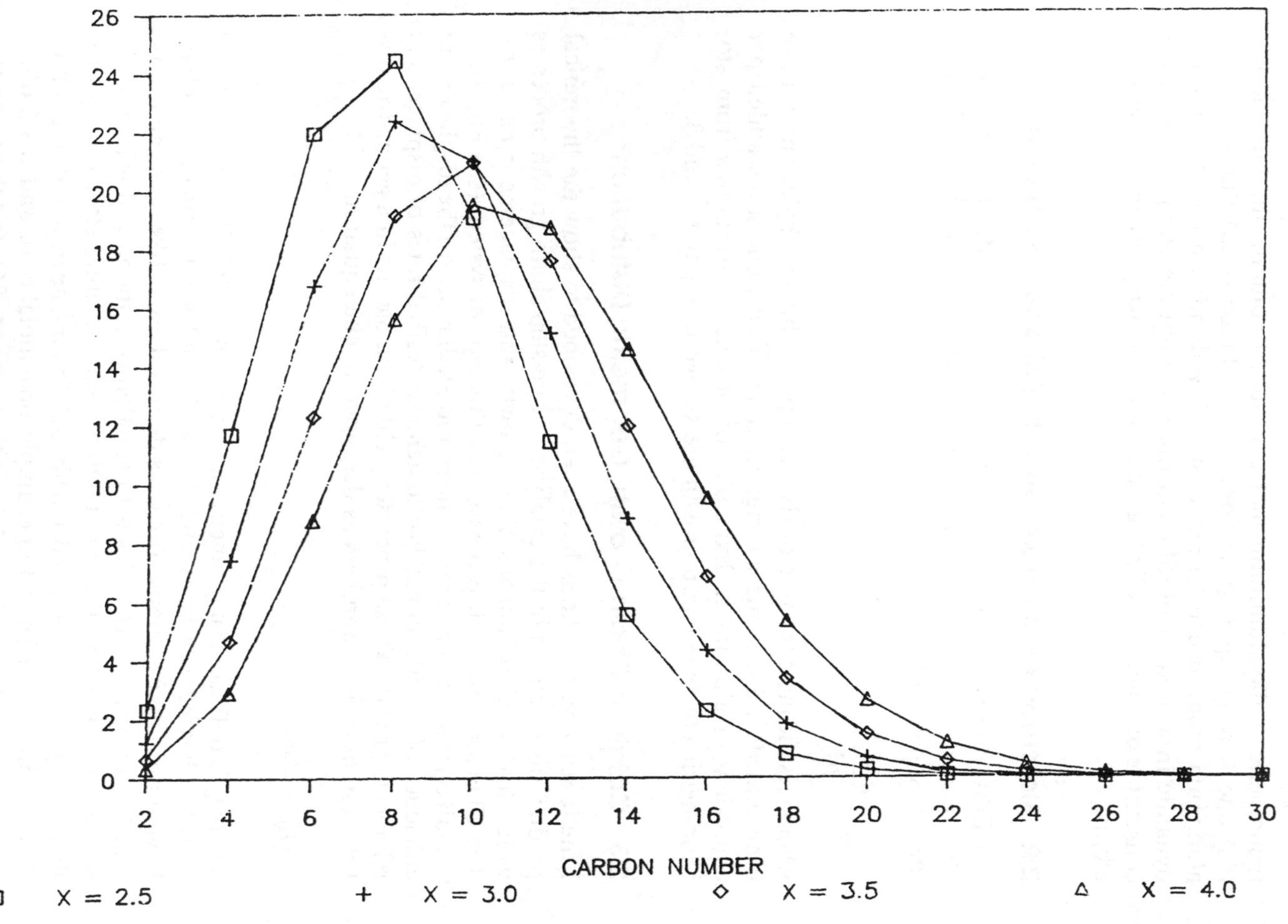

Figure 2. Theoretical stoichiometric growth.

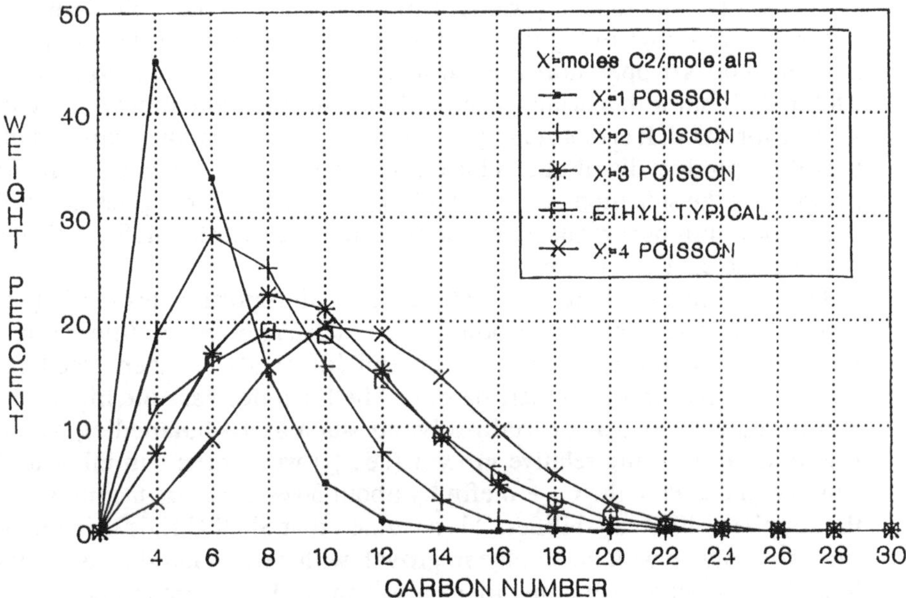

Figure 3. Stoichiometric growth: theoretical and an Ethyl typical.

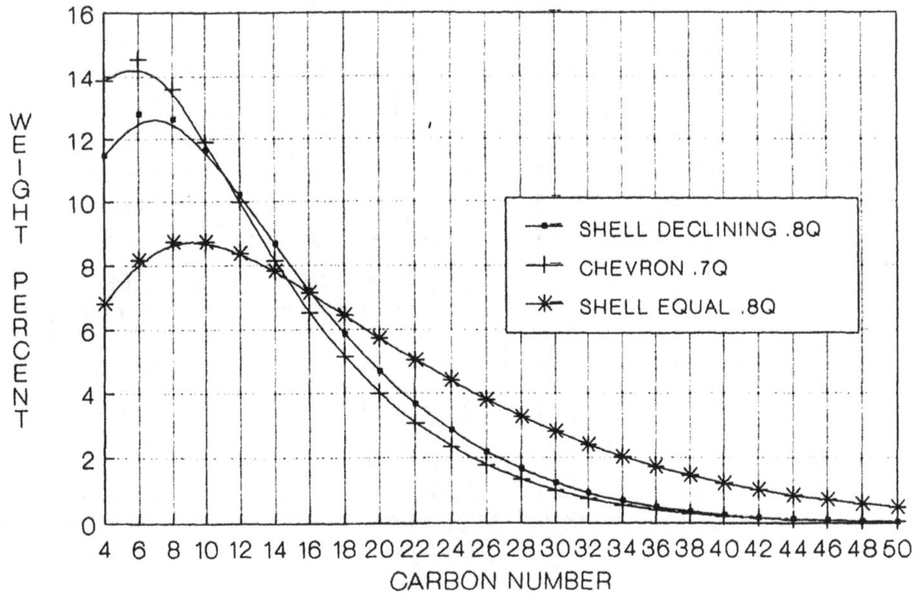

Figure 4. Comparison of catalytic chain growths: Shell declining, Chevron, Shell equal.

The relative amounts of product from the catalytic and stoichiometric processes are shown in Figure 5. In these plots, estimates have been made as to the modes of operation, including use of a Q of 0.7 for Chevron, a Q of 0.8 for Shell but with a declining growth rate with carbon number, and an X of 3.2 for Ethyl. It is reported by all three producers that they can and do vary their product distributions based on market requirements. From our perspective the stoichiometric process as practiced by Ethyl gives more of the carbon numbers required by current merchant alpha olefin markets than do the catalytic processes.

As part of its total SHOP operation, Shell diproportionates the higher olefins with lower olefins to produce C_{10}–C_{14} detergent range, linear internal olefins for internal oxo alcohol and linear alkylbenzene feed requirements. According to their patents, Shell can decrease the higher olefin production by using certain solvents, such as 1,4-butanediol, which results in a declining relative growth (i.e., Q) with carbon number and which can decrease the C_{20}+ olefin by about 30%. In Figure 6 comparing the suppliers' 1987 capabilities, it is assumed that Shell U.K. is using benzene as a solvent with no change in growth with carbon number, and that Shell U.S. is using 1,4-butanediol as a solvent with the corresponding decrease in growth rate with carbon number.

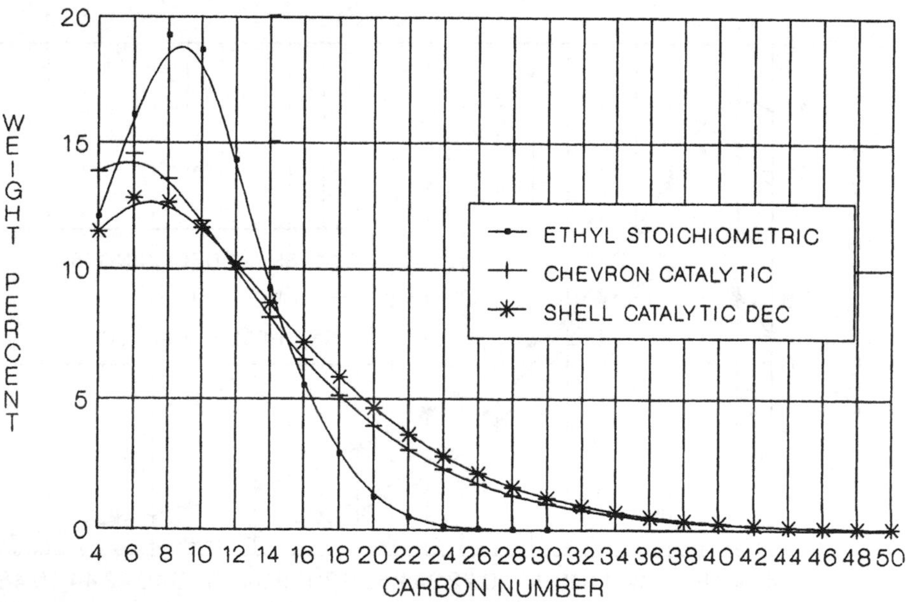

Figure 5. Comparison of Ethyl, Chevron, and Shell: estimated typical distributions.

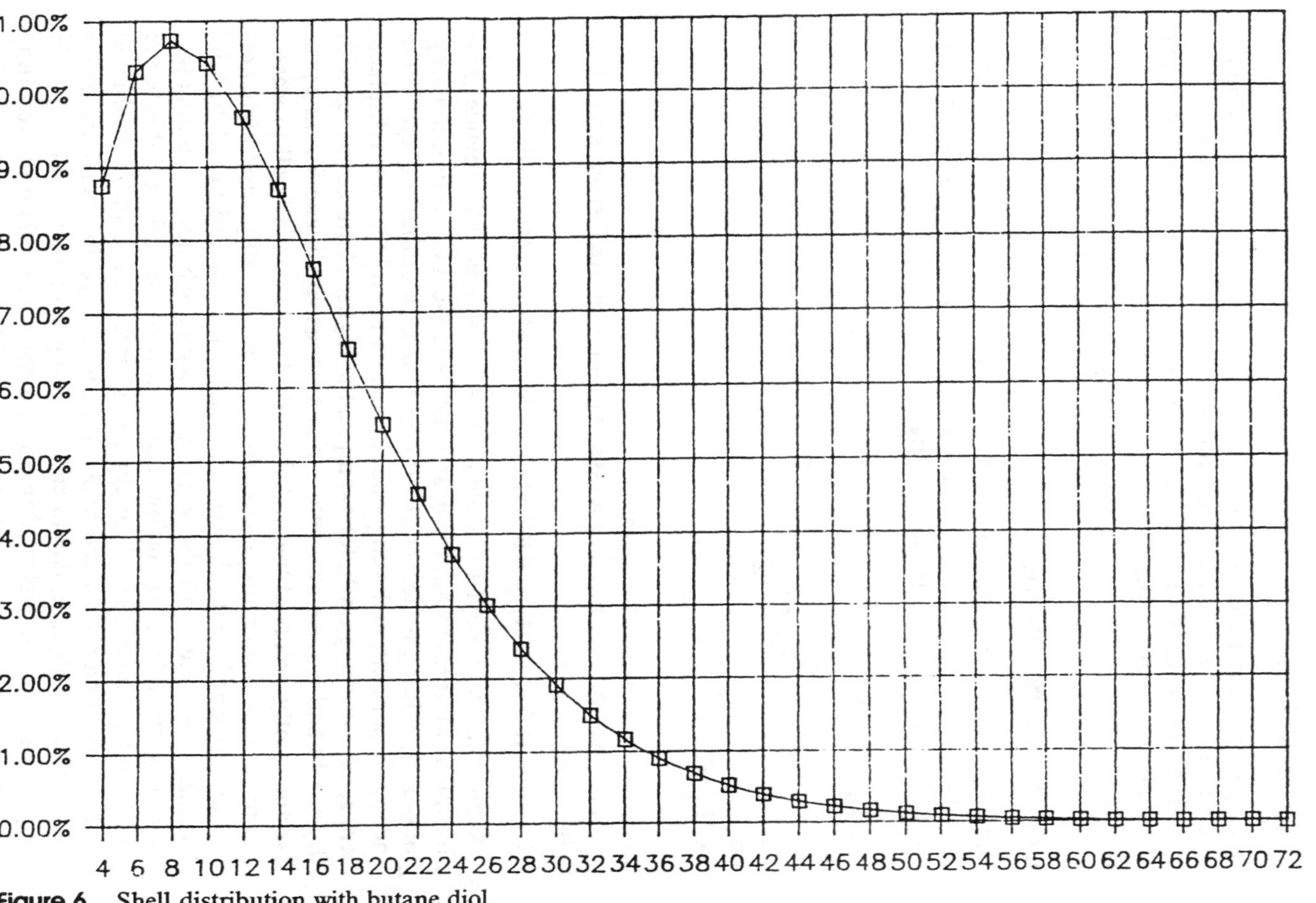

Figure 6. Shell distribution with butane diol.

3 BY-PRODUCTS

The types and quantities of nonlinear alpha olefins in the alpha olefin products depend on the reaction kinetics of the various chemical reactions that take place in the processes. Major by-products found in linear alpha olefins include vinylidene olefins, linear internal olefins, and paraffins.

3.1 Vinylidene Olefins

Vinylidene olefins are branched olefins in which the double bond is between the first and second carbon atoms and the branching alkyl group is attached to the second carbon (see Figs. 7 and 8).

$$\begin{array}{l} CH_3(CH_2)_m \\ \quad\quad\quad\quad \diagdown \\ \quad\quad\quad\quad\quad C{=}CH_2 \\ \quad\quad\quad\quad \diagup \\ CH_3(CH_2)_n \end{array}$$

where m and n are odd integers.

In the Chevron and Ethyl processes, vinylidenes are formed according to the scheme

$$R{-}Al + R'CH{=}CH_2 \rightarrow RR'CHCH_2{-}Al \rightleftharpoons RR'C{=}CH_2 + Al{-}H$$

As an example, 2-butyl-1-dodecene ($m = 9, n = 3$) is a C_{16} vinylidene olefin which can form by adding hexene ($R' = C_4$) to a C_{10} aluminum alkyl group or by adding dodecene ($R' = C_{10}$) to a C_4 aluminum alkyl. These reactions occur both in the ethylene chain growth and the olefin displacement steps. During growth, for example, the higher olefin competes with ethylene, but adds at a slower rate. An analogous vinylidene-forming process apparently occurs at the nickel catalyst atom in the Shell process.

In the organoaluminum process, since aluminum alkyls and higher olefins are present in both the ethylene chain growth and the olefin displacement steps, there will always be some degree of branched-chain alkyl group and vinylidene formation. Formation of vinylidenes affects the product distributions, as the branched alkyl groups are rapidly eliminated as vinylidenes and do not add back to the metal hydride as readily as do alpha olefins or ethylene. Olefin elimination gives an aluminum hydride group which tends to prefer addition to ethylene and then the resulting triethylaluminum group starts to grow again. Thus the actual distribution in a stoichiometric process will tend to be flatter than the theoretical

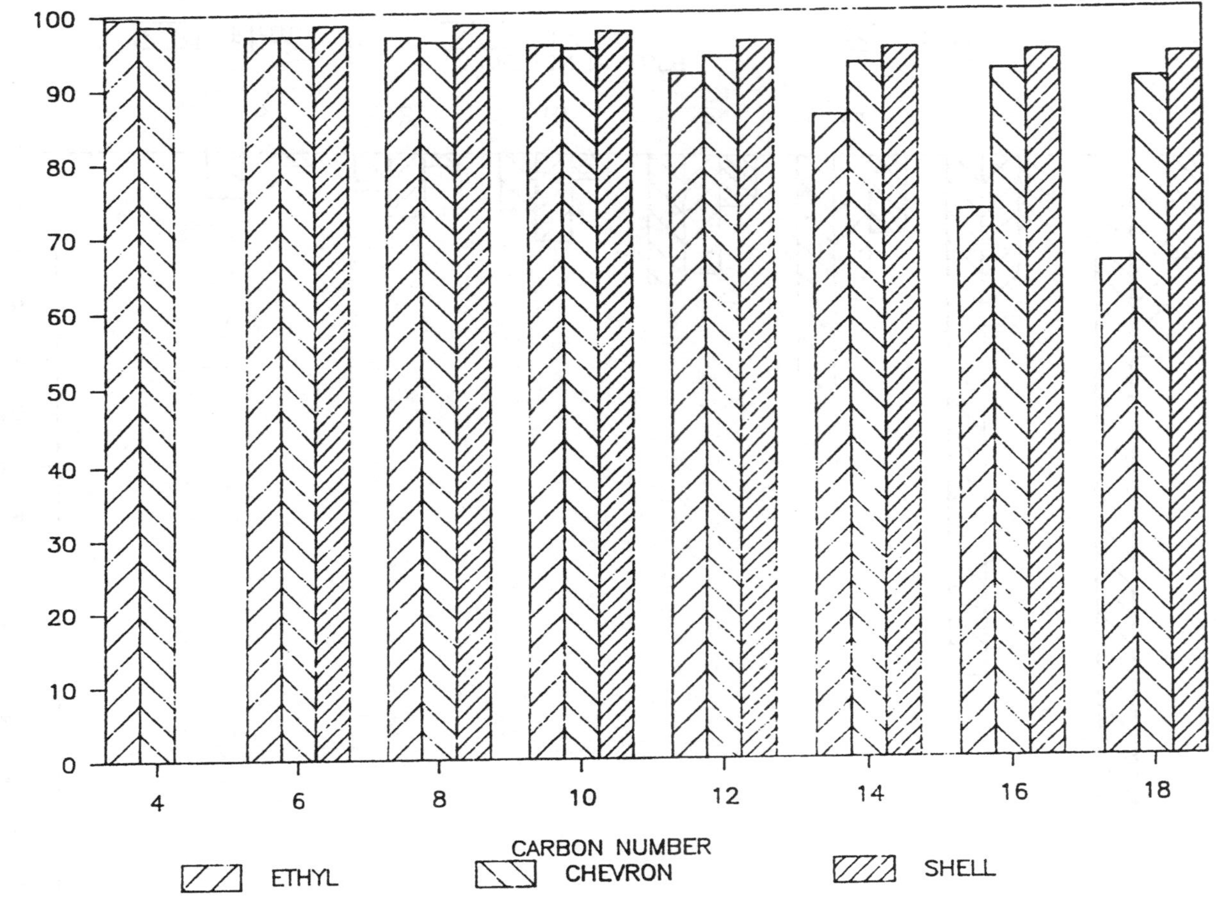

Figure 7. Typical vinyl olefin content (%). 1986 typicals are subject to change.

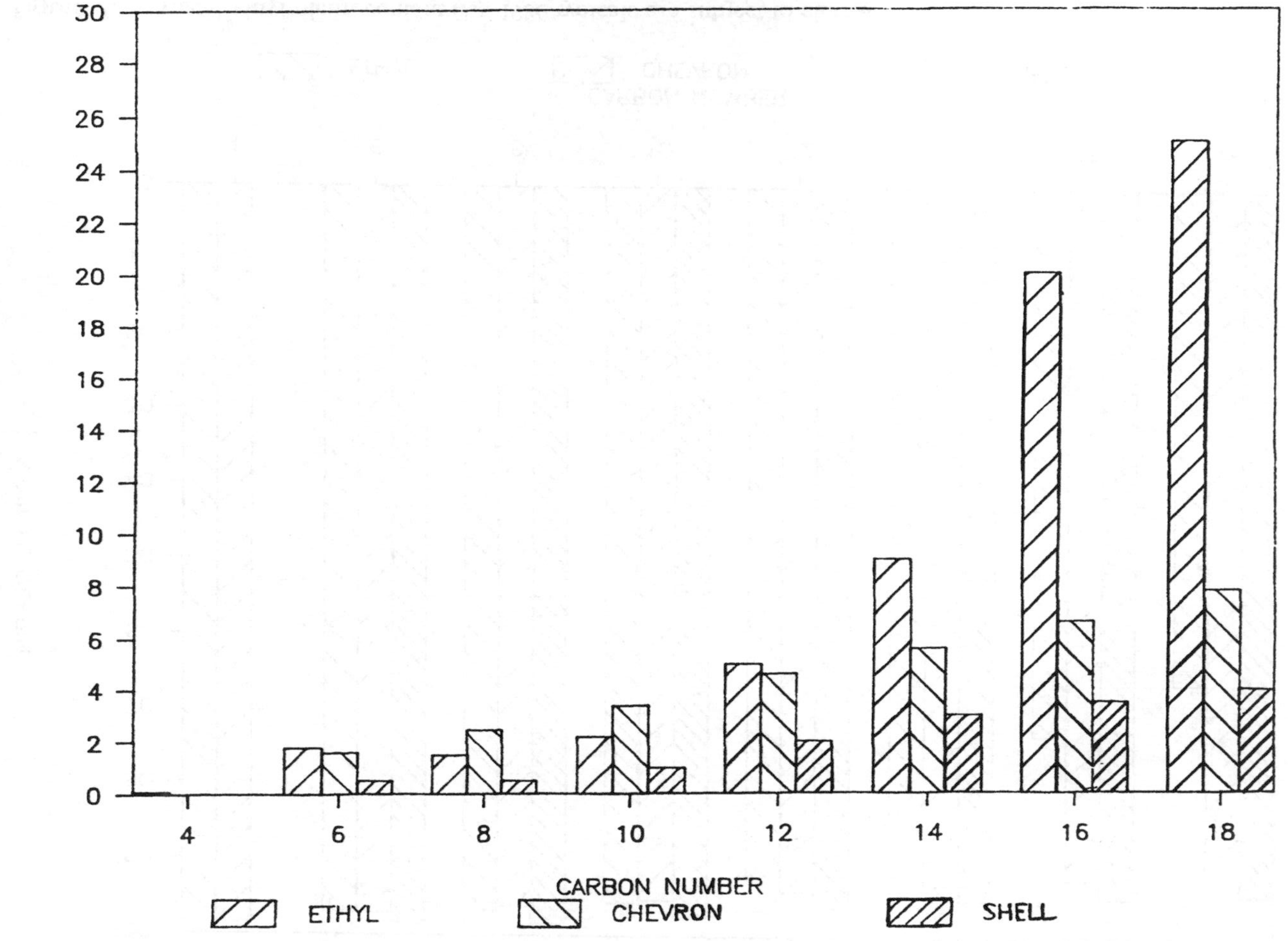

Figure 8. Typical vinylidene olefin content (%). 1986 typicals are subject to change.

Poisson curve. Under operating conditions, the actual carbon number distribution from "stoichiometric" chain growth may be like that from a combination of 10% catalytic and 90% theoretical stoichiometric processes.

3.2 Internal Olefins

Linear internal olefins in alpha olefin processes are formed by isomerization of linear alpha olefins. In these isomerizations, a catalyst usually causes the double bond to migrate toward the center of the molecule, with the degree of movement being a function of the catalyst, time, and temperature. An example of an internal olefin is 2-dodecene, which can be either cis (hydrogens on same side) or trans (hydrogens on opposite sides):

```
CH3(CH2)8       CH3
         \     /
          C=C                cis-2-dodecene
         /     \
        H       H


CH3(CH2)8       H
         \     /
          C=C                trans-2-dodecene
         /     \
        H       CH3
```

Products from the Shell and organoaluminum processes contain small amounts of internal olefins because conditions exist to cause the alpha olefins to be isomerized to internal olefins (Fig. 3.9). The nickel catalyst may cause this in the Shell process. In the organoaluminum processes some isomerization results from reverse addition of metal hydride to an alpha olefin followed by elimination of the 2-olefin.

3.3 Paraffins

Olefins from the Chevron/Gulf process contain about 1.4% paraffins which are formed during the hydrolysis step, in which the catalyst is deac-

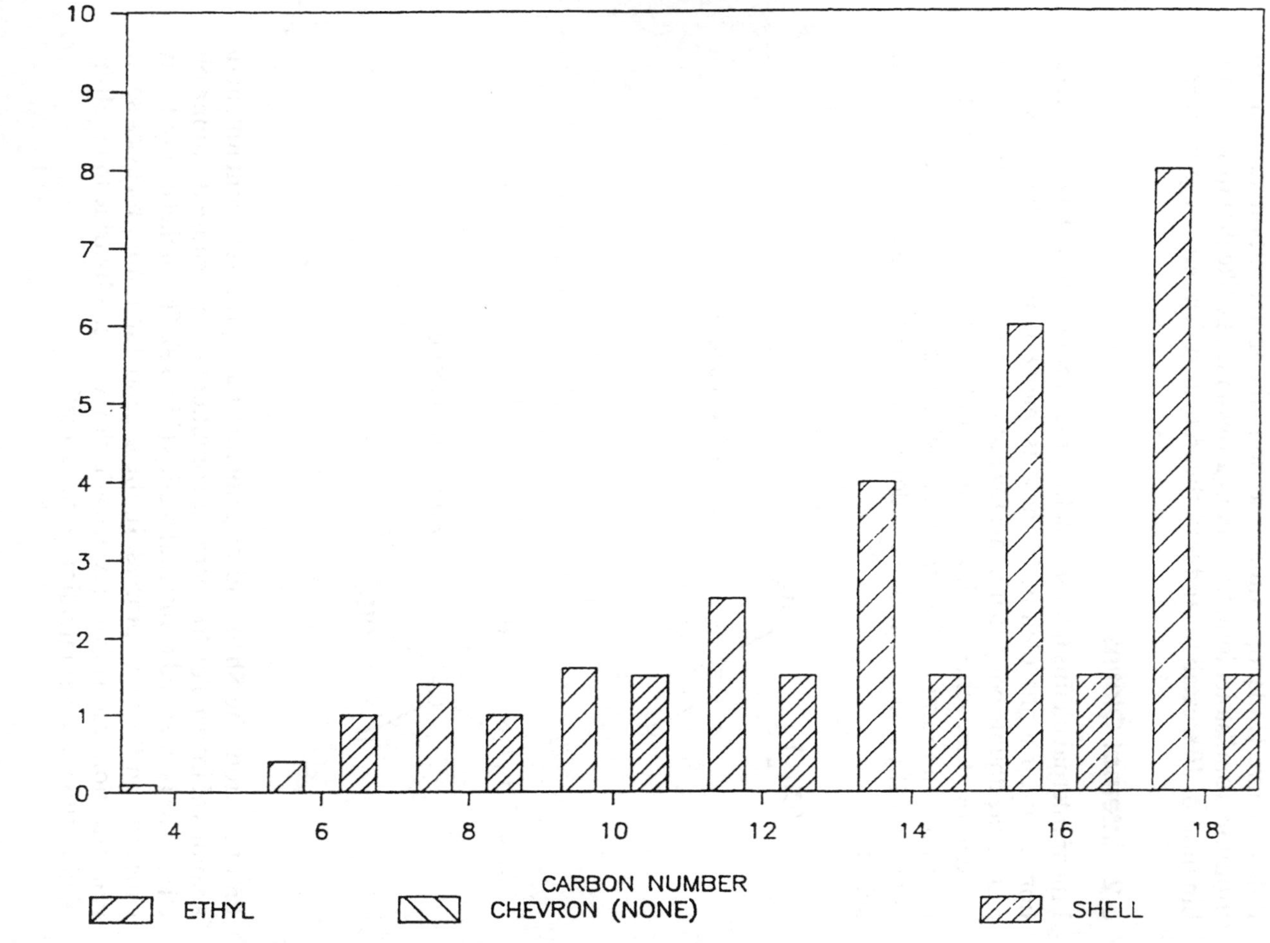

Figure 9. Typical internal olefin content (%). 1986 typicals are subject to change.

tivated at the end of the reaction. Ethyl's olefins contain up to 1% paraffins in the higher carbon numbers, with the paraffins resulting from one of the many steps in the process (Fig. 10).

3.4 Other Impurities

Because very high purity, polymer-grade ethylene is used, the alpha olefins have few impurities, such as sulfur, iron, and other metals that are sometimes found in hydrocarbons produced from petroleum fractions. Most of the impurities found in alpha olefins, such as peroxides and carbonyl compounds, are the result of contamination with oxygen in the atmosphere during handling. Avoidance of these impurities is discussed further in Chapter 18. All three U.S. suppliers have occasionally relied on toll processors to produce specific fractions, and these operations must be carefully carried out to avoid contamination.

4 OLIGOMERIZATION/CHAIN GROWTH EXAMPLE

The buildup of ethylene into long-chain alkylaluminum groups is similar to ethylene polymerization and ethoxylation of alcohol reactions, as the distribution of alkyl group chain lengths is based on the random chance of the various molecules meeting. In this example we define AlC_2 as being the ethyl derivative of one-third of an aluminum atom. If we start with 100 molecules of AlC_2 (e.g., as triethylaluminum), the first reaction of an ethylene molecule with an AlC_2 molecule gives an AlC_4 molecule. There are now 99 AlC_2 molecules and one AlC_4 molecule, so that the next reaction of an ethylene molecule will probably be with another AlC_2. As the population of AlC_4 molecules increases, the odds increase that an AlC_4 will react with an ethylene molecule to form an AlC_6. The changes in population (on a molal basis) will occur about as follows in stoichiometric chain growth:

Time	0	1	2	3	4	5
C_2	100	30	14	4	1	0
C_4	0	38	27	14	7	3
C_6	0	20	28	23	14	8
C_8	0	6	10	23	20	14
C_{10}	0	1	9	18	20	18

This reaction behavior is similar to that of polyethylene formation chemistry except that the number of moles of ethylene linked together are fewer in alpha olefin products and the mechanisms are different.

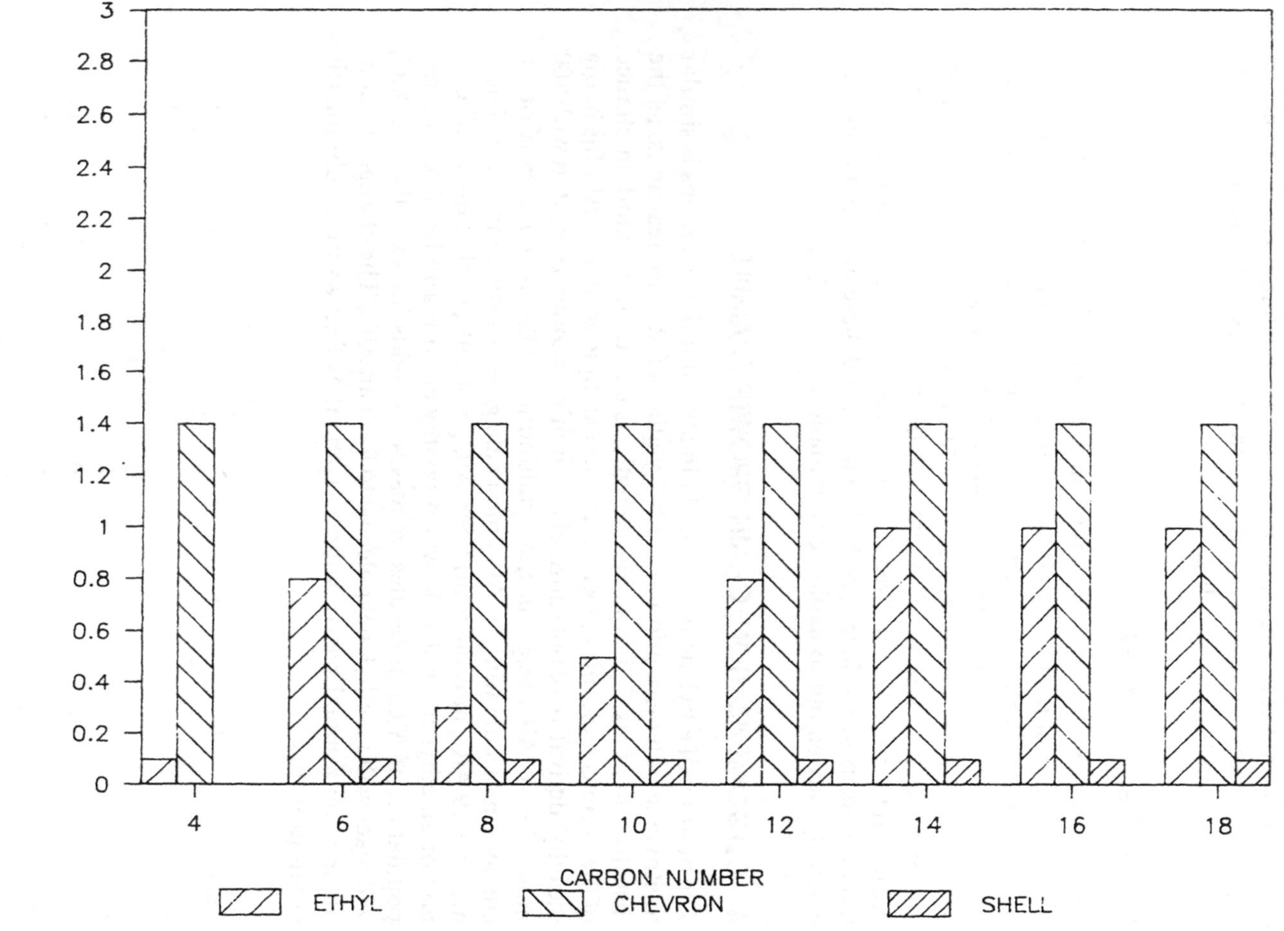

Figure 10. Typical paraffin content (%). 1986 typicals are subject to change.

Growth of inordinately long chains on the aluminum atom may contribute to pluggage of the alpha olefin process equipment. The polymer formation problem is one of the key differences between theory and practice. Once pluggage occurs, it is difficult to clear the equipment since pyrophoric aluminum alkyls may be found behind the pluggage. Over the more than 20 years of practice of Ziegler ethylene chain growth chemistry, techniques have been developed to reduce the severity of this operating problem.

The final product distribution is controlled by limiting the time and temperature of the reaction in batch processes and by controlling the temperature and throughput rate in continuous reactors (all of the commercial reactors are believed to be continuous). Again, the product distribution is influenced by side reactions, including the reaction of the olefins with the aluminum alkyls to form branched alkyls which are eliminated as vinylidene olefins.

By using a nickel catalyst and by having most of the olefin product in a separate phase, vinylidene formation is minimized in the Shell process. Some isomerization probably occurs in the reactor due to the nickel catalyst.

5 PROCESSES BY THE VARIOUS PRODUCERS

5.1 Gulf/Chevron/Mitsubishi

Chevron shut down a wax cracking plant in 1984 after they acquired Gulf with its ethylene-based alpha olefin plant. As stated in the introduction, the original Gulf plant is being referred to as the Chevron plant or the Chevron/Gulf plant, and the process originally developed by Gulf is being called either the Gulf process or the Chevron/Gulf process. References prior to 1984 show Chevron with a wax-cracking olefin plant in Richmond, California, and Gulf with an ethylene-based plant at Cedar Bayou, Texas. Originally placed in operation in 1966 with a capacity of 120 million pounds per year, the Chevron/Gulf facility at Cedar Bayou has been expanded several times to reach 250 million pounds per year in late 1986. The Chevron/Gulf capitalization is estimated at \$50 million. Gulf deserves much credit for pioneering the development of high-quality alpha olefins based on ethylene. Chevron announced plans in late 1986 to add 200 million pounds per year of capacity by early 1990. In their third quarter 1987 report to stockholders, they said that the investment will be \$87.5 million and the project will be expandable to 750 million pounds a year. They estimated that the project would increase their capacity to 450 million pounds annually, and projected that the market would grow at an annual rate of 4.5 to 7% into the 1990s.

Table 1 Annual Volume (Million Pounds per Year)

Carbon number	Degree of growth	
	Heavy	Light
4	33.4	54.9
6	35.1	49.4
8	32.8	39.5
10	28.7	29.6
12	24.1	21.3
14	19.7	14.9
16	15.7	10.2
18	12.4	6.9
20–28	30.3	11.8
30–40	5.5	0.9
Total	240.0	240.0

As stated above, Chevron/Gulf and Mitsubishi chain grow ethylene in a one-step system using catalytic amounts of aluminum alkyls. The operating conditions in the Chevron/Gulf reactors are thought to be in the vicinity of 400°F (200°C) and between 200 and 400 atm. Two possible distributions that might result are shown in Table 1 for heavy growth (at a higher Q) and for lighter growth (at a lower Q). Again this distribution is relatively flat.

Mitsubishi in Japan licensed Chevron's process and produces a similar distribution. Mitsubishi's total production is about 66 million pounds per year, or 25% of Chevron's capacity. Entry by Mitsubishi did much to develop alpha olefin markets in Japan.

5.2 Ethyl

Ethyl has acknowledged 950 million pounds per year of C_4-C_{26} capacity for alpha olefins and about 250 million pounds per year of primary alcohols capacity. Ethyl's alpha olefin process is often described as a two-step process. Ethyl's distribution has been described as follows:

	Capacity	
	%	MM lb/yr
C_4	13	124
C_6	55	522
C_{12}–C_{18}	30	285
C_{20}+	2	19
	100	950

This distribution is more peaked toward the C_6–C_{10} region. Additional distributions are given in the figures.

Huang [3] shows many different carbon number distributions. Huang states that "the process operated by Ethyl is basically a two-step process." In the first step, "long chain trialkylaluminum is grown at a low temperature to a desired average length and then sent to a high temperature displacement reactor," and in the second step an excess of ethylene is added to displace the alkyl groups. The second step was originally developed by Zosel in Ziegler's laboratory. How Ethyl performs the difficult separation of triethylaluminum from dodecene has not been reported.

Ethyl's olefin plant is interconnected with its alcohol plant and other sections of its Pasadena, Texas, complex (see Fig. 11). Ethyl reports many improvements on their original process. With the close integration of the two complexes, both olefins and alcohols may be peaked in the same or similar equipment. The complex includes recycle operations, many chain growth reactors, and various transalkylation (olefin displacement) reactors. These many steps allow Ethyl to tailor its alpha olefin and primary alcohol carbon number distributions to meet the changing demands of the marketplace. Optimization is effected through extensive computer models. Ethyl's entry into the market in 1971 did much to enhance the growth of markets in the United States, Europe, and other regions of the world, as Ethyl provided a substantial source of supply for alpha olefins.

5.3 USSR

A plant based on Ethyl technology is being constructed in the USSR. The plant capacity is reportedly in the range of 425 million pounds per year. Startup will probably be in about 1990.

Figure 11. Alpha olefin plant with chain growth reactors in the foreground and distillation columns in the background.

5.4 Shell

Shell has ethylene-based plants at Geismar, Louisiana, and at Stanlow, United Kingdom. The original SHOP (Shell Higher Olefin Process) plant was placed in operation at Geismar in 1977. As appears to be common among new olefin processes, it had operational problems that limited capacity until modifications were completed in about 1980, and once the modifications were completed it had capacity in excess of its initial rating. The original Geismar plant was built mainly to supply oxo alcohol markets. It had an original capacity rating of 425 million pounds per year. With additional investment it was rated in mid-1986 at 650 million pounds per year of total C_4-C_{40} and higher olefins. The original SHOP design was based on 300 million pounds per year of olefins allocated for internal use in Shell oxo alcohols and 125 million pounds per year of C_{10} through C_{20} alpha olefins allocated for merchant olefin sales. Capability was later added at Geismar to produce C_6 and C_8 single-carbon cuts. The Stanlow plant was built mainly to supply Shell's internal requirements for C_{10}-C_{18} olefins used in producing linear alkylbenzene, alcohols, and lube oils. Like the original Geismar plant, it was designed for modest merchant alpha olefin sales in the C_{10}-C_{20} range. The Stanlow plant has a capacity of 375 million pounds per year and reportedly is being upgraded to 450 million pounds per year. Shell has invested an estimated $120 million at Geismar and a reported $200 million at Stanlow. In November 1986, Shell announced plans to build a plant of capacity 525 million pounds per year at a cost of $150 million to begin operation at Geismar in 1989. Shell reported their process in detail in a *Chemical Engineering Progress* article by Freitas and Gum in 1979 [9]. The article agrees well with patents issued to Shell.

Shell had operated two other processes to produce olefins for detergents, including three wax-cracked units in Europe with a total capacity of 660 million pounds per year and a chlorination-dehydrochlorination (CDC) of paraffin plant located at Geismar, Louisiana, with a capacity of 160 million pounds per year. They could have built more of these types of plants but chose in the early to mid-1970s to invest in the new SHOP technology based on ethylene. One report had them shutting down the last of the wax-cracked units in the spring of 1987, but a later report suggested that the last one at Pernis may still be operating with a capacity of less than 100 million pounds per year. They restarted the CDC unit when U.S. demands required more capacity. It is likely that Shell will operate only ethylene-based plants when the second Geismar SHOP plant begins operation in 1989.

Shell's SHOP has two basic steps:

1. *Oligomerization or catalytic chain growth.* As described above, this step involves chain buildup from ethylene using a proprietary nickel catalyst dissolved in a solvent that produces a catalytic carbon number distribution similar to Chevron's. Shell produces a higher average carbon number than do Chevron/Gulf or Ethyl because Shell supplies alpha and internal olefins in the C_{11}–C_{14} range to their detergent derivatives, including oxo alcohols and linear alkylbenzene (LAB). They also use a C_8–C_{10} stream for C_9–C_{11} oxo alcohols and acids and a C_{15}–C_{19} stream for lube oil additives. Shell can decrease the relative amount of heavier carbon numbers by choice of solvent. Efforts at modeling the end uses and other sources suggest that Shell runs at a Q of about 0.8 as compared to a Q of about 0.7 for Chevron.

   Butene from the Shell process contains significant quantities of butene-2 apparently because a large portion of the butene-1 remains in the nickel-containing phase, giving it time to isomerize. A lesser percentage of the hexene-1 apparently remains in the catalyst phase, as the hexene-1 contains less internal olefin than does the butene.

   According to the Freitas-Gum article, the reaction takes place in a series of time tanks. A three-phase mixture is circulated through the reactors, and the heat of reaction is removed by water-cooled heat exchangers between the reactors. As they are formed, the alpha olefins separate from the solvent-catalyst phase and enter the hydrocarbon phase. Thus degradation of the alpha olefins is avoided, a key reason for the high linearity of Shell's higher-carbon-number olefins. Gaseous ethylene is the third phase. The alpha olefin product is scrubbed with fresh solvent to remove the catalyst. Some residual amounts of catalyst may remain in the higher (C_{16}+) alpha olefins.

The reactor operates at 80 to 120°C and 1000 to 2000 psi (70 to 140 atm). Temperature and catalyst concentration control the rate of reaction and the product carbon number distribution. High pressure is required to dissolve sufficient ethylene to obtain a high linearity of the product.

According to Shell patents, the choice of solvent affects the distribution of olefins, as solvents such as 1,4-butanediol result in a decreasing rate of growth with carbon number. Some sources believe that Shell uses a solvent like 1,4-butanediol at Geismar and a solvent like benzene at Stanlow. In preparing the graphs of distributions, it was assumed that 1,4-butanediol is used at Geismar to effect a declining rate of growth with carbon number, and it was assumed that benezene is used at Stanlow to effect a constant rate of growth with carbon number.

2. *Isomerization-disproportionation (I/D).* The I/D portion of the process results in rearrangements of the carbon numbers; for example, disproportionation of a C_4 molecule with a C_{20} molecule could yield two C_{12}'s, a C_{10} and a C_{14}, a C_{11}, and a C_{13}, or any combination adding to 24 carbon atoms, depending on positions of the double bonds in the starting molecules. In this step, the alpha olefins are first isomerized over a catalyst to predominantly internal olefins and then are sent to the disproportionation reactor. The catalyst for the isomerization is believed by some sources to be potassium carbonate. The main purpose of this section is to produce detergent range olefins (i.e., C_{11}–C_{14}) for feed to oxo and linear alkylbenzene reactors. The I/D section absorbs excess alpha olefins.

Before isomerization it is necessary to adsorb impurities from the feed olefins. Two beds are used, with one being regenerated while the other is being used. Impurities can include materials used in the oligomerization section, including the nickel catalyst, the solvent, the water used to kill the catalyst, and other materials used in preparation of the catalyst, including hydrofuran.

The product olefins from the disproportionation reactor are odd and even carbon number, internal linear olefins from C_4 through C_{40}+. The detergent range (C_{11}–C_{14}) internal linear olefins are used in Shell's alcohol plants in the United States and United Kingdom and in Shell's linear alkylbenzene plants in Australia, Europe, and South Africa. Most or all of the remainder (C_4-C_9 and C_{15}-C_{40}+) is recycled to the disproportionation reactor, but some of the C_8-C_{10} may go to C_9-C_{11} oxo alcohols and acids and some of the C_{15}-C_{19} may go to lube oil additives. The recycle loops are purged to avoid extensive buildup of undesirable components. Since the desired carbon number range may represent only 10 to 15% of the I/D products, there is a very large recycle of the lighter and heavier olefinic fractions. The recycle may approach 6 to 10 times the product olefin make.

In the disproportionation reaction, 1 mol of C_6 plus 1 mol of C_{24} will give an average of C_{15}; 2 mol of C_6 and 1 mol of C_{24} will give an average of 12, which is the desired carbon number. Thus 12 lb of C_6 and 24 lb of C_4 will give an average of C_{12}. The key principle is that the moles in equal the moles out and the pounds in equal the pounds out, so the average carbon number does not change. Otherwise, the distribution of the effluent from disproportionation is statistical.

Shell reported several reasons for going to the ethylene-based route instead of using a paraffin- or wax-based route. These included improved costs, lower energy consumption, reduced environmental impact, increased flexibility, use of ethylene as a feedstock (instead of one that may be less readily available), and the opportunity to enter the alpha olefin markets.

Disproportionation plus the ability to shift the carbon number distribution of the growth reactor, and thus of their detergent products, result in great flexibility for Shell. The SHOP plants must be of relatively large size in order to be economical because they consist of a large number of pieces of equipment. Labor and utility requirements of the SHOP process are estimated to be the highest of the commercial processes.

Entry by Shell into the alpha olefin markets assisted in the development of alpha olefin markets worldwide.

5.5 Shell and Chevron Wax Cracking Processes

Alpha olefins were produced in wax cracking plants by Chevron in the United States from 1962 through 1984 and by Shell in Europe from 1941 through 1987. In 1984, Chevron shut down its plant of capacity 90 million pounds per year. Shell had a capacity of 660 million pounds per year in three plants in the 1960s and 1970s. Shell was reported to have shut down the last of these plants in 1987, but another report suggested that they are still operating one unit. The wax cracking process requires special types of highly linear hydrocarbon waxes which come from certain crude oils such as those found in Libya or Indonesia. The wax cracking processes produced many carbon numbers for which there was no demand, and these olefins were probably sent back to the refinery for fuel value. It is believed that the large excesses of coproducts were the key factor causing these plants to be replaced by ethylene-based alpha olefin plants. Other factors may have been the lower quality of the olefins, the relative small size of the individual units, the age of the units, and difficulties in obtaining feedstocks.

5.6 Shell Paraffin Process

Shell produced linear internal olefins from paraffins using chlorination-dehydrochlorination (CDC) at Geismar, Louisiana, starting in about 1967. They shut down the CDC unit in 1981 and restarted it in mid-1985. These C_{11}–C_{14} odd and even, linear internal olefins are similar to those produced in the isomerization/disproportionation section of their SHOP unit and may be used in their oxo alcohol and LAB plants. They are generally unsuitable for other alpha olefin derivatives. It is likely that this plant of capacity 160 million pounds per year will be shut down when the new SHOP plant based on ethylene is in operation in 1989.

5.7 Other Announced Plants

At various times alpha olefin plants reportedly have been considered for Australia, Canada, India, Japan, Mexico, New Zealand, Saudi Arabia, the United States, Yugoslavia, Czechoslovakia, and perhaps elsewhere, citing among other things low-cost ethylene and captive markets as incentives. To date, many more have been considered and even announced than have been built. Present-day alpha olefin plants are complex and capital intensive (Figs. 12 and 13). History has shown that there is good reason to purchase the technology from those who know how to make the units operate if they will sell it. To be economical, they must be built very large, and this means that only a few can be justified, as the total market is much smaller than those of the larger commodity-type petrochemicals. Access to multiple markets for the many different products appears to be more important than low-cost feedstocks. It is difficult to find plant sites with good combinatons of low-cost ethylene and sufficient markets to justify world-scale alpha olefin plants.

5.8 Comparison of Processes

The ethylene-based plants continue to be expanded, while paraffin- and wax-based plants are being shut down, indicating that the ethylene processes are superior in today's environment. However, no single ethylene-

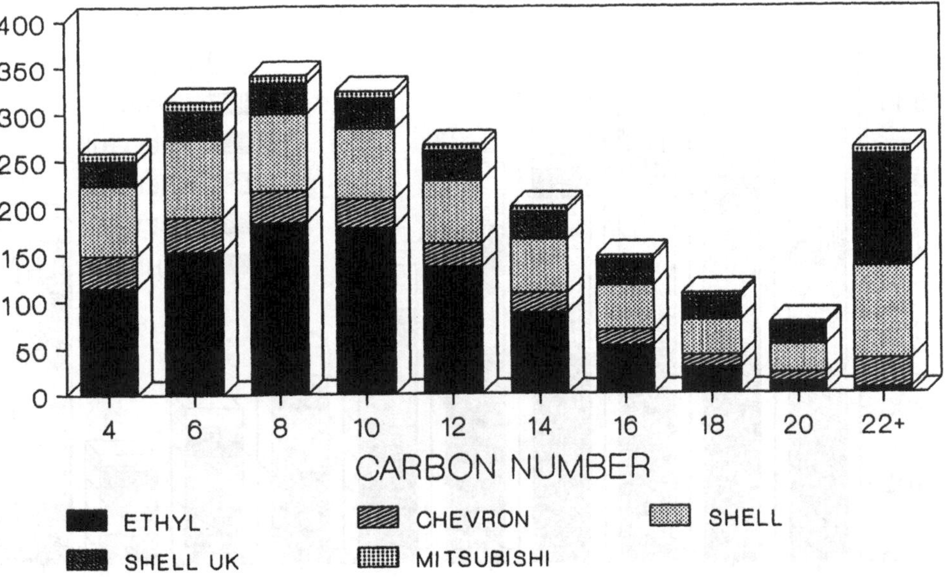

Figure 12. Estimated 1987 alpha olefin capacities (millions of pounds per year).

based process is obviously superior in all categories. Chevron's process appears to be the simplest and may have the lowest capital cost for a given volume. Shell's process has great flexibility in absorbing carbon numbers. Ethyl's allows increased production of specific carbon numbers while suppressing others. Size appears to have been a significant asset for Shell and Ethyl. All have hardware limitations that do not allow them as much flexibility as might be deduced from theory.

Shell's process gives a higher linear alpha olefin content at the higher carbon numbers, but its butene is of lower quality. The Chevron and Shell processes produce higher alpha olefin content at the higher carbon numbers (C_{14}+) than does the Ethyl process, and the Chevron and Shell higher alpha olefin content is preferred for some applications. The higher branching in the Ethyl C_{14}–C_{18} olefins results in their being preferred in several applications. For many applications, the difference in alpha olefin type is not important. Only Chevron distills and stores a linear olefin above C_{20}.

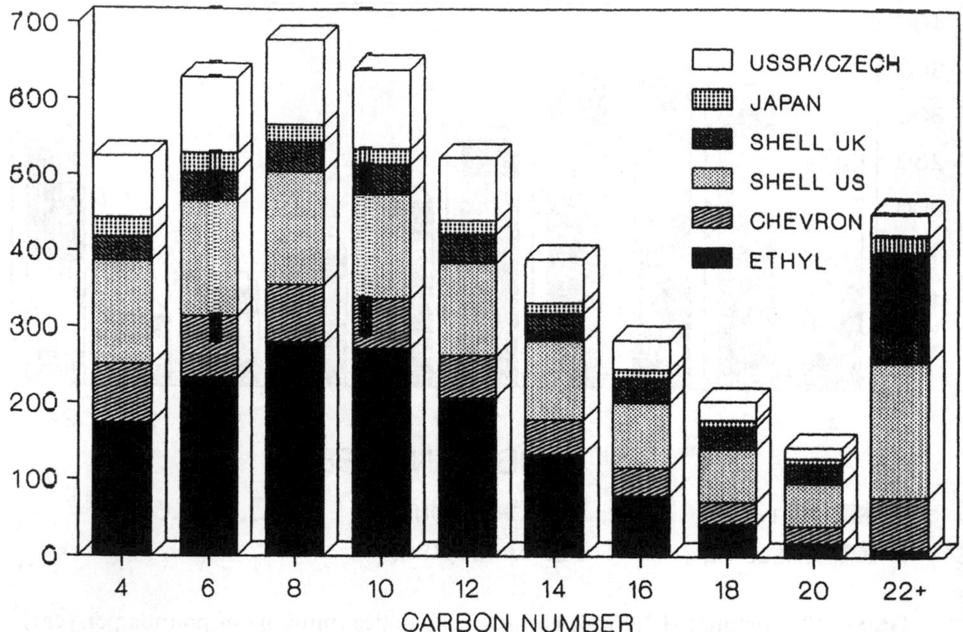

Figure 13. Estimated 1993 alpha olefin capacities (millions of pounds per year).

5.9 Other Routes

A number of other processes have been reported in the literature. In the United States, Exxon and Conoco (now Vista) piloted processes in the early to mid 1960s and Dow piloted a process in the late 1970s. These appear to be variations on the ethylene-based processes discussed above. It is believed that the current producers continue to study new processes or modifications to their present processes. Undoubtedly, other routes or processes will continue to be investigated.

6 MARKET MANAGEMENT

Access to markets for all the carbon numbers that are produced is important. Ethyl and Shell can recycle the excesses. Chevron utilizes large, long-term storage of excesses until the products can be marketed; this approach is less efficient than having direct outlets for the materials as they are produced. Like Chevron, Mitsubishi cannot balance carbon numbers, but in recent years Mitsubishi has produced only approximately half of the alpha olefins needed in Japan. All producers encourage development of new markets, particularly for those carbon numbers in excess at a given time.

An oil refinery is a good analogy to an alpha olefin plant. In general, a producer must manage markets for each carbon number and blend according to its availability and demand. The overriding consideration is that all markets sum to a profitable operation. Thus the price of a specific product may at times be higher or lower than the average of all carbon numbers. Long-term planning for an alpha olefin-consuming project should assume the average alpha olefin price rather than the current price of a specific product.

REFERENCES

General

1. Zietz, J.R., Jr., Robinson, G.C., and Lindsay, K.L., (Chap. 46.) in *Comprehensive Organometallic Chemistry* (G. Wilkinson, F.G.A. Stone, and E.W. Abel, eds.), Pergamon Press, Elmsford, New York, 1982.
2. Ziegler, K., et al., *Ann. Chem., 629:*53–89, 121–160, 172–198, 198–206 (1960).
3. Huang, C.-S.W., 8th Annual Science, Engineering, and Technology Seminars, Houston, May 24–25, 1986.
4. Albright, L.F., and Smith, C.S., *AICE J., 14:*325–330 (1968).
5. Demianiw, D.G., *Kirk-Othmer Encylopedia of Chemical Technology,* Vol. 16, third ed., p. 480–498 (1981).

Ethyl Process

6. Davis, W.T., and Gautreaux, M.F., Ethyl Corporation, U.S. patent 3,391,291 (July 2, 1968).
7. Lanier, C.W., U.S. patent 3,663,647 (May 16, 1972).

Chevron (Gulf) Process

8. Fernald, H.B., et al., Gulf Research and Development, U.S. patent 3,482,000 (Dec. 2, 1969).

Shell Process

9. Freitas, E.R., and Gum, C.R., *Chem. Eng. Prog., 73* (Jan. 1979).
10. Berger, A.J., U.S. patent 3,726,938 (Apr. 10, 1973).
11. Turner, A.H., *J. Am. Oil Chem. Soc., 60*(3): 623 (Mar. 1983).

CHAPTER 4
Polyethylene

GAYLON L. DIGHTON Dow Chemical U.S.A., Plaquemine, Louisiana

1 INTRODUCTION

The use of alpha olefins as comonomers with ethylene in the production of polyethlene in the mid-1980s is one of the largest applications for alpha olefins and is expected to be the fastest-growing market for alpha olefins through the early 1990s. Because of this relationship, in this chapter we include a glimpse of the history and development of the polyethylene family, including a brief description of the different processes for manufacturing "polyethylene," the utilization of various higher alpha olefins in polyethylene manufacture, the effect of the various alpha olefins on the polymer properties, and the relationship between the end-use applications and polymer properties.

It is intended as a broad overview of the polyethlene family and the relationship of alpha olefins to that family for those whose main expertise is outside the field of polyethylene. Most of the information presented here has been documented several times in papers, journals, and textbooks; therefore, specific individual references will be minimized or eliminated. Topical references at the end of the chapter direct those interested in more detail to some of the voluminous literature available on polyethylene.

2 THE POLYETHYLENE FAMILY

The three major divisions of the polyethylene family are (1) conventional or low-density polyethylene (LDPE), (2) linear or high-density polyethylene (HDPE), and (3) linear low-density polyethylene (LLDPE). The various types of polyethlene result from different manufacturing processes. LDPE is made by a high-pressure process in either an autoclave or tubular reactor. HDPE is manufactured by various low-pressure processes. Commercially, these processes are based mainly on Ziegler-Natta or Phillips catalyst technology. Although there are several other catalyst technologies available for producing HDPE, they are not significant commercial factors at this time. LLDPE is produced by various low-pressure processes that originally produced HDPE and that have been modified to have the capability to produce LLDPE. Some LDPE processes have also been adapted to produce LLDPE; however, these are not significant factors commercially.

2.1 Low-Density Polyethylene

LDPE is by far the oldest member of the polyethylene family. In 1933, Imperial Chemical Industries, Ltd. (ICI) started a systematic study of the high-pressure chemistry of organic compounds. In March of that year, polyethylene was discovered as a trace of white powder in a reactor vessel. Due to the high pressures and temperatures required, development of a commercial process for production of polyethylene was difficult. By 1937, ICI had a continuously running pilot plant in operation.

During World War II, the necessity for a suitable insulation material for radar installations stimulated the growth of LDPE. ICI provided all the information for production of LDPE to a U.S. government delegation. DuPont and Union Carbide became the first two producers of polyethylene in the United States.

The ICI process uses a stirred reactor operating at 15,000 to 30,000 psi and is commonly referred to as the autoclave process. Badische Anilin, Soda-Fabrik, A.G. (BASF), and others developed high-pressure polyethylene processes in which the reactor is a tube approximately 1 mile in length. A simplified process diagram for both LDPE processes is shown in Figure 1.

2.2 High-Density Polyethylene

HDPE came into existence approximately 20 years after the beginnings of LDPE. In the early 1950s, Karl Ziegler and Giulio Natta, working independently, were able to polymerize ethylene in laboratory glassware

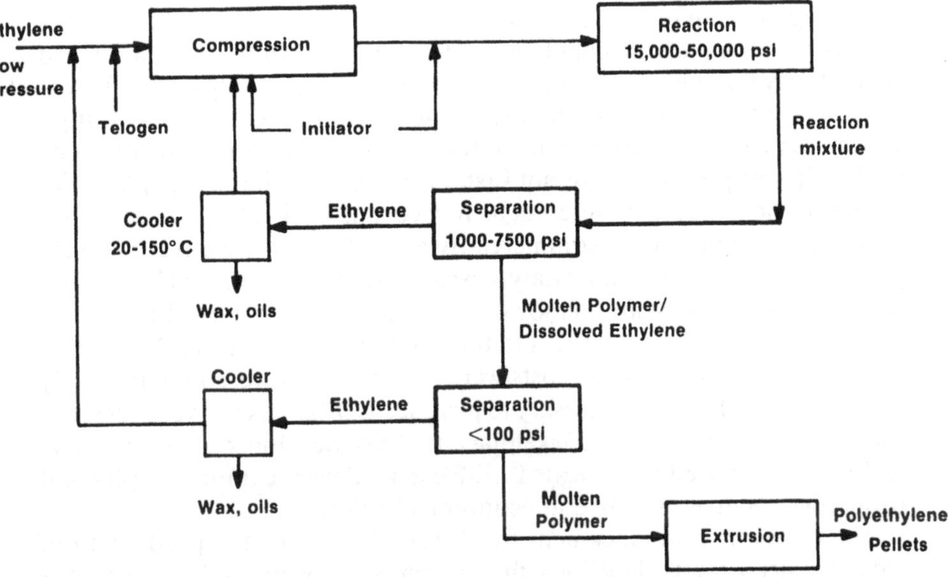

Figure 1. Typical high-pressure process for production of low-density polyethylene. (From U.S. patents.)

using aluminum alkyl-transition metal halide catalyst complexes. In addition to the Ziegler-Natta catalysts, other catalytic systems for low-pressure polymerization of ethylene were developed. Two of the most significant were developed by Phillips and Standard Oil of Indiana. Although all three of these catalyst systems were commercialized, the Phillips- and Ziegler-Natta-based processes are dominant in the industry.

2.3 Linear Low-Density Polyethylene

LLDPE was in the market development stages as early as the 1960s; however, the outstanding properties that contributed to the rapid growth of the product in the late 1970s and early 1980s were not the motives for the development of the LLDPE processes. The major driving forces were the desire to reduce the capital required to construct a high-pressure plant to produce LDPE, and to reduce the costs associated with the maintenance and operation of complex mechanical equipment at pressures in the range 15,000 to 40,000 psi.

Most of the processes developed by the various Ziegler-Natta and Phillips licensees for HDPE have the potential to produce some type of LLDPE. In 1958, DuPont had a Ziegler-Natta-based commercial plant operating in Canada capable of producing medium and low-density ethylene-1-alpha olefin copolymers. In the mid-1960s, Dow produced market development quantities of ethylene-1 C_6–C_7 alpha olefin copolymers via a Ziegler-Natta-based solution process. Various Phillip's processes produced a range of medium-density products during this same period. These products were not instant winners in the marketplace, for several reasons: (1) they were not as low cost as LDPE because alpha olefins were just becoming commercially available and were relatively expensive; (2) high-efficiency catalyst systems had not been developed, and therefore catalyst cost and removal both contributed to make LLDPE expensive compared to LDPE; (3) the Arab oil embargo and resulting dramatic increase in energy costs were still in the future, so the relatively large amount of energy necessary to produce LDPE was not yet a factor in relative costs; and (4) the marketplace had not developed the processing technology required to fabricate LLDPE to achieve the improved physical properties relative to LDPE at economical rates.

The dramatic announcement by Union Carbide of the production of "low-density polyethylene" via the gas-phase process and the development of high-efficiency catalyst in the Dow solution process radically changed the polyethylene world. LLDPE could be made by these two processes at costs that made the superior properties of the fabricated end products commercially attractive. LLDPE from Dow's solution process was recognized by *Industrial Research* magazine as one of the top industrial research developments for 1978. Since these events, several other manufacturers have also developed LLDPE processes (Table 1).

LLDPE Processes

As stated earlier, most HDPE processes, regardless of catalyst, have the potential to be modified to produce LLDPE, and some LDPE processes have produced LLDPE. Commercially, the most significant LLDPE processes have been developed from the solution and slurry HDPE processes and Union Carbide's gas-phase process. These processes will be described briefly.

Ziegler-Natta and Phillips Processes. Each of these processes is based on polymerizing ethylene in a liquid carrier using the different catalyst systems. Depending on whether the temperature at which the polymerization is carried out is below or above the polyethylene melting point, the process will be a slurry or solution process. Following polymerization in each process, the catalyst is "killed" and/or separated from the polymerization mass and the polymer then separated from the liquid carrier.

In the slurry processes, the polymer is separated as a powder or crumb by either one or a combination of centrifugation, filtration, or steam strip-

Table 1 Companies with LLDPE Processes[a]

Type process	Companies
Slurry	Phillips Solvay U.S.I. El Paso[b]
Solution	DuPont, Canada Dow DSM
Gas phase	Union Carbide BP (Naphthchimie)
Converted LDPE	CdF Chimie Dow[b] Arco[b]

Source: Based on B. H. Pickover, Linear low density polyethylene: an overview, material originally presented at Society of Plastics Engineers Regional Technical Conference, Akron, Ohio, 1982.
[a]Does not include licensees (i.e., Exxon, Mobil, Gulf, Northern, etc.).
[b]Not major commercial producers.

ping. The powder, thus separated, is usually dried and then melted, extruded, and pelletized to a bead or granule about ⅛ × ¼ in. In solution processes, the polymer is separated from the carrier as a molten mass by either one or a combination of distillation or flashing of the carrier, and devolatilizing extrusion, followed by pelletization to beads or granules. From the physical appearance of the finished product it is impossible to tell which process produced it. The major process steps for a solution or slurry LLDPE/HDPE plant are shown in Figure 2.

In both the Ziegler-Natta and Phillips processes, the development of more efficient catalyst systems in the late 1960s and early 1970s greatly simplified the processes by eliminating the need for catalyst removal from the polymer. High-efficiency catalyst technology was also the basis for the development of the gas-phase process.

Gas-Phase Processes. The advent of highly efficient catalysts for the polymerization of ethylene eliminated the need for catalyst removal in solution and slurry processes. This was the basis for the development of gas-phase processes for polymerizing ethylene. The advantage of a gas-phase process is the elimination of the liquid carrier, and the unit operations necessary to separate, recover, and purify it. The gas-phase processes that were developed were either stirred or fluid bed. The most

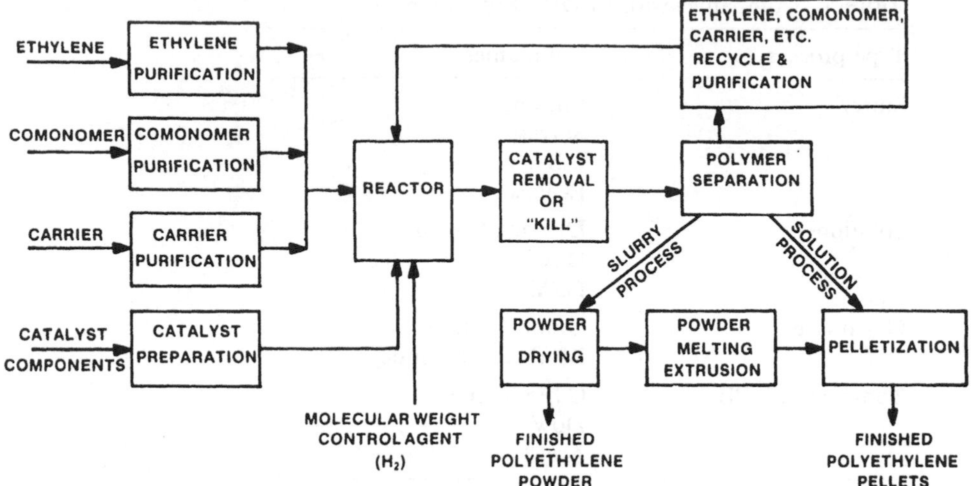

Figure 2. Typical low pressure slurry or solution process for production of polyethylene. (From U.S. patents.)

important commercially is the Unipol process developed by Union Carbide in the late 1960s (Fig. 3). In this process the hydrocarbon carrier for the reaction is the ethylene itself. Ethylene is circulated through a fluidized bed of polyethylene powder into which finely divided catalyst is introduced. From time to time, quantities of polyethylene powder, ethylene, and catalyst are removed from the reactor via a sequence of valves to a product discharge vessel. The ethylene is separated and recovered, and the polymer is obtained in dry powder form from the process. Because most fabricators prefer to handle beads or pellets instead of powder, the powder is then usually melted, extruded, and pelletized.

3 CHEMISTRY OF POLYETHYLENE

To explain the relationship of LDPE, HDPE, and LLDPE to the alpha olefin market, a few concepts of polymer chemistry will be touched on briefly.

3.1 Molecular Weight and Molecular Weight Distribution

When one manufactures a chemical such as caustic, it always has the same molecular structure. Concentration may vary, and the quantity of impurities increases or decreases, but the end product is, for all practical purposes, always the same, NaOH. Similarly, when one manufactures a

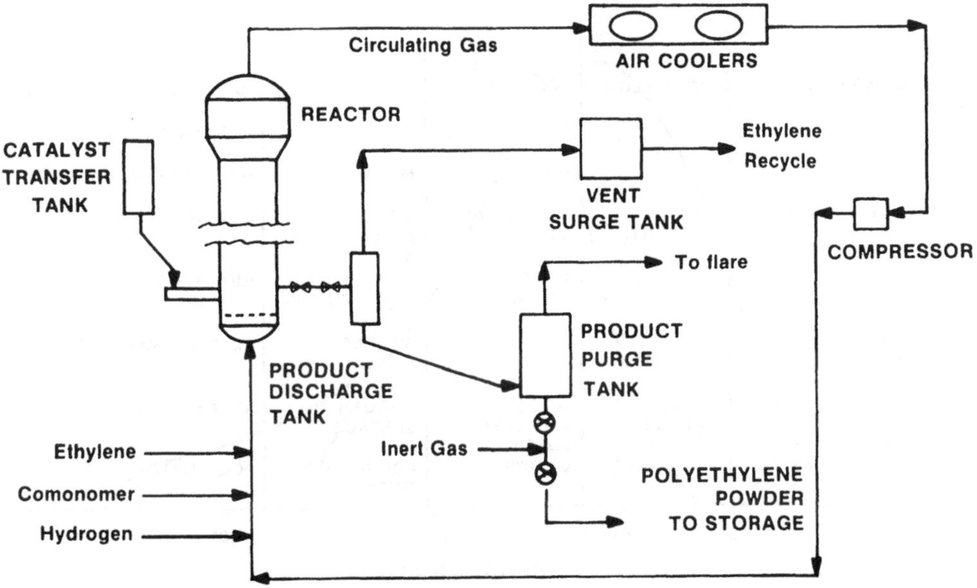

Figure 3. Union Carbide gas-phase process for production of polyethylene. (From U.S. patents.)

straight-chain alpha olefin, it always has the same number of carbon and hydrogen atoms in each molecule.

A commercial polyethylene is not composed of a single molecular weight species, but is a distribution of molecules having different molecular weights. Melt index or melt flow index is a criterion related to molecular weight used to categorize a commercial polyethylene. This is a measure of the viscosity of molten polyethylene and correlates roughly with the weight average of the different molecular weight molecules making up the polyethylene. The melt index is the grams of molten polyethylene that will flow through a 0.0825-in. orifice at 190°C under a pressure of 43.25 psi in 10 min. A higher melt index indicates a lower viscosity, hence lower average molecular weight for the molecules in the polyethylene.

Although melt index is generally thought of as representing some average molecular weight of polyethylene, it can be affected by the distribution of the individual molecular weight species in the polyethylene. The distribution of different-weight molecules also greatly affects the properties of the polyethylene. Figure 4 is a generalized molecular weight distribution curve. As seen from the curve, various molecular weight species contribute differently to polymer properties. For example, increasing the

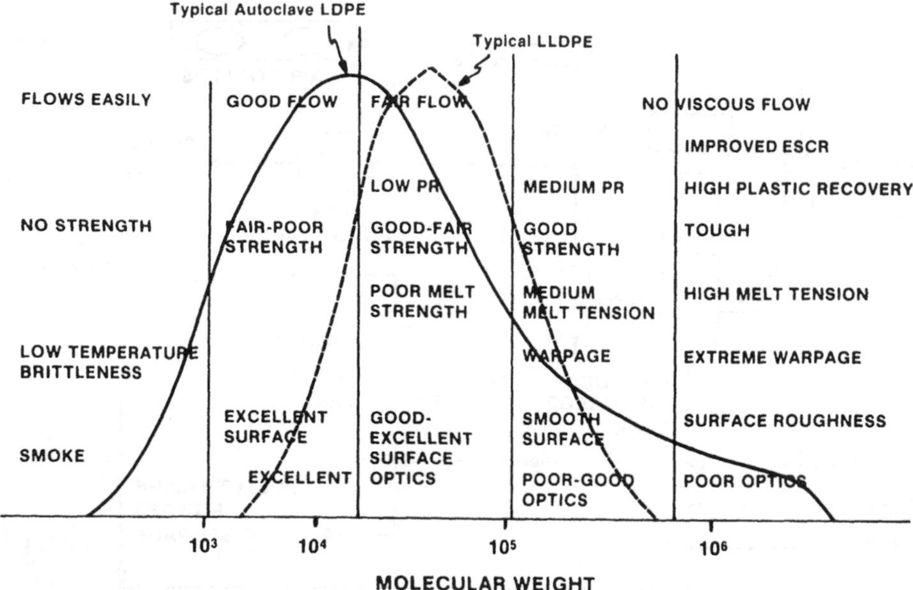

Figure 4. Effect of molecular weight on polymer properties. (From The Dow Chemical Company.)

amount of very high molecular weight molecules may improve environmental stress crack resistance; however, it may also cause extreme warpage, surface roughness, and poor optical properties. By increasing the number of molecules in the range of molecular weight from about 100,000 (usually written as 10^5) to 1,000,000 (i.e., 10^6), the polymer properties would have improved strength, less warpage, and the optical properties would be better than with more very high molecular weight material. By increasing the number of molecules in the range 10^4 to 10^5, processibility would improve, melt strength would be less, polymer strength properties would be decreased, film surface would become more glossy, and optical properties would become excellent.

Increasing the number of molecules with less than 10^3 molecular weight will greatly reduce strength. Low-temperature properties will be very poor, and the polymer will tend to smoke during fabrication at high temperature. Thus it is possible, based on the processing characteristics required and the end use of the polyethylene, to have an optimum molecular weight distribution for a given application. For example, for an overwrap film application where optical properties are most important, one would want to increase the amount of polymer in the range 10^4 to 10^5 and decrease the amount in the 10^6 range. However, for electrical cable jackets where environmental stress crack resistance is of greater significance, the amount of molecules in the 10^6 range should be increased. Therefore, the selection of an ideal molecular weight distribution for a given application will always be a compromise among the various properties desired.

The various manufacturing processes initially produced molecular weight distributions somewhat characteristic of the specific process. For example, the high-pressure LDPE autoclave tends to produce a broad molecular weight distribution with a high molecular weight tail compared to the high-pressure LDPE tubular process. In Ziegler-Natta- and Phillips-based processes, the solution process produces a very narrow molecular weight distribution compared to slurry processes using the same catalysts. The fluidized-bed gas-phase processes tend to produce a slightly broader molecular weight distribution than the solution process.

Because the polymer does not have to be handled in a molten state, the gas-phase and slurry processes can produce very high molecular weight powders. This is an advantage compared to the solution process; however, if the product must be supplied in pellet form, much of this advantage is lost.

Considerable research effort in each of the processes has been directed toward controlling molecular weight distribution. Through catalyst technology and reactor configuration, it is now possible to tailor the molecular weight distribution to the desired general shape in most low-pressure processes to a much greater extent than in the high-pressure processes. The low-pressure processes also have the capability to produce much more narrow molecular weight distributions than the high-pressure processes. Generally, the more narrow molecular weight distribution, the higher the strength-related properties of the fabricated article.

3.2 Chain Branching

An olefin having the same number of carbon and hydrogen atoms may have them arranged differently. For instance, the configuration about the double bond may be cis or trans, but that is the limit of different possible arrangements in a specific linear olefin. However, if one allows the olefin to be branched, the possible arrangement of carbon atoms becomes more complex. Similarly, in polyethylene, the introduction of branching and branch distribution, in addition to molecular weight distribution, increases the complexity of a commercial polyethlene by several orders of magnitude. To discuss in detail the effect of branching and branch distribution on the properties of polyethylene would require the rest of this book. However, since some branching is specifically related to the copolymerization of alpha olefins with ethylene, it is necessary to broach the subject.

Branching in polyethylene may be long-chain branching or short-chain branching. Either type individually, or both together, or no branching may be present in a given polyethylene.

Long-Chain Branching

Long-chain branching occurs during polymerization when a secondary active polymerization site occurs on the main polymer backbone and initiates polymerization of ethylene from that point. The resulting branch off the backbone may be from 20 to 30 carbons in length, to as many carbons in length as the main polymer backbone (i.e., thousands of carbons).

Long-chain branching affects the melt flow characteristics of a polyethylene. Long branches tend to become intertwined between molecules so that the molecules do not slip past one another as easily. Thus properties such as melt strength, melt elastic recovery, and die swell are related to the degree of long-chain branching in a polyethylene. As in molecular weight distribution, the optimal amount of long-chain branching is a compromise between processibility during fabrication and the properties of the finished article. For example, the increased melt strength of an LDPE from long branches makes fabrication of blown film easier than with LLDPE; however, the stresses in the film resulting when the long-chain branches are frozen, or crystallized, in a stretched configuration reduces the strength of LDPE film compared to LLDPE.

Short-Chain Branching

Short-chain branching occurs in two ways. In the high-pressure LDPE processes, side branches of less than five or six carbon atoms on the main polymer backbone or on long branches occur through a free-radical backbiting mechanism. A typical LDPE molecule may contain varying amounts of short branches, depending mainly on the reactor temperature and pressure at which the polymer was made. In all of the low-pressure processes, short-chain branching is obtained by the copolymerization of alpha olefins with ethylene.

Density of the polyethylene is a direct function of short-chain branching. Density may be illustrated by thinking of a bundle of straight, smooth sticks and a bundle of sticks which have had short pegs driven into them at intervals throughout their length. For the same dimension bundle, it is obvious that the bundle of straight, smooth sticks will contain more sticks, and weigh more. The density of a polyethylene affects such properties as melting point, stiffness, and barrier properties.

The question one might raise is: At what point does a short branch become a long branch? To answer that question, one must introduce the concept of chain folding and crystallinity. The study of polymer morphology and structure is a specialized field of polymer chemistry that we touch on briefly at this point.

As the length of a chain of carbon atoms grows longer, the potential of the chain folding back along itself becomes increasingly likely. It has been shown that the optimum fold length for polyethylene carbon atom chains is between 100 and 200Å. Thus the effect of the length of the short chains on the folding of the polyethylene chains into lamella (closely packed crystalline areas) like smooth sticks is affected by the length and number of short branches. Figure 5 is a very simplified diagram illustrating the different effects that propylene and octene have on the folding of the polyethylene chains. For example, the propylene branches may cause bulges and lumps in the folds, but the branches are not long enough to completely disrupt the folding mechanism. This effect may be shown graphically by plotting polymer density as a function of the number of short-chain branches.

Most branches in a polyethylene molecule terminate in a methyl group. The number of methyl groups for a series of ethylene-propylene and ethylene-1-octene copolymers of similar molecular weights and varying densities were determined. The data plotted in Figure 6 show the greater disruption of the chain folding mechanism by the octene comonomer than by propylene, confirming the concept shown in Figure 5. At some

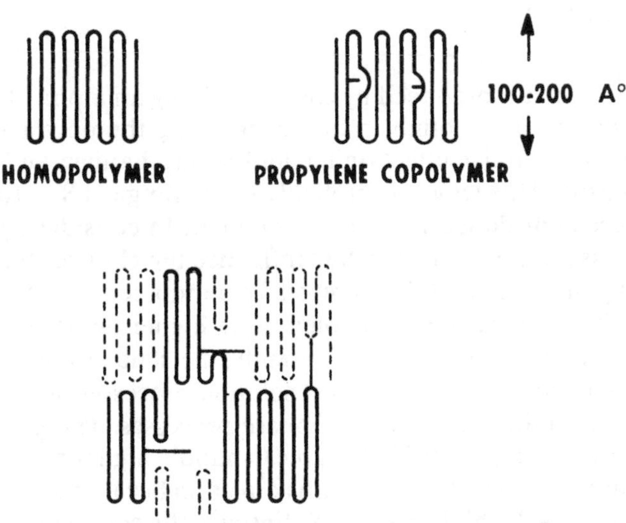

Figure 5. Relationship of short-chain branch length to main chain folding pattern. (From Dow Chemical Company.)

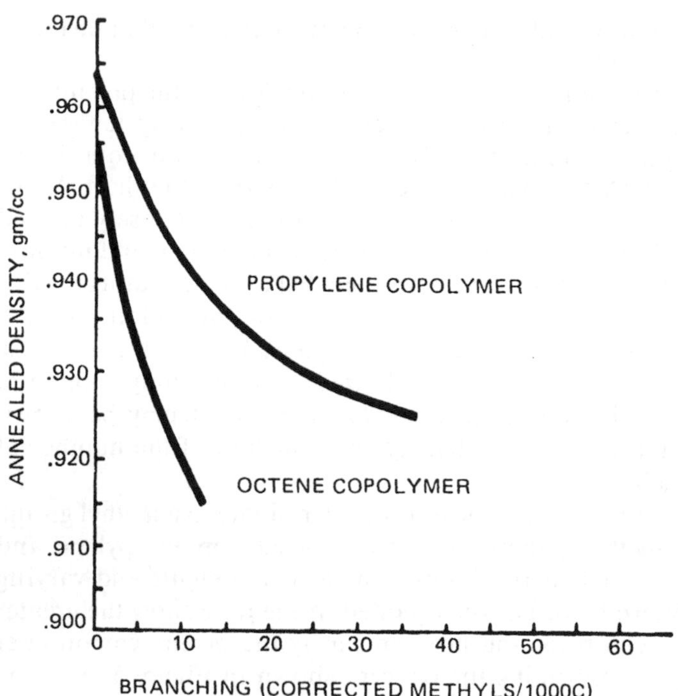

Figure 6. Relationship of short-chain branch length to polymer density. (From The Dow Chemical Company.)

point the length of a side chain produced by copolymerizing an alpha olefin with ethylene will be long enough to fold back along the polymer backbone, having a less disruptive effect on the folding mechanism and, therefore, polymer density. This length is probably in the range of 8 to 10 carbons, requiring decene or dodecene as a comonomer. In considering which comonomer to use, several other factors influence the choice. One is the relative difficulty in removal of the unreacted comonomer from the product and process. The longer the alpha olefin, the higher its boiling point and the more difficult it is to remove. Presently, the state of the art indicates that 1-octene is the optimum for most solution processes, while 1-butene is currently an optimum for the gas-phase processes. The gas-phase processes are progressing toward being able to handle higher alpha olefins and will probably continue to work in this direction because of the superior polymer properties the higher alpha olefins give the copolymers.

Beside the effect on polymer density caused by the length and number of short branches, there is also an effect on polymer rheology. Rheology is the study of flow, and there is much reference material available on polymer rheology. From a simplistic view, the longer the side branches are, the more they resist molecules slipping past each other freely. Going back to the illustration of smooth sticks and sticks with pegs along their length, it is intuitively obvious that it is easier to pull a smooth stick from a pile of sticks than one with many pegs. In addition, the longer the individual pegs, the more resistance is encountered. When branches are long enough to become entangled with nearby molecules, properties such as melt flow, melt strength, and plastic recovery are affected during processing, and properties such as "splittiness" (the tendency for the film to tear easily in the machine direction), tear strength, and puncture resistance are affected in finished products such as film. In some cases, the branches available from copolymerizing alpha olefins are not long enough to give the desired processing characteristics to the LLDPE and fabricators, especially in the area of blown film, blend conventional LDPE with LLDPE. The LDPE long-chain branches supply the increased melt strength and ease of processing, while the LLDPE provides increased toughness over LDPE alone.

3.3 Effect of Alpha Olefin on LLDPE

Regardless of the process by which an LLDPE is produced, the length of the short-chain branches will have essentially the same effect on overall polymer properties if the distribution of the branches is equivalent. Again, the optimal short branch length for a given application is usually a compromise. Longer-chain alpha olefins are usually more expensive, and in some processes, such as the gas phase, more difficult to utilize during polymerization. Thus LLDPE made using octene-1 is usually more expensive than LLDPE made using butene-1; however, the desirable physical properties of the octene-1 LLDPE may justify the increased costs for more critical applications.

4 POLYETHYLENE PROPERTIES AND END-USE APPLICATIONS

The wide variation in properties of the three different types of polyethylene allows them to be utilized in many different applications. Major end uses and recent trends in the U.S. market are shown in Table 2. These data show the concentration in various application areas of the different types of polyethylene.

Molding is dominated by HDPE. Film is largely LDPE, although LDPE volume is declining because of LLDPE growth in the market. Extrusion is mostly LDPE and HDPE, with HDPE showing a higher growth rate. Understanding why the different polyethylenes dominate the various

Table 2 U.S. Polyethylene Market Trends (Millions of Pounds)

	1982			1986			1987		
Market	HDPE	LDPE	LLDPE	HDPE	LDPE	LLDPE	HDPE	LDPE	LLDPE
Molding									
Blow	1714	38	4	2390	30	11	2690	22	14
Injection	1102	376	130	1550	340	250	1711	318	320
Roto	51	4	81	112	12	98	122	7	108
Total	2867	418	216	4052	382	359	4523	347	442
Extrusion									
Pipe and profile	537	33	55	682	72	61	770	110	66
Wire and cable	110	178	99	119	185	155	124	186	158
Coating	15	543		38	678		42	742	
Other	24	46		38	112		35	126	
Total	686	800	154	877	1047	216	971	1164	224
Film									
Total	288	3168	1041	496	2740	2075	586	2709	2415
Export	678	1045	99	920	795	156	915	882	110
Other[a]	506	680	66	670	760	266	830	894	302
Total	5025	6111	1575	7015	5724	3072	7825	5996	3493

Sources: U.S. International Trade Commission [22, 23] and SPI Committee on Resins Statistics as compiled by Ernst and Whinney [21].
[a]Includes material used for blending.

market segments requires some knowledge of the fabrication techniques and properties required in the finished article.

4.1 Blow Molding

Molding is the largest market for HDPE, and blow molding is the largest segment of the molding of HDPE. During the blow molding operation a molten parison (a relatively thick walled tube) is formed by forcing melted polyethylene through an annular orifice. A movable sectioned die then clamps around the parison, pinching the bottom end closed and squeezing the top around an inlet for pressurized air. Air is then blown into the parison, inflating it to fill the mold. Generally, the mold is cooled below its surface by circulating a cooling fluid. Thus when the molten polyethylene strikes the mold surface, it begins to freeze, forming a hollow thin-walled article shaped like the mold. When the polyethylene has solidified, the sectioned mold opens and the hollow article is ejected. Any flash at the seams where the mold sections come together is trimmed off and recycled.

A major market for blow-molded HDPE is small containers for the consumer market; for example, 1-gallon milk or bleach bottles, 1-quart oil "cans," and detergent bottles. The fabricator of blown containers is interested in making the container as thin as possible to minimize the amount of polyethylene used, and is also concerned with "top loading," the ability to stack full containers on top of each other. Therefore, one of the major properties desired in the polyethylene used is stiffness. As noted before, stiffness is a function of density; thus the majority of the blow molding market utilizes HDPE; although squeeze bottles and similar blow-molded items utilize LDPE or LLDPE for flexibility.

Homopolymer HDPE (i.e., polyethylene made from ethylene only) would be the material of choice based on stiffness. However, homopolymer HDPE is susceptible to environmental stress cracking in the presence of many liquids, such as household detergents and cooking oil. Because of this the blow-molded container market for HDPE can be broken into two main segments: detergent grade or stress crack resistant, and nondetergent grade. Detergent-grade HDPE is made by copolymerizing a small percentage (1 or 2%) of an alpha olefin with ethylene. Hexene is preferred in Phillips processes, while butene is preferred in the Unipol process.

Since copolymerizing the alpha olefin with ethylene lowers the polymer density and therefore the stiffness, the choice of a polyethylene for a given container has to be a compromise between wall thickness and stress crack resistance. For example, if the fabricator requires high resistance to stress cracking, as in detergent bottles, a lower-density HDPE must be used.

This, in turn, means that the walls of the container must be thicker to have the same rigidity as a homopolymer HDPE container. To keep costs down, the fabricator wants the thinnest-wall, stiffest container possible, with adequate environmental stress crack resistance.

Another important property in blow molding is melt strength. The polyethylene must have sufficient strength in the molten state to maintain a stable parison. If the melt strength is too low, the parison will stretch or sag and thin out in some sections, causing uneven wall thickness in the blow-molded article. Melt strength is a function of intanglement among polymer chains. For example, long-chain branching in LDPE increases melt strength. In HDPE, where no significant quantity of long-chain branching occurs, melt strength is obtained by using an HDPE with a large amount of very high molecular weight molecules. These molecules are long enough to become entangled with several other molecules, making it more difficult for individual molecules to slip past each other in the melt. Thus most HDPE used in blow molding has a melt index of less than 1. Because of this high melt viscosity requirement, polyethylene for blow molding is most easily produced in a slurry or gas-phase process.

Blow molding is not limited to small containers for the consumer market. Large-part blow molding has become a significant part of the market as fabricators learn to use HDPE of lower and lower melt index rating. Fifty-five-gallon HDPE blow-molded drums meeting DOT requirements are being manufactured. Blow-molded HDPE is also being used for vehicle fuel tanks. The trend toward more large blow-molded articles will probably continue as fabrication techniques improve. Choice of alpha olefin has little effect on the properties that the blow molder wants in polyethylene. Most of the HDPE blow-molding market will probably continue to be butene copolymers.

4.2 Injection Molding

About 70% of the polyethylene used in the injection molding market is HDPE. This is mainly heavy-walled items such as crates and tote bins. Toys and similar items overlap HDPE and LDPE. A significant portion of the remaining market is for thin-walled items such as housewares and lids. In this area LLDPE has been growing rapidly at the expense of LDPE.

In injection molding, the molten polyethylene under high pressure and considerably higher temperature than in blow molding is forced through a small nozzle at high shear rates into a cool mold. When the polyethylene has solidified, the mold is opened and the finished part ejected. Thus, in injection molding, the fabricator wants a polyethylene that flows easily during molding in addition to whatever properties are required in the finished article. For crates, the polyethylene must be stiff, resistant to impact, and have good low-temperature brittleness. For lids, flexibility, low warpage, and environmental stress-crack resistance are required.

In the United States, the HDPE portion of the injection molding market has grown at about 8% per year from 1982 to 1984. LLDPE has grown 36% per year during the same time. Some of this rapid growth has come at the expense of LDPE injection molding, which has declined slightly over the same period. This growth rate should not be expected to continue indefinitely when LLDPE substitution for LDPE is accomplished.

LLDPE polymers provide exceptional improvement in injection-molded articles when compared to LDPE. In injection molding, it is important that the melt viscosity of the polymer be low (high melt index) so that the polymer can rapidly fill the mold cavity and minimize cycle time. LLDPE is inherently tougher than LDPE and retains excellent physical properties at much higher melt indexes, compared to LDPE. This allows the injection molder to make a tougher part at the same production rate, or to down-gauge wall thickness and make an equivalent part strength-wise with less material. In addition to processibility and increased toughness, the narrow molecular weight distribution of the LLDPEs results in less warpage in the molded part. The lower melt viscosity and easier flow in the mold cavity contribute to a superior surface finish for LLDPE compared to LDPE.

The advantages of LLDPE over LDPE are most apparent in items with thin sections, such as thin-walled containers or lids. In lids, LLDPE has been shown to require 10 to 25% lower molding temperature, 10 to 50% shorter cycle times, and the lids can be 15 to 30% thinner. As a result, the fabricator will pay a premium for LLDPE. Within the LLDPE family, the type of alpha olefin used to produce the LLDPE will have some effect on properties of the finished article. Table 3 is a comparison of some commercially available injection-molding lid resins.

Regardless of comonomer, the LLDPEs have low-temperature brittle points well below conventional LDPE at a significantly higher melt index. Therefore, when molded into a container, such as a 1-gallon ice cream bucket, the LLDPE container will be much less likely to break when a housewife drops the cold container full of ice cream at the supermarket. The data also show the tremendous superiority of LLDPE over LDPE in resistance to stress cracking in the presence of vegetable oils. Although not as dramatic, the effect of comonomer in the LLDPE can also be seen. Resistance of the octene-1 LLDPE polymer is about three times as long as that of butene-1 LLDPE which is of similar melt index and density. This improved stress crack resistance of LLDPEs to vegetable oils is of significant value in lids and containers for products such as shortening, cooking oils, fatty food stuffs, and paints. Another difference seen from the data is

Table 3 Physical Properties of Thin-Walled Injection-Molding Polyethylenes

Property	ASTM test	LDPE	LLDPE alpha olefine	
			Butene	Octene
Melt index (g/10 min)	D-1238	35	50	50
Density (g/cm^3)	D-792	0.923	0.926	0.925
Tensile yield (psi)	D-638	1740	1600	1700
Flexural modulus 2% secant (psi)	D-790	34,000	50,000	47,000
Low-temperature brittle point (°C)	D-746	−70	<−76	<−76
Tear resistance (psi)	D-1004	13.0	17.5	20.0
Vicat softening point (°C)	D-1525	96.5	89.0	99.0
Vegetable oil stress crack, F_{50}	—	10 min	30 hr	100 hr

Source: The Dow Chemical Company.

in tear resistance. Although both LLDPEs are superior to LDPE, the octene-1 LLDPE is superior to the butene-1 LLDPE.

For the same density, LLDPE is stiffer than LDPE. The effect of different alpha olefin comonomers on the density stiffness relationship for LLDPE is insignificant. Therefore, if stiffness is a critical property, the butene-1 LLDPE may be favored over a higher olefin LLDPE because of cost; however, if resistance to stress cracking is of paramount importance, the choice will be an octene-1 or hexene-1 LLDPE.

To sum up the situation in injection molding of thin-walled parts and lids, the advantages of LLDPE are savings in cycle time, molding energy, and resin usage, plus superior physical properties in the finished part. LLDPE will continue to grow in market share in lids and thin-walled containers.

4.3 Rotational Molding

Rotational molding is another fabrication technique used to produce hollow articles. Although parts similar to some blow-molded items are made by rotational molding, articles with thin-walled sections, such as gallon bottles, are not. Rotational molding of polyethylene is used to produce large unsymmetrical hollow articles with relatively thick walled sections. For example, vehicle gas tanks may be rotationally molded in shapes to fit odd-shaped spaces in an automobile. Large children's toys, such as hobby horses and large recreational items, are also rotationally molded from polyethylene. Compared to the injection- and blow-molding markets, rotational molding is very small; however, the growth in the United States over the past five years is above 74% compared to around 52% for the total molding market in polyethylene.

In rotational molding a hollow mold is partially filled with powdered polyethylene. The mold is then heated while being rotated in multiple plains. In some cases, after a few minutes of heated rotation the mold is stopped and the excess polyethylene withdrawn. The heating and rotational cycle are started again to continue to fuse the polyethylene and allow the material remaining inside the mold to flow out and coat the mold uniformly.

The main advantages of the rotational molding process are the low cost of the mold in comparison to blow or injection molds and the ability to make very large items. The polyethylene used for rotational molding must be in a finely divided form; therefore, all LDPE must be ground for this application. LLDPE and HDPE from gas-phase processes and certain slurry processes have a potential economic advantage because those processes produce a finely divided product prior to pelletization, although most of these materials are currently pelletized and ground to obtain adequate stabilization and particle size.

For example, large articles for outdoor use must be protected from ultraviolet (UV) light by the addition of UV stabilizers. To get adequate protection, UV stabilizers are usually added by compounding the stabilizers with molten polyethylene. The stabilized product is then ground to powder for rotational molding. Stiffness is an important property in large rotational molded parts; therefore, the growth in this market is mostly HDPE.

4.4 Film

For many years the largest market for polyethylene has been film. In the film market the lion's share historically belonged to LDPE; however, LLDPE has been replacing existing LDPE volume and taking most of the growth in the new LDPE film applications. LDPE volume has actually declined since 1983. The HDPE film market has been only about 15% the size of LDPE; however, HDPE use is growing with improved fabrication techniques for HDPE films of very low melt index. The film business is divided by fabrication technique between blown and chill-roll cast film. Each of these fabrication methods has slightly different polymer requirements.

Blown Film

Blown film is the largest segment of the market. Because blown film is produced by stretching a bubble of molten polymer film around an entrapped quantity of air, polyethylenes used in this process must have superior melt strength. This melt strength must be balanced against the ability of the molten polymer to flow adequately, so that the film may be "drawn" or stretched to the required thickness. LDPE is the preferred material for ease of fabrication into blown film because of the melt strength associated with long-chain branching. HDPE was very difficult to fabricate into blown film because the low melt strength at the required melt conditions for rapid production would not allow the fabricator to easily maintain a stable air bubble. Early copolymers of ethylene-propylene and ethylene/butene-1 also lacked bubble stability during fabrication. The development of higher-alpha-olefin LLDPEs gave some improvement in processibility over ethylene-propylene copolymers and HDPE. In addition, the LLDPEs gave some significant improvements in the finished film properties. The HDPE and shorter short-chain branched copolymers produced films that had high tear strength across machine direction (CD), but the tear strength along the machine direction (MD) was very poor. Tear strength, as a function of alpha olefin comonomer, is plotted in Figure 7.

Another dramatic change in the property of the finished film related to the alpha olefin used to make the LLDPE is impact strength. As the short-chain branches become longer and can interact more with nearby molecules in the film, they increase the impact strength of the film. Impact strength is measured by dropping a weighted dart from a given height onto a taught film. For the consumer, improved strength means that the film is not as easily punctured by odd-shaped objects. For example, when a bag full of garbage is dropped, the bag will not break and the garbage spill out. The effect of various alpha olefins (length of short-chain branches) on the impact strength of films fabricated at two sets of conditions from LLDPEs with similar melt viscosity and density is shown in Figure 8. These data also show the tremendous effect that fabrication conditions can have on the properties of a film from the same LLDPE.

In addition to impact strength, a film should have adequate tensile strength so that the film will not break when a full bag is picked up by grabbing the top. The effect of increasing alpha olefin length on film tensile is not as dramatic as the effect on impact strength. The data in Figure 9 show the superiority of LLDPE to a LDPE of comparable melt index and density.

For garbage bags and heavy-duty industrial bags, the optical properties of the film are not important. However, for soft goods and food packaging, the consumer is interested in film clarity and gloss. A glossy or shiny pack-

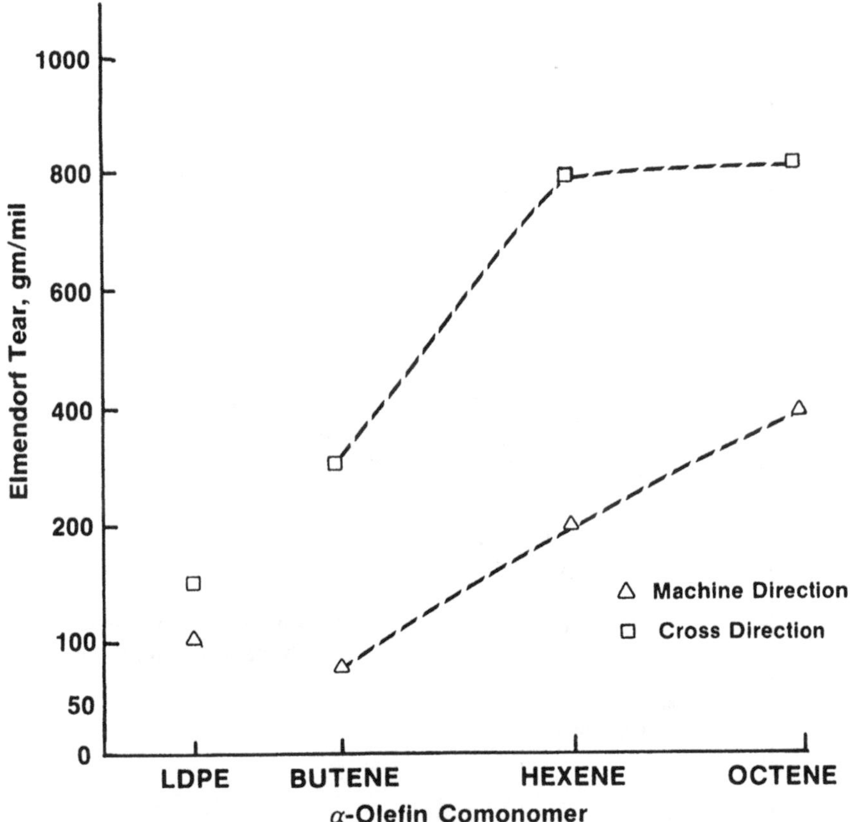

Figure 7. Relationship of short-chain branch length to polymer film tear strength. (From The Dow Chemical Company.)

age is more attractive on the shelf, and the consumer wants to be able to see what is inside the package. The effect of different alpha olefins on the optical properties of blown film are shown in Figure 10. Gloss (the shiny surface) improves from butene-1 to hexene-1 to octene-1. Haze (an inverse measure of see-through clarity) drops greatly from butene-1 to hexene-1 but does not change much from hexene-1 to octene-1. Data for a typical LDPE film of the same thickness indicate that the LLDPE optical properties are not quite as good. However, it is important to remember the difference in film strength and that the haze as seen by the ultimate customer would be reduced by using a thinner-gauge film. Also, many overwrap applications use cast film, which can be fabricated at conditions to optimize the optical properties of LLDPE.

Several other important film properties do not seem to be greatly influ-

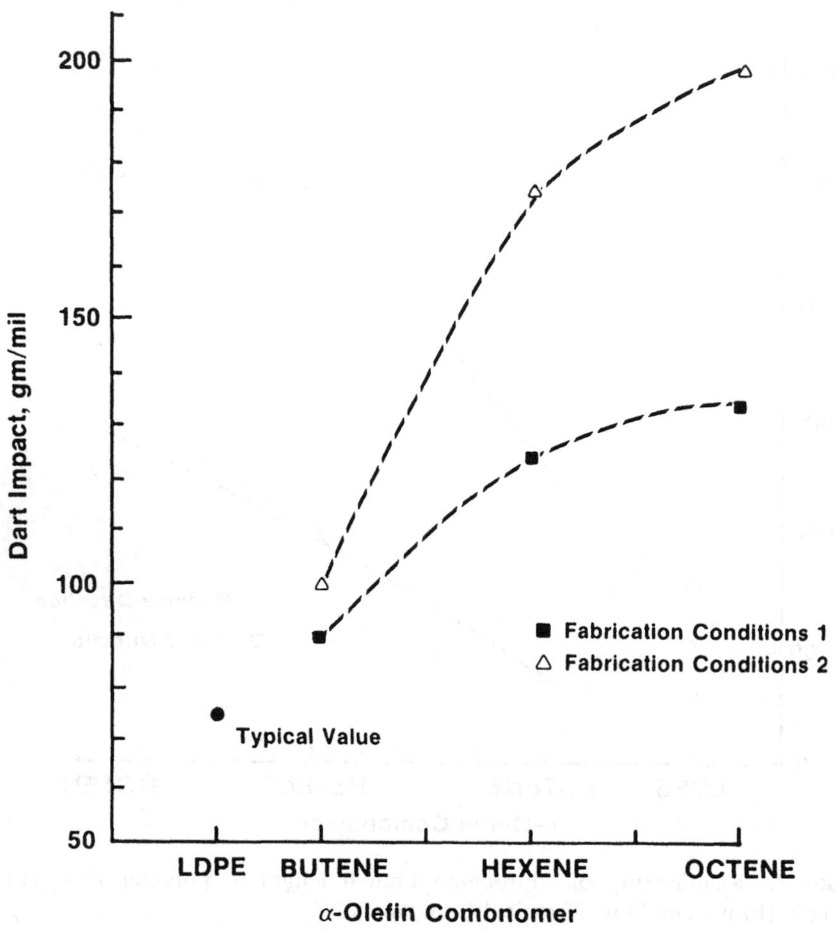

Figure 8. Relationship of short-chain branch length to polymer film impact strength. (From The Dow Chemical Company.)

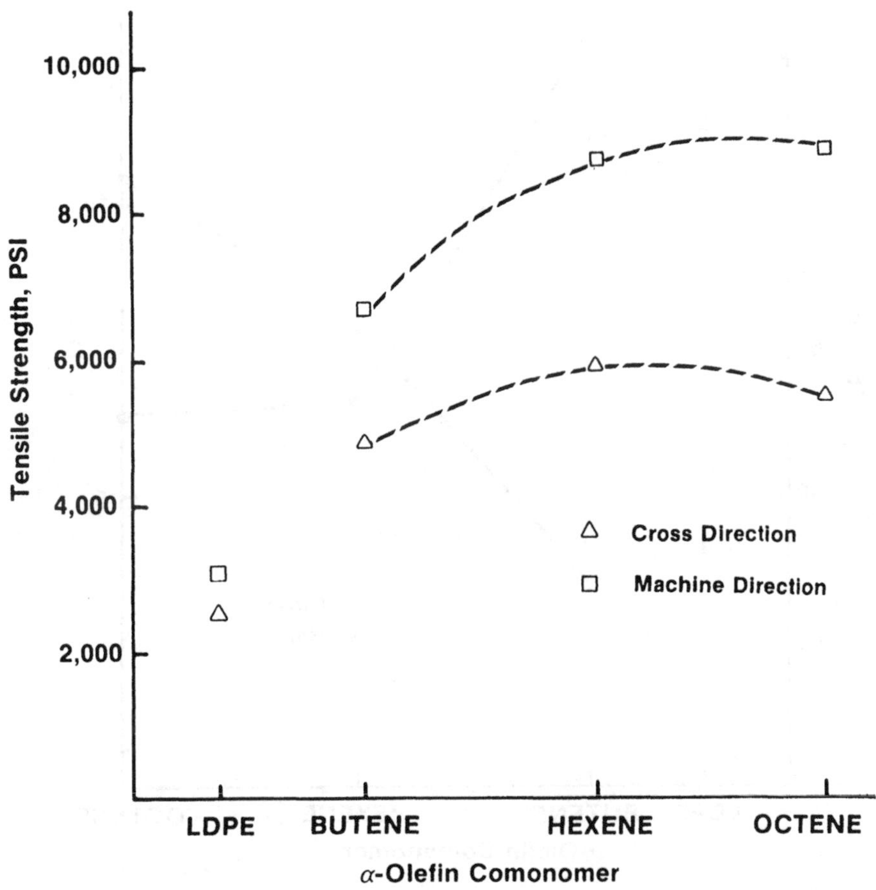

Figure 9. Relationship of short-chain branch length to polymer film tensile strength. (From The Dow Chemical Company.)

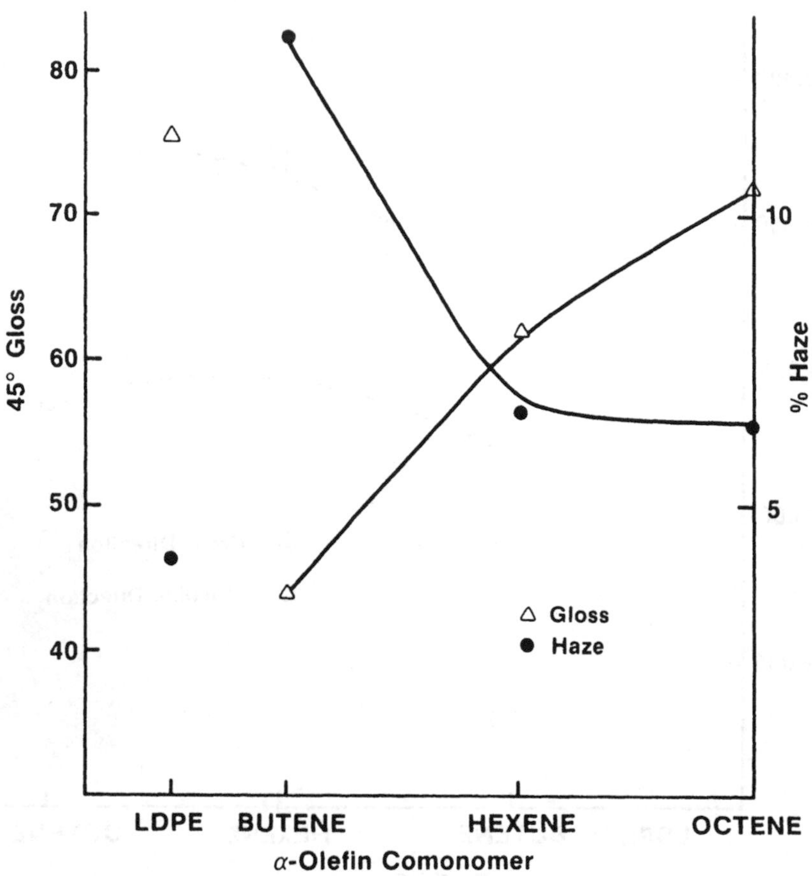

Figure 10. Relationship of short-chain branch length to optical properties of polymer films. (From The Dow Chemical Company.)

enced by the type of alpha olefin in the LLDPE. Film stiffness is a function of density and shows little difference among butene, hexene, and octene LLDPEs of the same density. Percent elongation, or the stretchiness of a polyethlene film, is related to molecular weight distribution and long-chain branching. There is little difference in elongation of LLDPE made with the various alpha olefins. However, there is a significant difference between LLDPE and an LDPE of similar melt viscosity and density. A market that takes advantages of the elongation properties of LLDPE is that of the stretch wrapping of pallets of various products for shipment.

Another specialized film market is the ice bag market. Film in this market must have good low-temperature properties. Historically, ordinary

Table 4 Comparison of Film Properties LLDPE and EVA

Propery	Octent-1 LLDPE		EVA	
Temperature	Ambient	0°C	Ambient	0°C
Thickness (mils)	1.75		2.0	
Tensile yield (lb/in.)				
MD	2.8	5.3	2.2	5.4
TD	3.0	6.5	2.2	5.3
Ultimate tensile, (lb/in.)				
MD	10.1	>11.8	6.9	10.4
TD	8.8	>10.3	6.5	11.3
Elongation (%)				
MD	700	>1219	480	751
TD	755	>1219	565	998
Impact (g)	235	475	375	425
Puncture resistance (lb)	16.2	—	9.9	—
Heat seal strength (lb/in.)	7.3	—	6.5	—

Source: The Dow Chemical Company.

LDPE was not adequate, and most of the market was satisfied by ethylene vinyl acetate (EVA) copolymers or similar products. The good low-temperature brittleness and strength properties of LLDPE allow the film fabricators to make LLDPE ice bags with superior properties to the EVA bags. Table 4 is a comparison of a 1.75-mil octene-1 LLDPE film with a 2.0-mil EVA ice bag film. Even though it is 12% thinner, the LLDPE is equivalent to the EVA copolymer.

Cast Film

Many of the comments about blown film also apply to cast film. However, since the molten film is extruded onto a highly polished chill roll, melt strength is not as important. Also, melt temperature can be higher. Increasing the melt temperature reduces polymer melt viscosity, which produces a smoother film surface with fewer defects. The smoother the film surface, the more glossy the film. In addition, internal optical properties of polyethylene films are improved the faster the films are cooled from the melt. Thus for a given polyethylene, it is usually possible to make a film with better optical properties on a chill-roll cast film unit than on a blown film unit. The effect of melt temperature on film gloss and haze for LLDPE is shown in Figure 11.

Increases in the amount of alpha olefin in the LLDPE improve the opti-

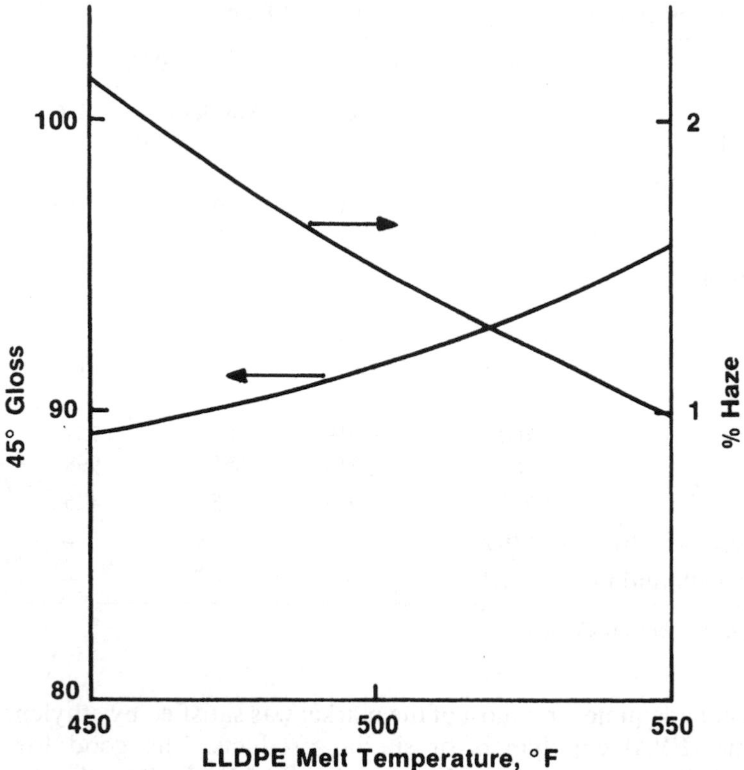

Figure 11. Relationship of film fabrication temperature to gloss and haze of LLDPE film. (From The Dow Chemical Company.)

cal properties of the film as shown in Figure 12. The typical LLDPE film resin is around 10 wt % alpha olefin.

Chill-roll cast film strength properties for LLDPE are clearly superior to LDPE for the same melt index and density. Therefore, the cast film fabricator can select a higher melt index LLDPE and improve output rate, compared to LDPE, with no loss in strength properties. Alternatively, the fabricator may down-gauge film thickness and still maintain strength, while improving see-through clarity because of the thinner film.

4.5 Extrusion and Other Processes

In addition to blown film and cast film, which are extrusion processes, there are several other extrusion processes that use significant quantities of polyethylene. The largest of these are electrical wire and cable jacketing, pipe and profile, and coating of various substrates.

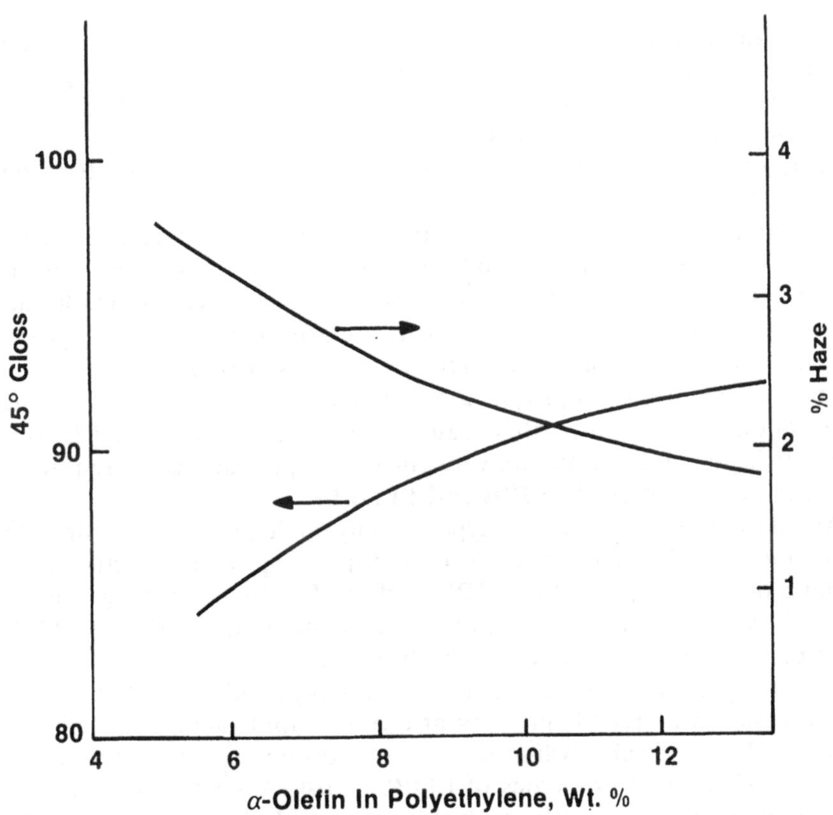

Figure 12. Relationship of octene-1 content to gloss and haze of LLDPE film. (From The Dow Chemical Company.)

In a typical extrusion process molten polyethylene is continuously forced through a die at relatively low shear rates (compared to injection molding), then cooled until frozen in the shape determined by the extrusion die. The finished product may be wound into rolls as with electrical wire flexible tubing and some pipe or cut into appropriate lengths for less flexible products.

LDPE consumption in extrusion applications has been growing faster percentagewise than either HDPE or LLDPE. In absolute pounds the growth in LDPE is greater than either HDPE or LLDPE individually and only slightly less than their combined pounds. In the United States, HDPE, LDPE, and LLDPE are concentrated in different segments of the extrusion markets (Table 2). For example, HDPE is largest in pipe and profile, where stiffness is an important factor. Because of superior melt processing characteristics LDPE is largest in substrate coating (i.e., coating of paper and cardboard to make items such as pouches and milk cartons). Wire and cable is a mix of all these types of polyethlene, with specific applications taking advantage of the different strong points of the three types of polyethylene.

For example, the outer jacket on power cable may be HDPE for resistance to abrasion. In colder climates LLDPE may be used because of superior low-temperature brittleness properties. Since wire and cable applications such as power and utility lines are expected to last many years, penetration of these markets by new materials is fairly slow until aging studies and regulatory approvals are completed. Partly as a result of this application history and partly because of easier processing, LDPE continues to be the major choice for wire and cable applications not requiring the specific properties of HDPE and LLDPE.

An additional use of all three types of polyethylene is in blending with each other or with other plastics to achieve specific properties. This area is a static or declining use for LDPE and HDPE but a growing use for LLDPE. Many film fabricators now routinely blend LLDPE with LDPE to improve strength properties of the finished film.

To sum up, the extrusion markets will probably continue to show mixed performance with HDPE growing at a more rapid rate than LDPE or LLDPE. Wire and cable will more or less follow new construction trends. Because of the melt properties of LDPE, substrate coating will remain largely the domain of LDPE, with some new LLDPE products entering the marketplace. LLDPE use as an additive to LDPE for improving finished film properties will probably increase.

5 POLYETHYLENE PRODUCTION CAPACITY

Polyethylene production capacity has grown from a laboratory experiment to today's 65 billion pounds in approximately 50 years. In the early stages there was only LDPE to consider. Later LDPE and HDPE were easily identifiable. However, LLDPE has blurred the picture considerably. Many plants that were originally designed to produce HDPE can manufacture LLDPE with minor modification, and some LDPE processes have been adapted to make LLDPE.

The distinction between processes is least clear between HDPE and LLDPE. Several plants and in some cases trains within a single plant are "swing" plants or trains. They can produce either HDPE or LLDPE, depending on the demand and profitability. Thus when one tries to determine the relative capacity available for the different polyethylenes, there is considerable variability in the data.

Probably the most reliable data for actual polyethylene use are the data from U.S. International Trade Commission, the Society of Plastics Industry, and *Modern Plastics* magazine. The historical data from these sources are in reasonable agreement, with only a 3.9% spread for 1984. The problem one has with these data in relating them to alpha olefin demand is that there is no breakdown between LDPE and LLDPE. This is a very significant factor. While the HDPE and LLDPE processes can and do overlap in ability to make either HDPE or LLDPE, the product overlap is largely between LDPE and LLDPE. Thus a manufacturer switching a train from HDPE to LLDPE will be reducing the availability of HDPE but increasing the availability of LDPE since LLDPE will probably displace LDPE in some end use.

The U.S. Industrial Trade Commission data split HDPE/LDPE at 0.940 g/cm^3 density. Therefore, some quantity of alpha olefin is also included in the HDPE since a homopolymer of ethylene alone will have a density in the range of 0.96 g/cm^3 or above. As discussed in the section on blow molding, manufacturers frequently copolymerize some quantity of alpha olefin (usually butene-1 or hexene-1) during the production of HDPE to improve stress crack resistance.

With these caveats in mind, one can still see the impact of LLDPE in the U.S. polyethylene market in Figure 13. From less than 500 million pounds in 1979, LLDPE production has grown to 3.5 billion pounds in 1987, representing 25% of the combined LLDPE and LDPE production. A similar trend is apparent in the global production capacity of LLDPE shown in Figure 14. While LLDPE demand and capacity have increased, LDPE demand has stabilized or decreased. Because of the product and process advantages of LLDPE, no new LDPE capacity is anticipated.

The main reason that LDPE volume has not declined further is because of the relative incremental costs. Most LDPE producers have older plants that are largely depreciated so that the cost to manufacture LDPE is basically raw materials and conversion cost. As these plants continue to age and maintenance costs increase, more of them will be shut down. Producers needing additional polyethylene capacity will choose LLDPE or, more likely, LLDPE/HDPE swing-type processes for their new plants. Construction of a new high-pressure LDPE plant will be a rare occurrence.

The historical growth trend for polyethylene will probably continue into the 1990s; however, the growth rate of the various types will differ within the total trend. HDPE will continue to grow at a rate above the total polyethylene growth rate, LDPE will probably show a small growth or decline, and LLDPE will pick up the growth that historically belonged to LDPE. These trends mean that more alpha olefins will be required in the polyethylene business in the 1990s.

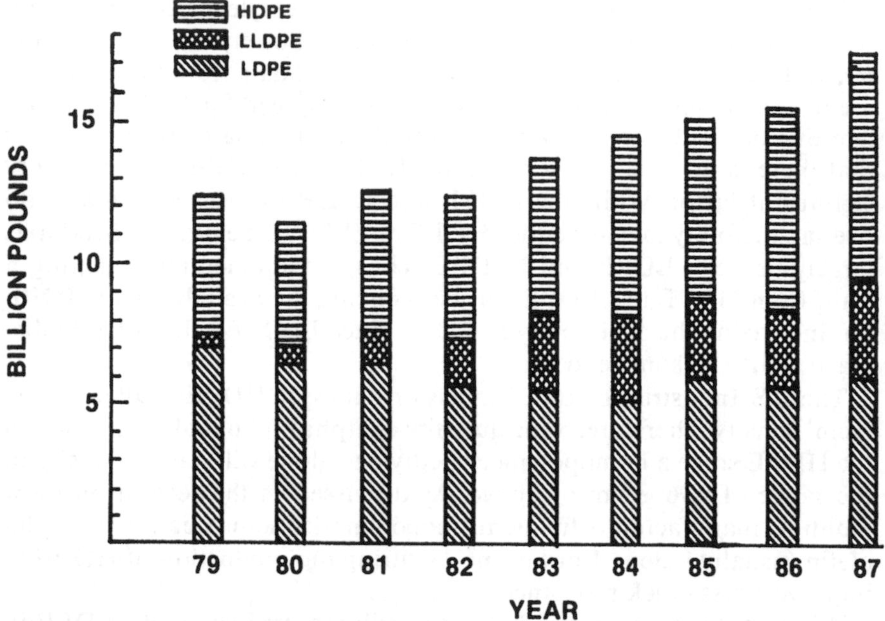

Figure 13. Trend of U.S. production of various types of polyethylene from 1979 to 1984. (From U.S. Trade Commission.)

6 ALPHA OLEFIN DEMAND

6.1 Past and Present

In many HDPE products the properties relative to short-chain branch length of from two to six carbons (butene-1 to octene-1) are not different enough to justify the use of the more expensive comonomers. Since by definition HDPE is 0.940 g/cm^3 and above in density, HDPE products have relatively little alpha olefin content. Butene-1 is a significant portion of this demand. However, because of process limitations or preferences and for critical applications in film and molding, a HDPE manufacturer may select alpha olefins other than butene-1 for copolymerization. The apparent increase in the use of hexene-1 in HDPE may reflect this. This complexity of processes and applications makes it impossible to say with certainty the individual alpha olefin requirements for HDPE. It is possible to estimate the total alpha olefin requirements reasonably well by assuming a 1% concentration of alpha olefin in HDPE as shown in Figure 15.

The alpha olefin requirements for LLDPE are currently divided among butene-1, hexene-1, and octene-1 with a small amount of 4-methyl-

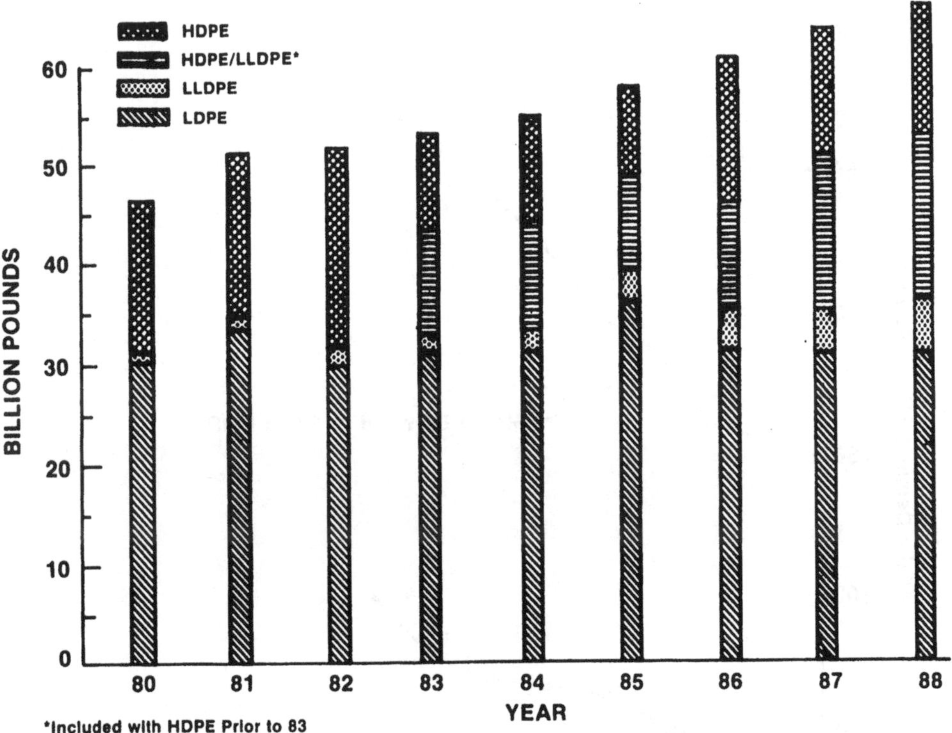

Figure 14. Trend of global production capacity for various types of polyethylene from 1980 to 1988. (From The Dow Chemical Company.)

pentene-1 as a new entry. The individual alpha olefin used in a specific instance depends on the end product requirements and process capability. Because of superior product properties, the trend is toward hexene-1 and octene-1. Assuming an average of 10% concentration in the polyethylene, the total alpha olefin demand for LLDPE is shown in Figure 16.

6.2 Future Possibilities

The growth of polyethylene has been the result of continuous development in both production techniques and product properties. As polyethylene producers have learned to control processes to tailor molecular weights and branch length, product properties have improved. Fabricators have developed techniques to take advantage of the new product properties. These trends will continue, and if current research efforts at controlling the distribution of branching prove successful, implementation of this technology in production plants may alter the need for specific

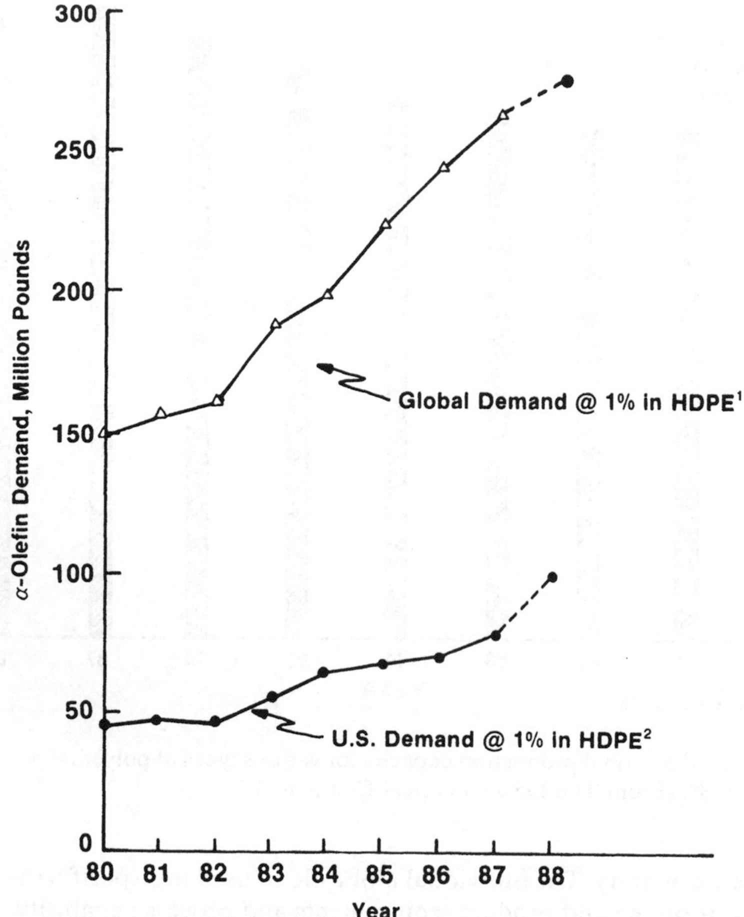

[1]Based on capacity assuming 20% HDPE/LLDPE swing capacity used to produce LLDPE
[2]Based on sales

Figure 15. Trend of alpha olefin utilization in the manufacture of HDPE from 1980 to 1988. (From The Dow Chemical Company.)

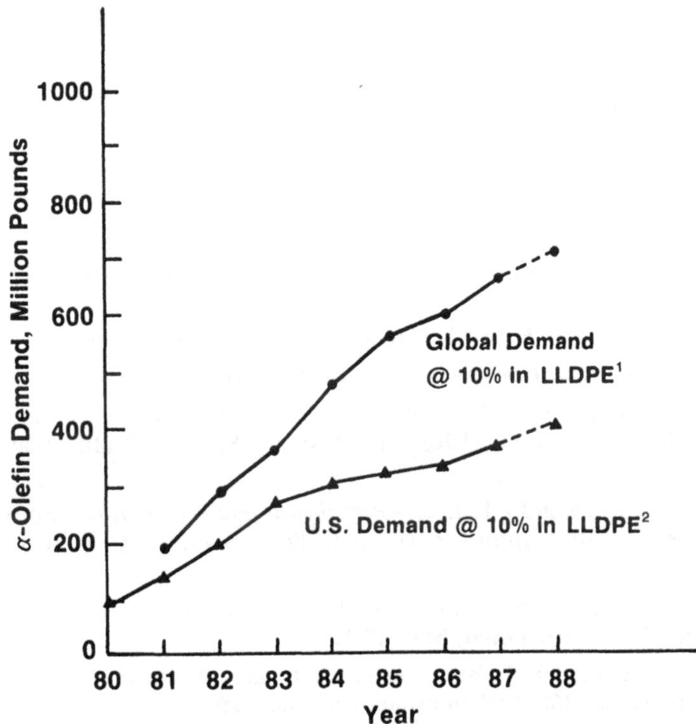

[1]Based on capacity assuming 20% HDPE/LLDPE swing capacity used to produce LLDPE
[2]Based on sales

Figure 16. Trend of alpha olefin utilization in the manufacture of LLDPE from 1980 to 1988. (From The Dow Chemical Company.)

alpha olefins for a given process or property. Successful development of in situ production of alpha olefin in the polyethylene polymerization process could also significantly affect alpha olefin use in polyethylene. Another potential change in the alpha olefin use pattern could result from the successful development of terpolymers in which propylene is used to replace a portion of the higher alpha olefins.

These and other research and development efforts will contribute to continued activity, growth, and new developments in the alpha olefin/polyethylene relationship. This should make the next 10 years even more challenging than the last 10 years.

ACKNOWLEDGMENT

Many of the polymer data for this chapter were supplied by Polyethylene Research and Development, Texas Division, Dow Chemical U.S.A.

REFERENCES

History and Process

1. Raff, R. A. V., and Allison, J. B., *Polyethylene* Vol. 11, Interscience, New York, 1956.
2. Renfrew, A., and Morgan, P. (eds.), *Polyethylene*, Iliffe, London, 1960.
3. Sittig, M., *Polyolefin Resin Processes,* Gulf Publishing, Houston, 1961.
4. Macgovern, R. L., Gianella, F. M., and Doles, G. E., *Linear Polyethylene and Polypropylene,* Report 19, Stanford Research Institute, Stanford, Calif., Nov. 1966.
5. Macgovern, R. L., and Kondo, T., *Linear Polyethylene and Polypropylene,* Supplement B, Report 19B, Stanford Research Institute, Stanford, Calif., Feb. 1974.
6. Chadwick, J. L., and Magovern, R. L., *High Density Polyethylene,* Supplement C, Report 19C, SRI International, Mar. 1979.
7. Chen, J. C. F., Magovern, R. L., and Sinclair, K. B., *Low Density Polyethylene* Supplement B, Report 36B, SRI International, Aug. 1980.

Chemistry

8. Gaylord, N. G., and Mark, H. F., *Linear and Stereoregular Addition Polymers,* Interscience, New York, 1959.
9. Raff, R. A. V., and Doak, K. W. (eds.), *Crystalline Olefin Polymers,* Part II, Vol. 20, Wiley, New York, 1965.
10. Raff, R. A. V., and Doak, K. W. (eds.), *Crystalline Olefin Polymers,* Part I, Vol. 20, Wiley, New York, 1965.
11. Rudolph, D. D., *Polymers Structure, Properties and Applications,* Cahners Books, Boston, 1972.
12. Vandenberg, E. J., and Repka, B. C., Ziegler-type polymerizations, in *Polymerization Processes,* Vol. 29, (C. E. Schildknecht and I. Skeist, eds.), Wiley, New York, 1977.

Applications

13. Bernhardt, E. C. (ed.), *Processing of Thermoplastic Materials,* Reinhold, New York, 1963.
14. Tadmor, Z., and Gogos, C. G., *Principles of Polymer Processing,* Wiley, New York, 1979.

15. Martino, R., Here come new-breed linear low density polyethylenes that are really tough, *Mod. Plas.*, 126–128 (Apr. 1984).

Demand and Capacity

16. *LLDPE Opportunity or Threat,* a multi-subscriber study, Chem Systems, Inc., June 1980.
17. *High-Density Polyethylene,* Report 80-4, Chem Systems, Inc., May 1981.
18. The statistical story of industry recovery, *Mod. Plast.*, 57–67 (Jan. 1984).
19. Sinclair, K. B., *Polyolefins: the Experience of the Last 10 Years and Prospects for the Coming Decade,* paper presented at the SRI Chemical Industries Division Symposium, 1984.
20. Statistics say another good year, *Mod. Plast.*, 61–71 (Jan. 1985).
21. *Monthly Statistical Report,* Society of Plastics Industry, Apr. 1985.
22. *Preliminary Report on U.S. Production of Selected Synthetic Organic Chemicals (Including Synthetic Plastics and Resin Materials),* U.S. International Trade Commission, Washington, D.C., published monthly.
23. Section VIII, *Synthetic Organic Chemicals—United States Production and Sales, 19XX,* U.S. International Trade Commission, Washington, D.C., published annually.

Miscellaneous

24. *What's New in Polyolefins,* SPE Retec, Houston, 1975.
25. *Linear Low Density Polyethylene Materials, Processing, Applications,* SPE Retec, Akron, Ohio, 1982.
26. *Polyolefins IV . . . Innovations in Processes, Products, Processing and Additives,* SPE Retec, Houston, 1984.

CHAPTER 5
Plasticizers

DAVID PAUL Monsanto Chemical Company, St. Louis, Missouri

1 INTRODUCTION

Plasticizers are materials which are added to high-molecular weight polymers to make them softer and more flexible. Plasticizers also improve workability by lowering polymer melting temperatures and improving melt flow characteristics, both of which facilitate the production of finished goods. Substances commonly called plasticizers are more precisely "external" plasticizers; they form stable molecular-level physical mixtures with polymers. Plasticization can also be achieved "internally" by building flexibilizing segments into the "backbone" of the polymer molecule. A hybrid approach is also used, wherein the flexibilizing agent is mixed into the polymer and then chemically appended to the backbone by grafting. This discussion focuses on external plasticizers (hereafter simply called "plasticizers"), with particular emphasis on those derived from linear alpha olefins.

Of the hundreds of different polymers in use today, relatively few employ large commercial volumes of plasticizers. Flexible poly(vinyl chloride) (PVC) is by far the largest outlet for alpha olefin-based plasticizers. Many other polymers will readily accept plasticizer but yield no commercially useful products. Polystyrene, for instance, becomes a viscous, sticky, semisolid at plasticizer levels of only 10 to 15%. Even at lower plasticizer levels, properties such as heat distortion temperature and creep resistance are seriously impaired. When plasticizers are used at all in these plastics, it is at very low levels (½ to 2%) to enhance melt flow. In these cases the additive is more properly called a "processing aid" than a plasticizer, since modification of the end-use properties of the polymer is not the objective. In general, amorphous, noncross-linked polymers do not form useful compositions when plasticized; a three-dimensional structure, either some form of cross-linking or a moderate level of crystallinity, is present in all commercially important plasticized systems.

To obtain a useful plasticized system, it is important to have the right "chemistry" (good compatibility) between polymer and plasticizer. Because of this, different polymer systems usually have different preferred plasticizers or plasticizer families. In many ways plasticizers behave like solvents, albeit inefficient ones, and much of the theory developed to explain solvent action can be applied to plasticizers. The old concept that "like dissolves like" can serve as a rough indicator of plasticizer effectiveness; there are, however, a number of exceptions to this rule, and many times this concept will not predict the optimum plasticizer for a given resin. For example, poly(vinyl chloride) resin is not soluble in its own monomer, and chlorinated paraffins are far from being the best plasticizers for chlorinated polyethylene. A complete discussion of plasticizer/polymer interactions would involve solubility parameters, hydrogen bonding indices, dipole moments, dielectric constants, and interaction parameters. Such a discussion falls outside the scope of this book but is thoroughly covered in the principal reference, *The Technology of Plasticizers,* by Sears and Darby [1]. Much of the material in this chapter is drawn from their book; it is highly recommended as a source of further details on the theory and uses of plasticizers.

2 PLASTICIZER USE IN PVC ("VINYL")

By far the predominant use for plasticizers derived from linear alpha olefins is in plasticized poly(vinyl chloride) compounds, commonly called "flexible PVC" or "flexible vinyl." By convention, formulations for these mixtures are always based on 100 parts by weight PVC resin, with the relative weights of the other ingredients expressed in "parts per hundred parts resin" (phr). Table 1 shows the most common ingredients used in flexible PVC compounds, along with the normal range of usage levels. Of all the additives, the plasticizer has the greatest influence on the physical properties of the composition. As the plasticizer level is increased, the compound becomes softer, more extensible, and flexible at lower temperatures (Fig. 1). At very low plasticizer levels, typically 3 to 10 phr, hardness is increased and extensibility reduced even beyond the values for the pure resin. This

Table 1 Common Components of Flexible Poly(vinyl chloride) Formulations

Relative amount by weight	Ingredient	Function/comments
100	PVC resin	Strength; holds the compound together. Formulations are usually based on 100 parts by weight resin.
15–80 (melt process) 25–100 (plastisol)	Plasticizer	Gives the system softness and flexibility.
½–10	Stabilizer	Retards decomposition of the resin due to heat. Various organometallic compounds are used, frequently in combination with epoxidized oils.
0–1	Antioxidant	Protects branched alkyl plasticizers against oxidative chain scission. Bisphenol A is commonly used; typically some is added to branched alkyl plasticizers at time of manufacture.
0–1	UV adsorber	Protects compound against degradation by ultraviolet light.
0–50	Filler	Reduces cost, can improve processing. Frequently, finely ground calcium carbonate is used. Alumina trihydrate can be used to reduce flammability of compound.
0–10	Pigments	Color, UV protection, flame retardance.
0–½	Lubricant	Improves processing. Stearic acid or metal stearates often used; linear alkyl plasticizers require less lubricant than other types.

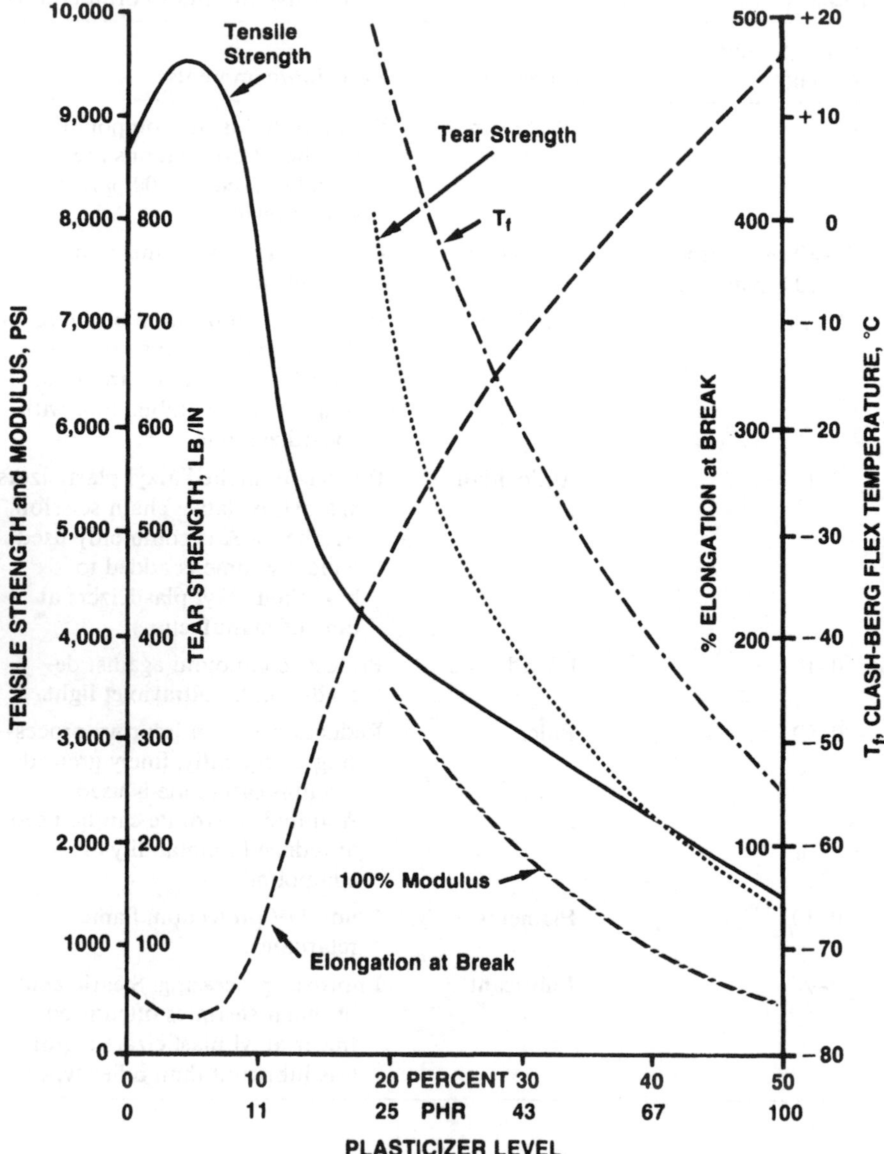

Figure 1. Effect of plasticizer concentration on the physical properties of PVC. As the plasticizer level is increased, the compound becomes softer, more extensible, and flexible at lower temperatures, while ultimate tensile and tear strength are reduced.

"antiplasticization" phenomenon occurs only in polymers having an appreciable level of crystallinity, such as PVC. When low levels of plasticizer are added to such resins, the mobility of the resin molecules is slightly increased. This allows some small segments that are frozen in amorphous configurations to reposition themselves into the preferred crystalline orientations (Fig. 2b). The higher level of crystallinity thus produced causes the system to be harder and less extensible than the pure resin. Further addition of plasticizer solvates and partially "dissolves" the crystalline regions, producing the familiar softening action of plasticizers discussed previously (Fig. 2c, d).

Finished vinyl goods are manufactured from flexible PVC compounds by many different techniques, most of which are variations on two basic technologies: melt processing and plastisol. These two processes employ PVC resins that are quite different in particle size and porosity; other formulation ingredients are similar.

The "calender grade" or "suspension" resin used in melt processes is made with relatively large grains, in the range 50 to 200 μm, and has a high degree of internal porosity and surface irregularity. This type of resin can easily absorb its own weight in liquid plasticizer and still remain a "dry" free-flowing powder. Preparation of such a "dry blend" (containing all the formulation ingredients) is the first step in most melt processes. Finished goods are made by melting, or "fusing," this dry blend by one of several mechanical/thermal processes, shaping the hot molten vinyl into the desired form, and then cooling the product.

Plasticol technology is based on PVC "dispersion" resin made by the emulsion process. This resin has a much smaller particle size than suspension resin (typically, 0.2 to 1.5 μm), no internal porosity, and does not readily absorb plasticizer at room temperature. When plasticizer is added to emulsion resin, the finely divided resin forms a dispersion in the plasticizer and the whole system remains a liquid. To produce finished goods, the liquid plastisol (containing all the formulation ingredients) is poured into molds or spread onto a fabric substrate or carrier paper at room temperature and then heated. As the plastisol reaches its "gelation temperature" the plasticizer begins to solvate the amorphous portions of the resin; the resin particles swell and the system solidifies into a gel. The material is heated further to the "fusion temperature" at which the gelled particles melt and fuse together to form a homogenous mass. When cooled, vinyl made by this method is practically indistinguishable from that produced by the melt process.

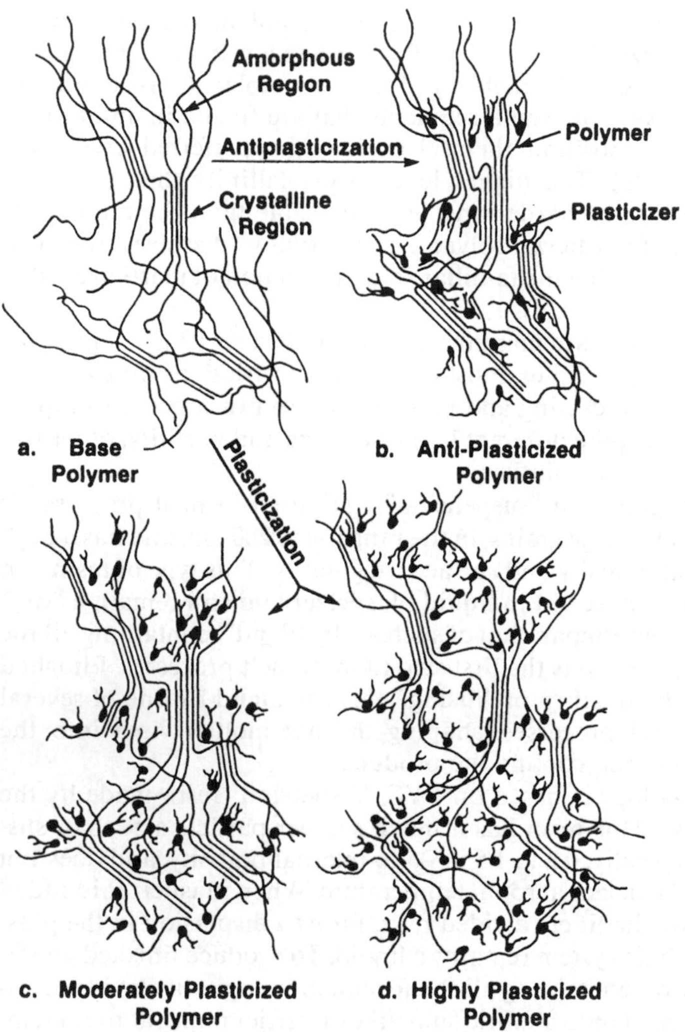

Figure 2. Stylized representation of semicrystalline polymer containing various amounts of plasticizer: (a) base polymer containing no plasticizer; (b) antiplasticized polymer in which a small amount of plasticizer allows additional crystallinity to form; (c and d) polymer containing higher levels of plasticizer that swell amorphous regions and partially dissolve crystalline regions.

3 PLASTICIZER PROPERTIES

To be commercially successful as a plasticizer, a material must possess certain attributes. These fall into two groups: per se properties and performance properties. As the name implies, the per se properties are measured on the plasticizer alone. Many of these, such as high purity, low odor, and freedom from residual acidity, are taken for granted today; no supplier could survive producing a material deficient in these areas. A few per se properties such as viscosity and vapor pressure do vary considerably among commercial plasticizers and are reflected in processing or end-product performance.

Performance properties are the functional characteristics that the plasticizer imparts to the compound in which it is used; they are measured by testing the complete compound after mixing and fusion. It is these qualities, along with cost, that primarily determine plasticizer choice for a given end use. Among commercial materials, performance properties vary widely, although members of the same chemical family have many characteristics in common. This is where plasticizers derived from linear alpha olefins show the greatest superiority over competitive materials. Some compound performance characteristics are influenced by several ingredients. Tear strength, for instance, is strongly influenced by resin molecular weight and filler level in addition to plasticizer level. The following discussion is confined to those properties that are determined primarily by plasticizer choice and are relatively independent of other formulation variables.

4 IMPORTANT PERFORMANCE PROPERTIES

4.1 Compatibility

The most fundamental quality needed in a plasticizer is good compatibility with the polymer being modified. Used in this context, "compatibility" denotes that the plasticizer will readily solvate the polymer and that molecular mixtures of the two are stable over a wide range of compositions and end-use conditions. Polymer/plasticizer combinations that are impossible to mix on a molecular level are said to be "incompatible." Plasticizers with marginal compatibility are difficult to mix with the polymer and tend to be fugitive; that is, they readily migrate out of the compound when subjected to mild stress of some form. The loop compatibility test [2] is a simple but highly effective way of measuring compatibility. In this test, strips of compound are folded into U shapes and held between parallel metal plates. The strips are removed at intervals and unfolded; incompatibility is shown by the appearance of plasticizer droplets (exudation) on the inside surface of the bend. A visual rating scale enables quantification of this effect. The ideal plasticizer would exhibit zero exudation in this test, as many commercial PVC plasticizers do. At the opposite end of the spectrum, aliphatic hydrocarbons exude very heavily from PVC; chlorinated paraffins perform somewhat better. Both of these "secondary plasticizers" can only be used in combination with a more compatible plasticizer or in compounds with high filler levels. Problems involving marginal plasticizer compatibility are aggravated by small amounts of incompatible lubricants or stabilizers in the compound. Conversely, the presence of a highly compatible "primary" plasticizer makes the compound more tolerant of secondaries.

4.2 Softening Efficiency

The main reason that plasticizers are included in PVC compounds is to impart softness and flexibility to the polymer system. In the vinyl industry, softness is quantified by several methods. Shore hardness is measured using a Durometer [3]; this device indents the sample with a small stylus linked to a scale and a calibrated spring. Several different scales based on various springs and indentor geometries cover the range of hardnesses encountered in the vinyl and rubber industries. In all cases, higher numbers indicate less indentation and therefore greater hardness. This test is quick and easy to run but imprecise since the results are very operator dependent. A more accurate measure of softness is extension modulus [4]. This procedure measures the force required to elongate a standard test specimen a given amount, usually 100%. Modulus measurements are less operator dependent than Durometer readings, but more time and equipment are required to prepare and test the samples. As in the hardness test, lower numbers indicate greater softness. Tensile (breaking) strength and tear strength usually go down as hardness is decreased; elongation at break goes up.

Not all plasticizers produce the same degree of softening in polymer systems. Those that produce a greater softening at a given usage level are said to be "more efficient." High softening efficiency is not always a desirable property; when the polymer is more costly than the plasticizer, inefficiency is preferred. This is because higher plasticizer levels are required to achieve the desired softness, thereby reducing the overall cost of the formulation.

4.3 Low-Temperature Performance

One very important characteristic of plasticizers is how much they enhance the low-temperature flexibility of the compound to which they are added. A number of tests to measure this are used in the vinyl industry. The Clash-Berg test [5] is a quick and precise method of quantifying low-temperature flexibility of plasticized PVC compounds. In this procedure a small strip of material is cooled to subfreezing temperatures and then slowly heated until it regains a predetermined degree of flexibility. The end point of the test (T_f or flexibility temperature) is the temperature at which the sample can be twisted through 200° of arc on the apparatus, corresponding to an apparent modulus of rigidity of 45,000 psi. This point roughly approximates the glass transition point (T_g) of the plasticized polymer.

Two other common low-temperature tests find the point at which a specimen breaks when suddenly bent. The solenoid brittleness point test (T_b) [6] is commonly run on heavy extruded products, either cut or remolded into specimens of a standard thickness. The SPI impact test [7], sometimes referred to as Masland impact, measures the brittleness point of thin films. In all three of these tests lower values (more negative) indicate better low-temperature performance.

4.4 Volatility

Volatility is another very important performance characteristic of commercial plasticizers. This is because plasticizers behave much like solvents when incorporated into resin systems. Only weak van der Waals forces hold the plasticizer molecules to the resin chains; these molecules are relatively free to move along the chains and from chain to chain. They are also in dynamic equilibrium with plasticizer molecules that are not associated with resin chains and with plasticizer at the surface of the article. With no strong chemical bonds to restrain them, plasticizer molecules at the surface are subject to loss by evaporation, the rate of which depends mostly on the ambient temperature and the vapor pressure—or volatility—of the plasticizer itself. As might be expected, the volatility of plasticizers is strongly dependent on molecular weight. Molecular structure, discussed later, is also important.

Volatility performance of plasticizers is measured different ways in different industries, but most methods test the plasticizer in the resin system of interest rather than by itself (per se). In the rubber industry, samples are often simply hung in a forced-draft oven and checked for weight loss. The deficiency of this technique is that some cross-contamination of samples can occur from plasticizer vapor present in the airstream; if plasticizers of widely differing volatility are involved, the results can be significantly affected. A common test in the vinyl industry is the activated carbon method [8]. In this procedure weighed disks of thin vinyl sheet are placed in closed jars and covered with granular activated carbon. The jars are placed in an oven for a specified time, after which the vinyl samples are removed from the carbon and reweighed. Because the carbon adsorbs all the evaporated plasticizer, there is no danger of cross contamination, even when samples of differing volatility are placed in the same jar.

5 MANUFACTURE OF PLASTICIZERS FOR PVC

Most plasticizers used in flexible PVC compounds are esters, made by reacting an alcohol with an acid or anhydride in the presence of a catalyst. The vast majority of commercial products are derivatives of organic difunctional acids or anhydrides and monohydric aliphatic alcohols (Fig. 3a,b). Several products are made from trifunctional acids or anhydrides, and a few are made from monobasic acids, most of which are reacted with dihydric or polyhydric alcohols (Fig. 3c); a few esters of monobasic acids and monofunctional alcohols are used. Glycol ethers, alicyclic alcohols, and aromatic alcohols are employed in some specialty plasticizers. The phosphates are the only family of plasticizers that are derivatives of an inorganic acid. As shown in Figure 3d, they are made from phosphorous oxychloride instead of phosphoric acid itself; here, phenols are more commonly used than alcohols.

6 SYNTHESIS OF PLASTICIZER ALCOHOLS

Alcohols used in most ester plasticizers are produced by several different routes. Most present-day commercial processes employ the oxo reaction (more properly called hydroformylation) at some point; the products of these processes are known as "oxo" alcohols. One of the first commercial petrochemical processes made "plasticizer range" alcohols (7 through 13 carbon atoms) from olefins generated in the "polymer gasoline" ("polygas") manufacturing process (Fig. 4). This route produces higher olefins by polymerizing propylene and butylenes in the presence of phosphoric acid supported on a solid substrate. The output of this "solid phosphoric acid unit" is fractionated to separate out olefins of the desired chain length (n), which are then hydroformylated to aldehydes having $n + 1$ carbons. The finished alcohols are generated by hydrogenation of the aldehydes. Because the oxo alcohols are isomeric mixtures of branched-chain molecules, they are known commercially as "iso" alcohols to distinguish them from unbranched "normal" alcohols. This nomenclature does not follow the IUPAC rules since typically there are several methyl branches per molecule, not just one in the beta position; also the numerical root used in naming these alcohols counts the total number of carbon atoms in the molecule, not just the carbons in the main chain.

Probably the single most widely manufactured plasticizer alcohol is 2-

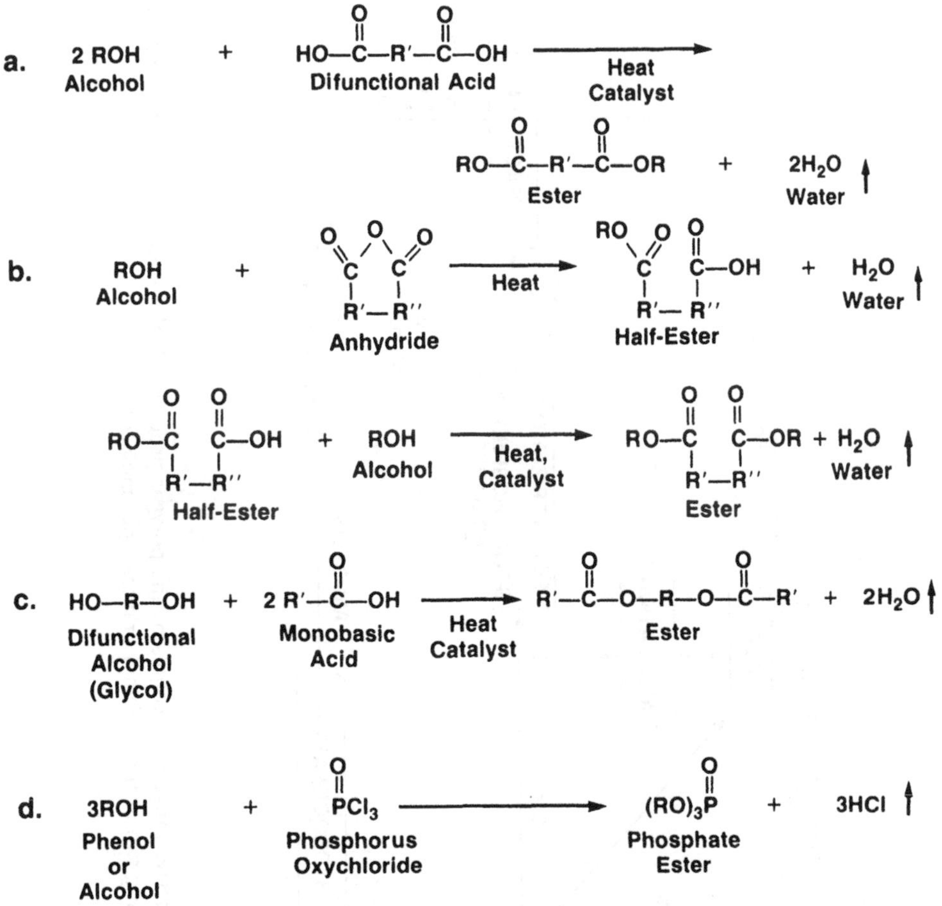

a. 2 ROH (Alcohol) + HO—C(=O)—R′—C(=O)—OH (Difunctional Acid) —Heat, Catalyst→ RO—C(=O)—R′—C(=O)—OR (Ester) + $2H_2O$↑ (Water)

b. ROH (Alcohol) + Anhydride (cyclic: O=C—O—C=O, with C—R′—R″—C) —Heat→ RO—C(=O)—R′—R″—C(=O)—OH (Half-Ester) + H_2O↑ (Water)

RO—C(=O)—R′—R″—C(=O)—OH (Half-Ester) + ROH (Alcohol) —Heat, Catalyst→ RO—C(=O)—R′—R″—C(=O)—OR (Ester) + H_2O↑ (Water)

c. HO—R—OH (Difunctional Alcohol (Glycol)) + 2 R′—C(=O)—OH (Monobasic Acid) —Heat, Catalyst→ R′—C(=O)—O—R—O—C(=O)—R′ (Ester) + $2H_2O$↑

d. 3ROH (Phenol or Alcohol) + O=PCl_3 (Phosphorus Oxychloride) → $(RO)_3P$=O (Phosphate Ester) + 3HCl↑

Figure 3. Chemistry of ester plasticizer production. The majority of commercial materials are made from monohydric alcohols and dibasic acids or anhydrides as in (a) and (b). A few products are derived from glycols and monobasic acids (c). Phosphate esters are produced from phosphorus oxychloride instead of phosphoric acid (d).

ethylhexanol, used to make di(2-ethylhexyl) phthalate (DEHP)—commonly but imprecisely known as dioctyl phthalate (DOP). Unlike the derivatives of the polygas process, this alcohol is not an isomeric mixture but a single species. In the "oxo/aldol" manufacturing process (Fig. 5), the first step is the oxonation of propylene to yield a mixture of butyraldehydes. An aldol condensation of the *n*-butyraldehyde followed by dehydration yields an eight-carbon unsaturated aldehyde which is readily hydrogenated to form 2-ethylhexanol (2EH).

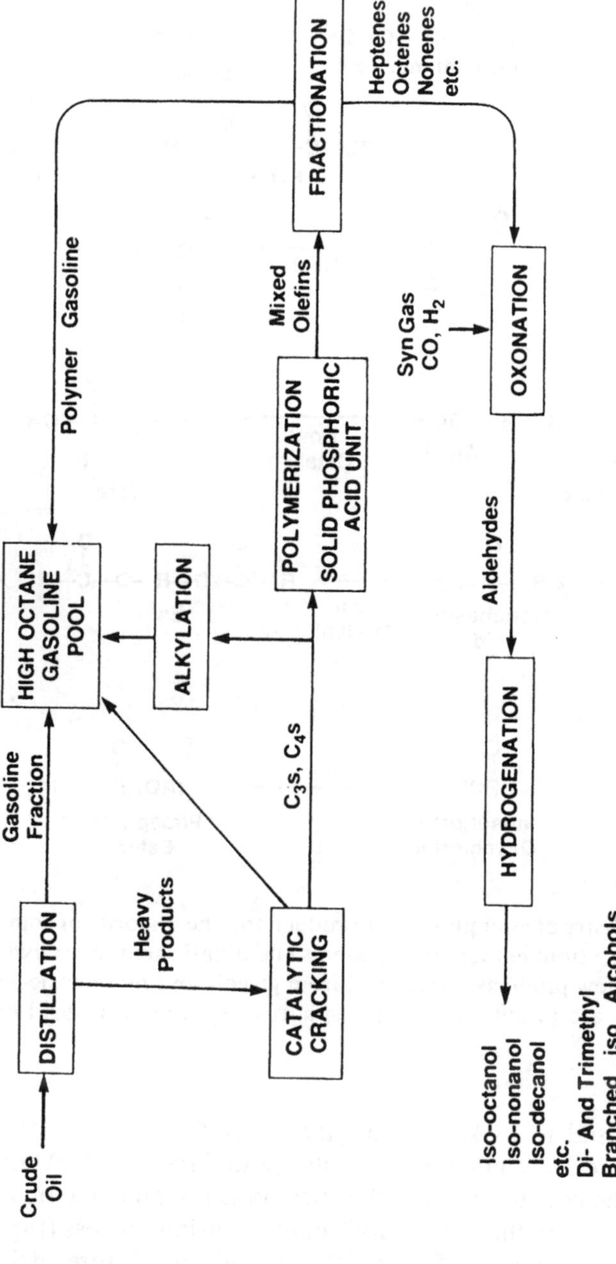

Figure 4. Production of plasticizer range iso alcohols by the polygas process. Alcohols made by this process are highly branched, typically having two or three methyl branches per molecule.

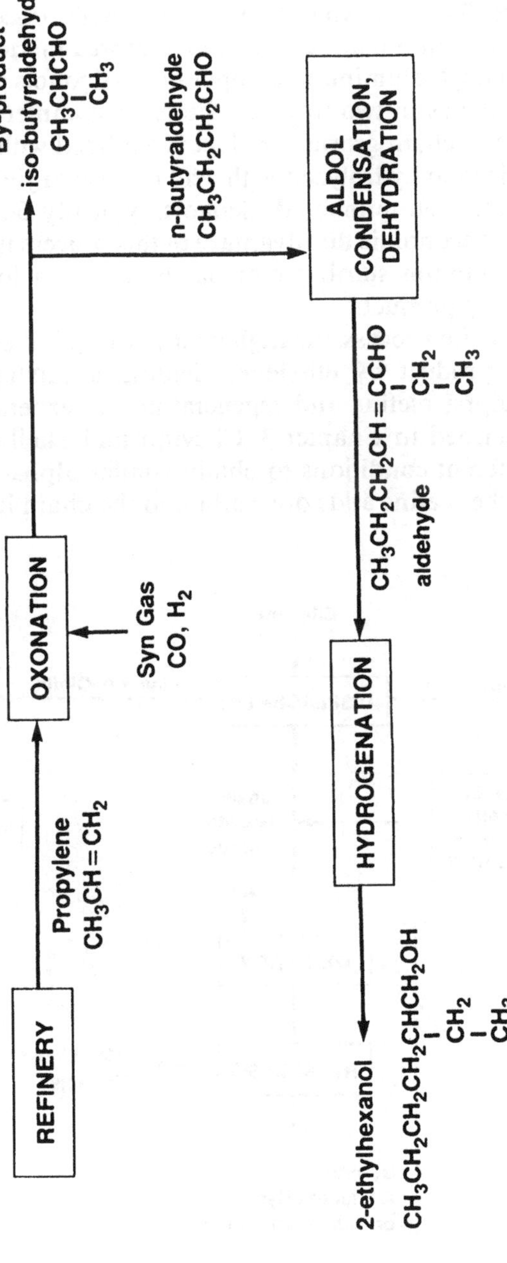

Figure 5. Oxo/aldol route for the production of 2-ethylhexanol. This process is the most common route to 2EH, the most widely produced plasticizer alcohol in the world.

Linear olefins for plasticizer alcohols are produced from ethylene as discussed in Chapter 3 and shown in Figure 6. In Ethyl's process, ethylene is added to triethyl aluminum under pressure to form a mixture of higher even-numbered trialkyl aluminum compounds. Alcohols are derived from these intermediates in two ways. In the older commercial process, the trialkyl aluminum chain growth product is oxidized with air to form the alkoxide. This is then hydrolyzed with acid or base to yield a mixture of completely linear, even-numbered alcohols, typically 6-, 8-, and 10-carbon. A significant economic disadvantage of this process is that the expensive trialkyl aluminum starting material is lost as a low-value inorganic aluminum by-product.

In Ethyl's alpha olefin process, the higher alkyl group is displaced from the chain growth product by ethylene, yielding a mixture of even-numbered linear alpha olefins and regenerating the expensive triethyl aluminum. As described in Chapter 3, Chevron and Shell oligomerize ethylene under different conditions to obtain similar alpha olefins. Hydroformylation of the olefins adds one carbon to the chain length, yield-

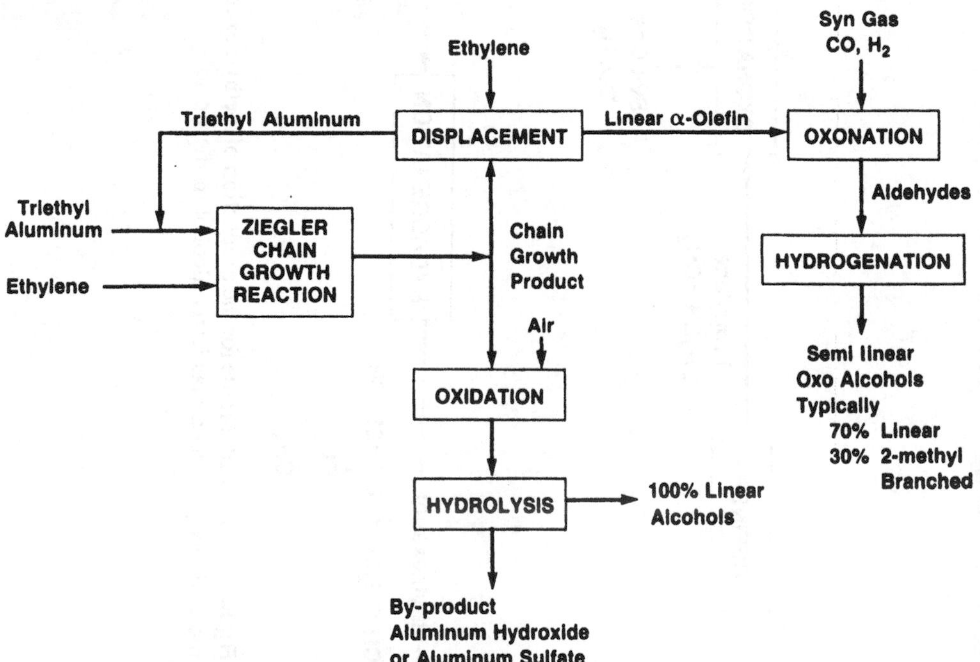

Figure 6. Synthesis of linear alcohols from ethylene. Depending on how the trialkyl aluminum chain growth product is treated, the alcohol produced can either be completely linear or semilinear.

ing the aldehydes. Depending on oxonation conditions and alkyl chain length, the carbon can add to either side of the olefinic double bond. In one commercial process, the addition is roughly 85% to the alpha carbon of hexene-1 but only 55% alpha for decene-1. Alcohols are generated by hydrogenation of the mixed odd-numbered aldehydes. The most widely used product of this route is roughly a 30/40/30 mixture of 7, 9, and 11 carbon alcohols; on the average, it is about 70% linear and 30% 2-methyl branched. Although the oxo alcohols produced in this process are not 100% straight chain, the esters of these materials are known as "linear plasticizers" or sometimes just as "linears."

The newest commercial route to olefins for plasticizer range alcohols is the Dimersol X [9] process (Fig. 7). This starts with *n*-butenes, which are by-products from the synthesis of methyl tertiary butyl ether (MTBE), an octane booster for unleaded gasoline. Dimerization of the *n*-butenes gives a mixture of predominantly monomethyl-branched eight-carbon olefins. Nine-carbon alcohols are made from these by conventional hydroformylation followed by hydrogenation. The oxo alcohol produced from Dimersol olefin is primarily monomethyl branched, with some dimethyl branching and small quantities of linear material.

7 RELATIONSHIP BETWEEN CHEMISTRY AND PERFORMANCE

The chemical characteristics of both the acid and alcohol precursors contribute to the performance of an ester plasticizer in the finished product. Although the influences of these two factors are not entirely independent of one another, they do have separate and predictable effects on performance. A good understanding of this structure/property relationship enables a formulator to select the best plasticizer to meet a given set of performance targets.

The acid is commonly thought of as the primary component of an ester molecule. Typically, several chemically distinct esters are commercially available from each of these acids; together, they are thought of as a family of plasticizers. Each of these families has a particular set of advantages and disadvantages which must be considered when formulating. Table 2 outlines the characteristics of the major ester families in comparison with the orthophthalates, the most widely used group and the de facto industry reference point for plasticizer performance.

8 EFFECT OF ALCOHOL STRUCTURE

Within a plasticizer family, the performance characteristics of the parent acid are modified by the alcohol with which it is esterified; Table 3 outlines the influence of several alcohol types. Of the four shown, only the

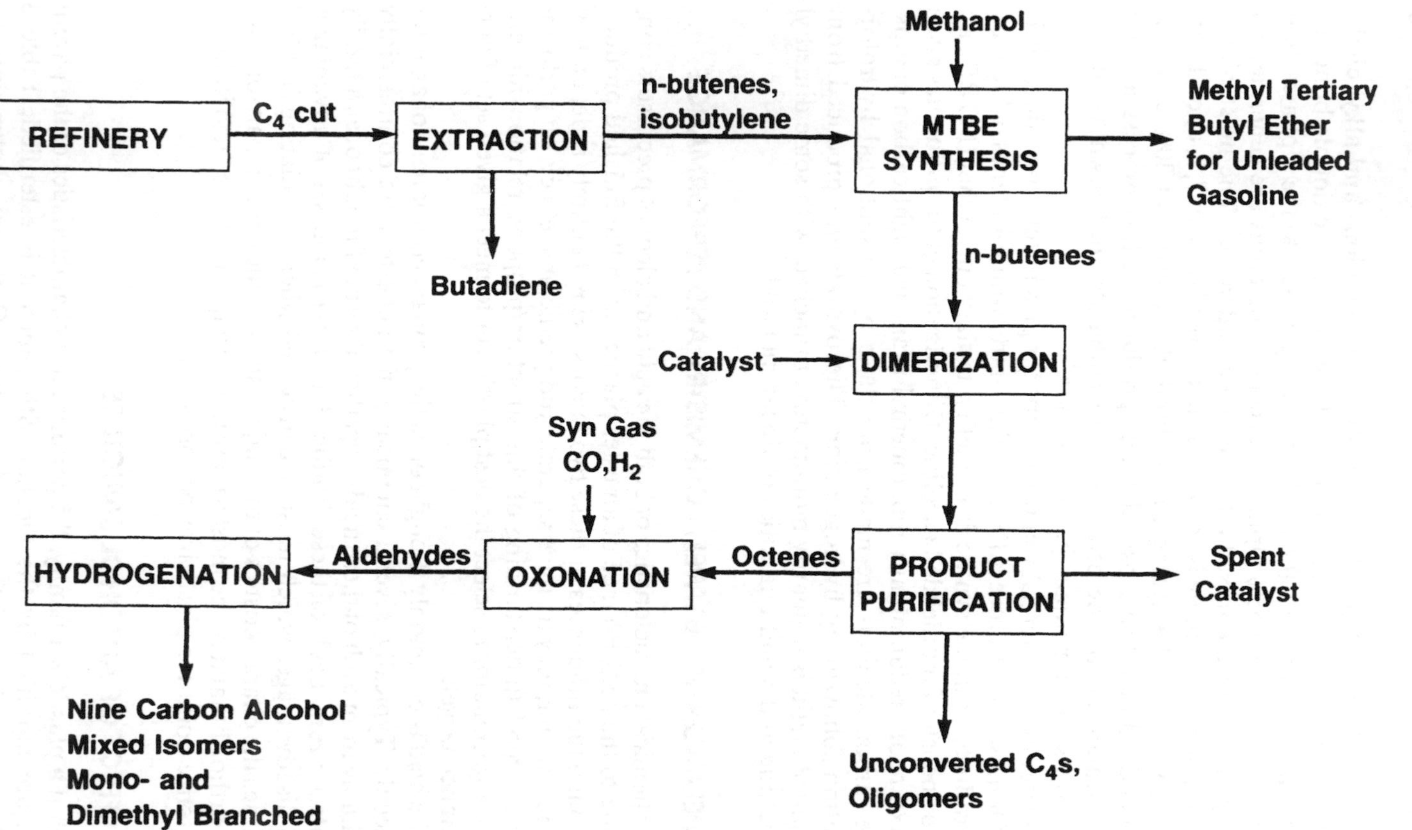

Figure 7. Dimersol X process for the manufacture of octenes for conversion to nine-carbon plasticizer alcohols. This scheme starts with the mixture of *n*-butenes that emerge unreacted from MTBE synthesis.

Table 2 Plasticizing Characteristics of Major Ester Families in PVC, Holding Alcohol Moiety Constant

Ester family	Advantages	Disadvantages	Comments
Phthalates (ortho isomer)	Good compatibility with PVC and other polar polymers, good balance of properties, low cost		By far the largest ester family commercially: industry workhorse; standard of performance against which esters are judged
Terephthalates (para isomer)	Better low temperature than ortho, lower volatility than ortho, nonmarring to lacquered surfaces, low cost	Less light stable in PVC than ortho isomer, less compatible with PVC than ortho isomer	Less efficient than ortho isomer, small commercial volume
Isophthalates (meta isomer)	Nonmarring to lacquered surfaces, slightly less volatile than ortho isomer	Cost, availability	Very little in commercial use today
Phosphates	Lower flammability than other ester plasticizers, excellent compatibility with PVC and other polar polymers	Cost, lower heat stability than phthalates in PVC	Other properties largely determined by alcohol moiety

(continued)

Table 2 *(Continued)*

Ester family	Advantages	Disadvantages	Comments
Dibasic acid esters			
Monomeric (C_n = 5,6,8,10)	Excellent low-temperature performance, good UV resistance	Cost (especially higher C_n), lower compatibility, greater hydrocarbons extractibility, high volatility at low C_n	Adipates (C_n = 6) are the largest of this group commercially
Polymeric	Very low volatility, good hydrocarbon solvent resistance, nonmigratory	Cost, hydrolytic instability, poor low-temperature performance	Most commercial products based on adipic acid; some on glutaric and azelaic acids
Trimellitates	Low migration, very low volatility	Cost	Good balance of other properties, similar to ortho phthalates, but less efficient
Benzoates	Good compatibility with PVC and other polar polymers, high solvating power	High volatility, poor low-temperature performance, support fungal growth	Used primarily in PVAc and acrylic latex caulks and adhesives
Citrates	Broad FDA coverage, naturally occurring acid	Cost	Small commercial volume

Table 3 Principal Characteristics of Ester Plasticizers Made from Various Alcohol Types[a]

Alcohol used in ester	Advantages	Disadvantages	Comments
Linear alkyl (paraffinic)	Excellent low-temperature performance, good (low) volatility, good light and oxidative stability, low plastisol viscosity and good rheology	Somewhat more migratory than branched alkyls	Derived from ethylene
Branched alkyl (paraffinic)	Slightly more compatible with PVC than linears, good electrical characteristics, low cost	Susceptible to oxidative chain scission, poor (high) volatility	Derived from propylene or butylene; largest commercial volumes, many producers
Alkoxy	Enhanced compatibility with some non-PVC resin systems	Migratory, water sensitive, poor electrical characteristics	Little used in PVC; specialty applications only
Aryl (aromatic)	Excellent (very low) volatility, excellent compatibility with PVC and other polar polymers, powerful solvent action, very fast fusion in PVC systems, imparts stain resistance to vinyl	Poor low-temperature performance, high specific gravity, poor plastisol rheology and viscosity stability	Derived from toluene: used primarily in vinyl flooring and specialty applications

linear and branched alkyls are widely used in the mainstream of the vinyl industry. In alkyl alcohols, molecular weight and the amount of chain branching are the two structural factors that determine plasticizer performance. Although it affects most per se properties only slightly, chain branching has a strong influence on performance in PVC. This is readily seen in low-temperature and volatility characteristics, but is equally significant, only less obvious, in the areas of oxidative stability and sunlight resistance.

8.1 Compatibility

Within a given plasticizer family, compatibility with PVC is strongly influenced by the chain length of the alcohol precursor and to a lesser extent by the amount of chain branching present. Below a carbon number of nine, dialkyl phthalates show excellent compatibility irrespective of chain branching. At a chain length of 10 carbons, the first evidence of decreasing compatibility appears. Compatibility with PVC drops sharply beyond this point; a linear 12-carbon phthalate is practically incompatible and is not used in commercial applications. Adding carbon to the alkyl chain in branches hurts compatibility less than a simple lengthening of a linear chain. Evidence of this may be seen by comparing di(2-methyldecyl) phthalate with di(*n*-undecyl) phthalate. Although these both have 11-carbon alkyl chains, the branched material exudes less in the loop test. Ditridecyl phthalate (DTDP), a highly branched 13-carbon ester, is compatible enough for use in wire insulation compounds containing moderate amounts of filler.

8.2 Softening Efficiency

The softening efficiency of a plasticizer is dependent on molecular weight and chemical structure. In a given chemical family, lower-molecular-weight materials are more efficient since in a given weight of plasticizer there are more molecules available to solvate sites on the resin chains. Figure 8 illustrates this for branched-chain phthalates; at any plasticizer level the eight-carbon ester gives a lower 100% modulus than the nine-carbon material, which is in turn more efficient than the 10-carbon ester. Branching in the alkyl moiety decreases the softening efficiency of a plasticizer. In Figure 9 the highly branched material gives higher modulus values than the lightly branched ester; it in turn is less efficient than the predominantly linear material. All three of these phthalates have nine-carbon alkyl groups.

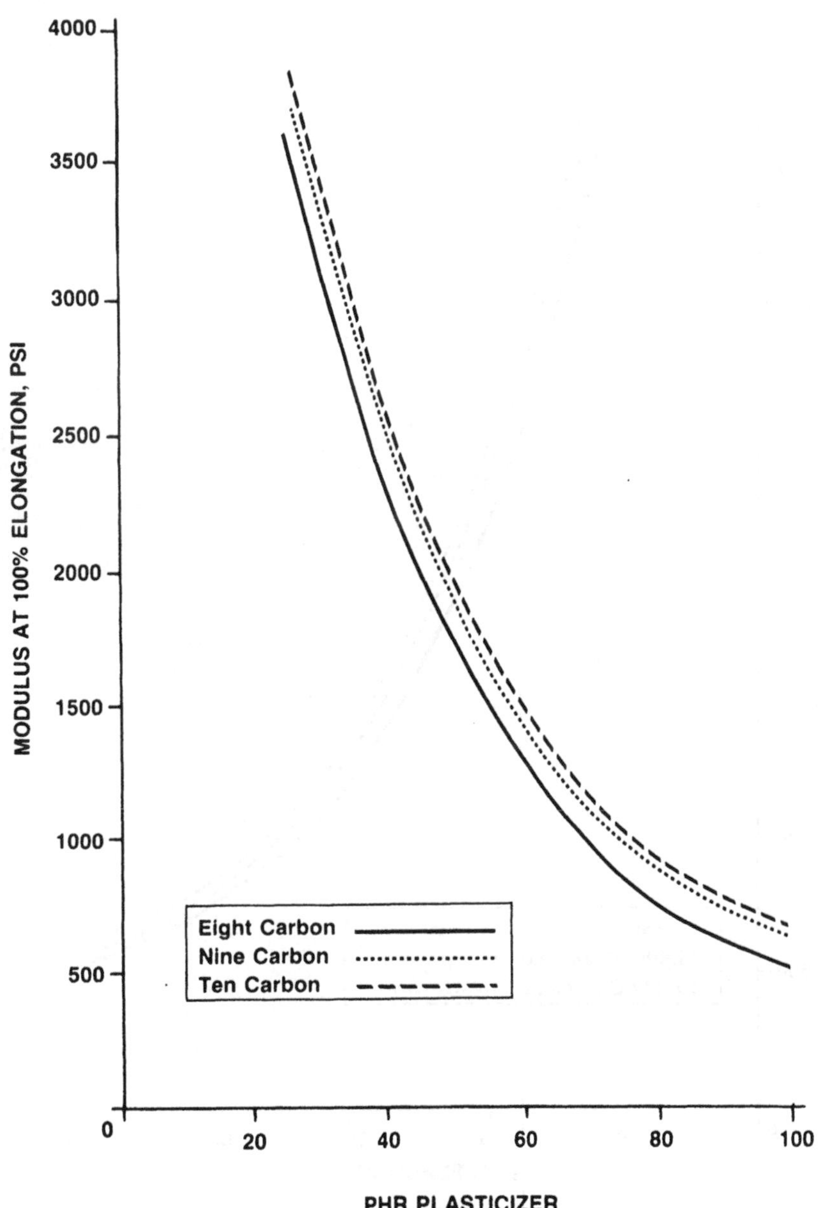

Figure 8. Influence of molecular weight on softening efficiency of branched-chain dialkyl phthalates in PVC. In a series of structurally similar plasticizers, softening efficiency increases as carbon number decreases. Throughout the practical range of plasticizer levels the eight-carbon ester gives the lowest modulus (greatest softening efficiency), while the ten-carbon material produces the highest modulus (least efficient).

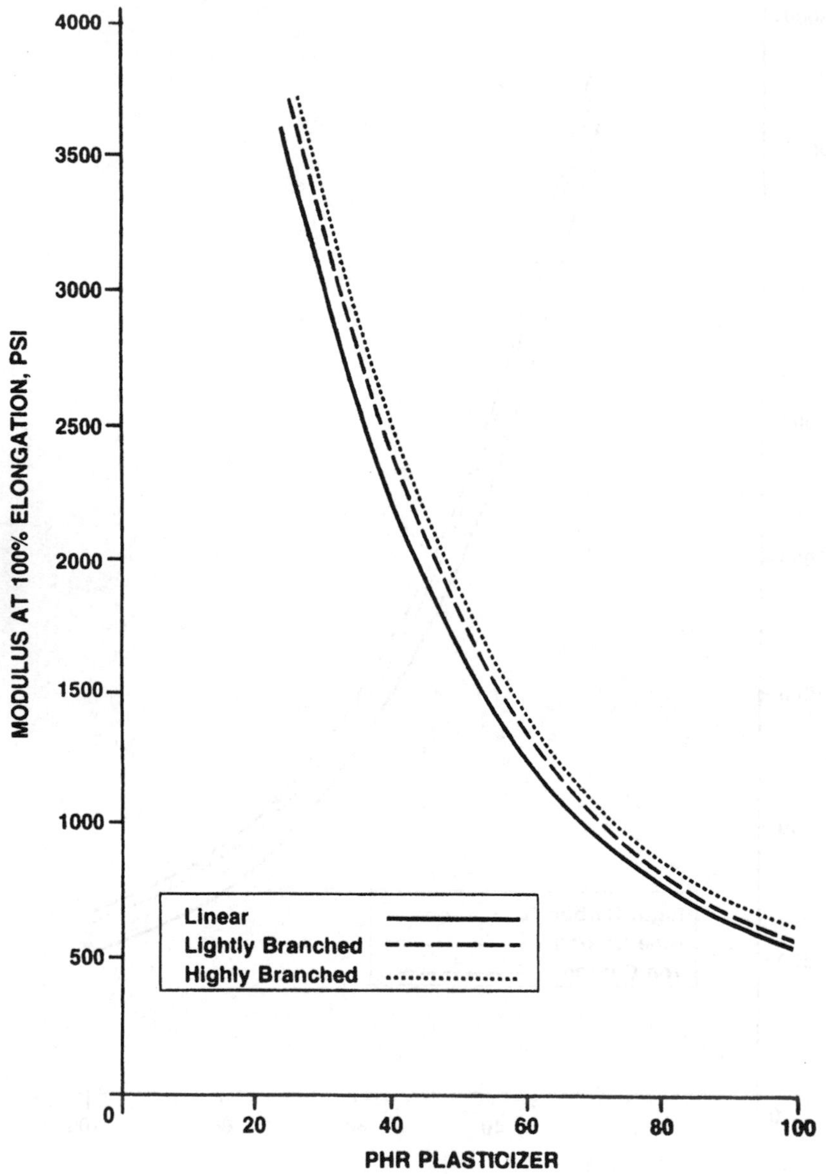

Figure 9. Effect of chain branching on softening efficiency of nine-carbon dialkyl phthalates in PVC. At any given plasticizer level the linear ester is the most efficient (lowest modulus), while the highly branched material is the least efficient (highest modulus).

8.3 Low-Temperature Flexibility

In Table 4, data from the Clash-Berg test illustrate how branching in the alcohol moiety detracts from the low-temperature performance of plasticizers. The ester of lab-prepared 2-methyl octanol, in which all the alcohol groups have a single methyl branch, gave an 8°C poorer T_f than that of the 100% linear material. The Monsanto and Gulf (Circa 1970) commercial nine-carbon alcohols, which contain different ratios of straight-chain and 2-methyl-branched materials, gave intermediate values closely proportionate to their isomeric compositions. Olefins produced by the Dimersol and polygas processes give alcohols that have several methyl branches per molecule. The esters of these materials have the poorest low-temperature properties of commercial nine-carbon dialkyl phthalate plasticizers.

Table 4 Effect of Methyl Branching on Low-Temperature Performance of Nine Carbon Dialkyl Phthalate Plasticizers

Plasticizer alcohol	Branching in alcohol	T_f of ester in PVC[a] (°C)
Normal (lab prep.)	Practically no branching Essentially 100% linear	−50.3
Monsanto process (nine-carbon fraction only)	70% linear 30% monomethyl branched	−48.5
Gulf process (ca. 1970) (nine-carbon fraction only)	55% linear 45% monomethyl branched	−46.5
2-Methyl octanol (lab prep.)	Essentially 100% monomethyl branched	−42.2
From Dimersol olefin (Monsanto pilot plant)	1–2% linear[b] 59% monomethyl branched 39% dimethyl branched 0 trimethyl branched	−40.4
From polygas olefin (primarily nine-carbon fraction)	1% linear[b] 11% monomethyl branched 75% dimethyl branched 13% trimethyl branched	−36.4

[a]Test formulation: 100 parts PVC resin, medium molecular weight
67 parts Dialkyl phthalate plasticizer
0.5 part Dibasic lead stearate (stabilizer)

[b]Branching of eight-carbon olefin from which nine-carbon alcohol was made. Branching in alcohol is somewhat higher than in parent olefin.

Interestingly, the size of the branch on the alcohol moiety exerts relatively little influence on low-temperature properties. Table 5 illustrates what happens as the length of the side chain is increased while the overall carbon number is held at nine. As these data show, the mere *presence* of a branch has a greater effect on low-temperature performance than the length of the branch. Other data [10] indicate that T_f is hurt more by chain branching close to the carboxyl group than by similar branching located farther away. Although the reasons for this behavior are not well documented, steric hindrance is probably the major factor involved. Branching, especially close to the carboxyl groups, is thought to restrict the motion of the alkyl chains relative to the ring and carboxyl structures. This apparently reduces the mobility and thus the "efficiency" of the free volume introduced into the polymer system by the plasticizer. It may also force the plasticizer molecules into less favorable orientations with respect to the resin chains.

Within limits, increasing the carbon number of structurally similar dialkyl phthalates improves their low-temperature properties. For 100% linear materials, the limit in PVC is at 11 carbons, beyond which com-

Table 5 Effect of Side Chain Length on Low-Temperature Performance of Nine-Carbon Dialkyl Phthalate Plasticizers

Alkyl moiety	Structure	T_f in PVC[a] (°C)
n-Nonyl	C–C–C–C–C–C–C–C–C–	−50.3
2-Methyl octyl	C | C–C–C–C–C–C–C–C–	−42.2
2-Ethyl heptyl	C | C | C–C–C–C–C–C–C–	−41.7
2-Propyl hexyl	C | C | C | C–C–C–C–C–C–	−40.1

[a]Test formulation: 100 parts PVC resin, medium molecular weight
67 parts Dialkyl phthalate plasticizer
0.5 part Dibasic lead stearate (stabilizer)

patibility with the resin decreases dramatically with a consequential loss of low-temperature flexibilizing efficiency. As shown in Table 6, branched esters show a greater response to lengthening the main chain than do linear materials. Taking a slightly different viewpoint, it could also be said that branching has less adverse effect on long-chain phthalates than on short-chain esters.

Table 7 shows brittleness point data for several commercial plasticizers over a range of concentrations in PVC. At all levels, the linear phthalate gives the best performance, while DINP, a highly branched nine-carbon ester, is the worst. DOP and DIDP are slightly better than DINP. In the case of DOP, there is less branching in the alcohol moiety, and DIDP has a longer alkyl chain length by one carbon. Figure 10 shows the results of SPI impact testing of 10-mil (0.010-in.)-thick smooth calendered films having various plasticizer levels. As in the previous tests, the linear ester gives the best performance of the phthalates evaluated. Dioctyl adipate (DOA), an excellent low-temperature plasticizer, is included on this chart for reference.

Occasionally, individuals attempt to judge the low-temperature performance of a plasticizer on the basis of the freezing point of the pure ester itself. Although superficially logical, this approach is unreliable. In PVC, compatible linear esters always give better T_f and T_b performance than their branched-chain counterparts, although in some cases the branched plasticizers have lower reported freezing points (Table 8).

Table 6 Influence of Alkyl Chain Length and Branching on the Low-Temperature Performance of Dialkyl Phthalate Plasticizers in PVC[a]

Side-chain carbon number[b]	Main-chain carbon number 6	7	8	9	10	11	12
0 (normal)	−48.1	−48.7	−49.6	−50.3	−52.2	−52.1[c]	−43.2[c]
1 (methyl)	−38.6	—	−42.2	—	−48.5	—	—
2 (ethyl)	−39.4	−41.7	—	—	—	—	—
3 (*n*-propyl)	−40.1	—	—	—	—	—	—

[a] Data reported are the Clash-Berg flex temperature, T_f (°C).

Test formulation: 100 parts PVC resin, medium molecular weight
67 parts Plasticizer
0.5 part Dibasic lead stearate (stabilizer)

[b] In all cases, the branching was at the 2-position of the main chain.

[c] Low-temperature performance deteriorates after the limit of compatibility with the polymer is reached.

Table 7 Low-Temperature Brittleness Performance of Commercial Dialkyl Phthalates in PVC[a]

	Solenoid brittle point, ASTM D-746			
Plasticizer level	Linear 7-, 9-, 11-carbon	DOP	DINP	DIDP
25 phr (20%)	−5°C	+1	+8	+4
43 phr (30%)	−31	−20	−19	−21
67 phr (40%)	−47	−40	−38	−41
100 phr (50%)	−60	−49	−47	−51

[a]Formulation: 100 parts PVC resin, medium high molecular weight
As shown Plasticizer
1 part Barium cadmium stabilizer

Table 8 Disparity Between Ester Freezing-Point and Low-Temperature Performance in PVC

	11-carbon linear phthalate	10-carbon branched phthalate	7- and 9-carbon linear adipate	8-carbon branched adipate
Plasticizer freezing point (°C)	+2	−53	−13	<−70
T_f of PVC compound at plasticizer level of:				
43 phr (30%)	−29	−19	−49	−46
67 phr (40%)	−52	−38	−73	−66
100 phr (50%)	−64	−53	—	—

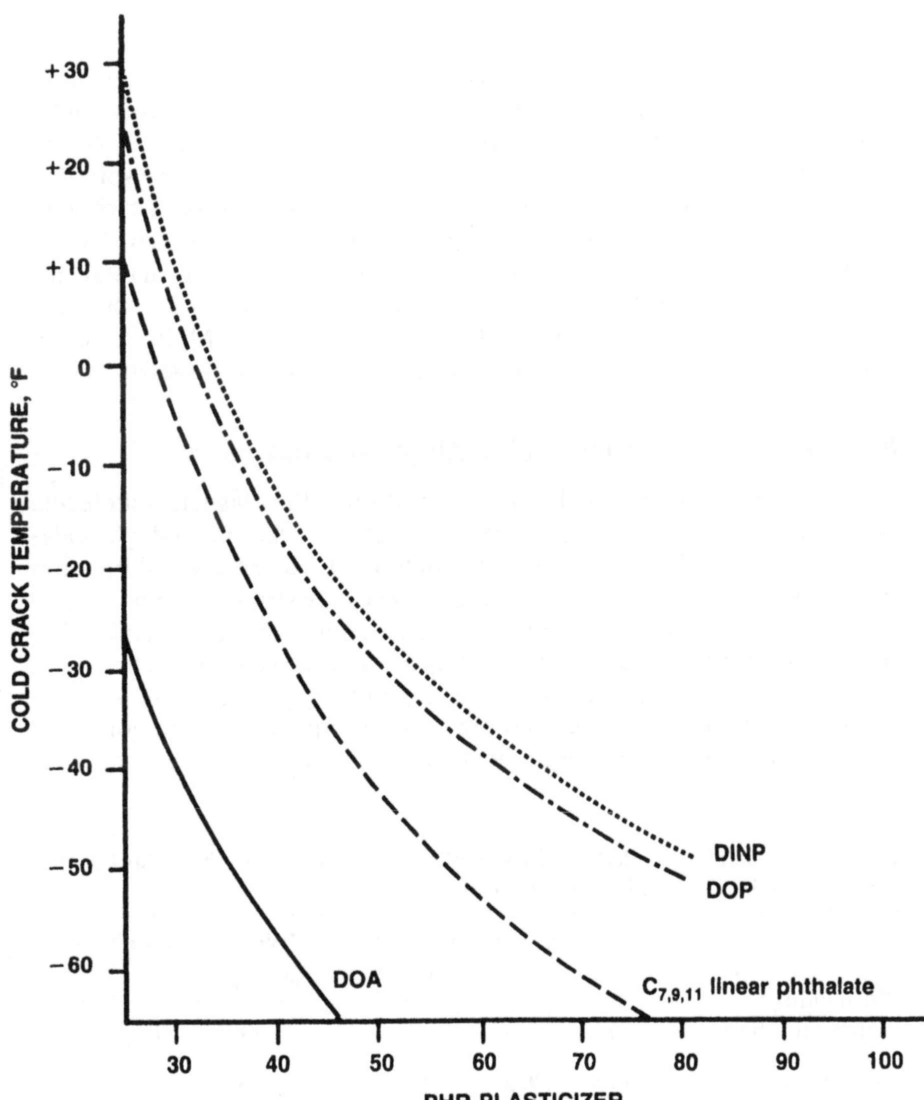

Figure 10. Low-temperature performance of plasticizers in PVC as measured by the SPI impact test on 10-mil films. In a given family such as the phthalates shown here, linear esters give better low-temperature performance than branched-chain materials. Adipates, represented here by DOA, give much better performance than phthalates (contrast directly with DOP).

8.4 Volatility

As might be expected, volatility performance of alkyl ester plasticizers is strongly dependent on molecular weight and thus alkyl chain length. Alkyl chain configuration is also important, as shown in Table 9. As the main-chain carbon number of the alkyl groups is increased, the volatility is reduced until the PVC compatibility limit is reached. Adding carbon to the chain in the form of side branches is much less effective; in fact, at higher carbon numbers it appears to detract. Clearly, for a given total carbon number the lowest-volatility chain configuration is linear, while the worst is branched, especially if the branch is large. Multiple small (methyl) branches are equally deleterious, as shown in Table 10.

8.5 Oxidative Stability and Sunlight Resistance

The degree of branching of the alcohol portion of the plasticizer molecule determines its susceptibility to oxidative chain scission and photodegradation. This is because at every branch there is a tertiary carbon atom in the alkyl chain; these tertiary carbons are easily stripped of their single hydrogen atom to form free radicals when energized by heat, light, radiation, or peroxides. The free radicals thus formed start chain reactions and rapidly multiply, especially in the presence of oxygen, and degrade both plasticizer and polymer. Because of their destructiveness, it is important to eliminate the sources of free radicals as much as possible. Since linear

Table 9 Influence of Alkyl Chain Length and Branching on the Volatility Performance of Dialkyl Phthalate Plasticizers[a]

Side-chain carbon number[b]	Percent plasticizer loss from PVC for main-chain carbon number:						
	6	7	8	9	10	11	12
0 (normal)	46.5	20.3	6.5	2.7	1.5	1.6	1.5
1 (methyl)	34.6	—	6.0	—	2.2	—	—
2 (ethyl)	18.2	7.7	—				
3 (*n*-propyl)	9.1	—	—				

[a]Test conditions: ASTM D-1203, 6 days at 87°C, specimens packed in activated carbon
Test specimens: Disks, 2 in. diameter, 0.040 in. thick
Test formulation: 100 parts PVC resin, medium molecular weight
67 parts Plasticizer
0.5 part Dibasic lead stearate (stabilizer)

[b]In all cases, the branching was at the 2-position of the main chain.

Table 10 Effect of Methyl Branching on Volatility Performance of Nine-Carbon Dialkyl Phthalate Plasticizers[a]

Plasticizer alcohol	Branching in alcohol	Percent plasticizer lost from PVC
Normal (lab prep.)	Practically no branching Essentially 100% linear	2.7
Monsanto process (nine-carbon fraction only)	70% linear 30% monomethyl branched	3.7
Gulf process (ca. 1970) (nine-carbon fraction only)	55% linear 45% monomethyl branched	4.5
2-Methyl octanol (lab prep.)	Essentially 100% Monomethyl branched	6.0
From Dimersol olefin (Monsanto pilot plant)	1–2% linear[b] 59% monomethyl branched 39% dimethyl branched 0 trimethyl branched	6.1
From polygas olefin (primarily nine-carbon fraction)	1% linear[b] 11% monomethyl branched 75% dimethyl branched 13% trimethyl branched	7.6

[a]Test conditions: ASTM D-1203, 6 days at 87°C, specimens in activated carbon
Test specimens: Disks, 2 in. diameter, 0.040 in. thick
Test formulation: 100 parts PVC resin, medium molecular weight
67 parts Dialkyl phthalate plasticizer
0.5 part Dibasic lead stearate (stabilizer)

[b]Branching of eight-carbon olefin from which nine-carbon alcohol was made. Branching in alcohol is somewhat higher than in parent olefin.

alkyl plasticizers contain few, if any, tertiary carbon atoms compared to branched alkyl esters (Fig. 11), they are the plasticizers of choice to minimize free-radical formation in applications involving oxidative aging conditions or sunlight exposure. The test results plotted in Figure 12 show how much the plasticizer can affect the durability of vinyl exposed to sunlight. In this experiment, two flexible vinyl sheets, identical except for plasticizer, were aged outdoors in Florida. At regular intervals small samples were checked for degradation using microscale version [11] of the SPI low-temperature brittleness test. The point at which degradation becomes significant is indicated by a sharp rise in the measured brittleness temperature. As expected, the formulation made with the linear phthalate resisted sunlight longer than the sheet containing the highly branched

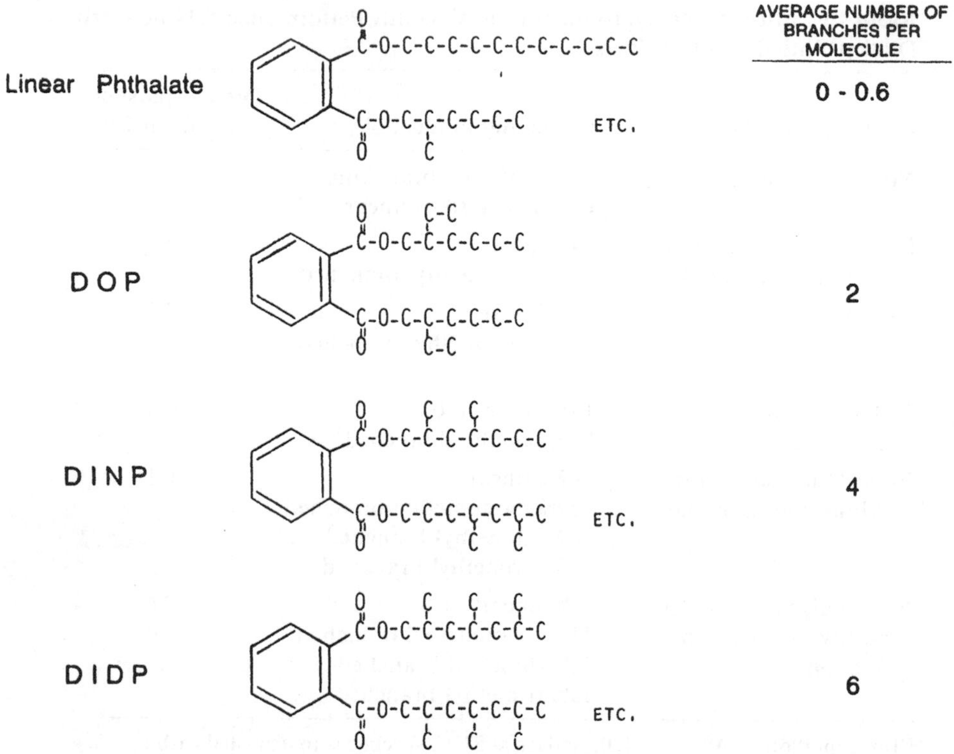

		AVERAGE NUMBER OF BRANCHES PER MOLECULE
Linear Phthalate	C-O-C-C-C-C-C-C-C-C-C-C-C / C-O-C-C-C-C-C-C ETC.	0 - 0.6
DOP	C-O-C-C-C-C-C-C / C-O-C-C-C-C-C-C	2
DINP	C-O-C-C-C-C-C-C-C / C-O-C-C-C-C-C-C-C ETC.	4
DIDP	C-O-C-C-C-C-C-C-C / C-O-C-C-C-C-C-C-C ETC.	6

Figure 11. Chain branching present in commercial dialkyl phthalate plasticizers. By creating tertiary carbon atoms where free radicals are easily formed, branches in the alkyl chain increase the plasticizer's vulnerability to degradation by heat, oxygen, and light.

DIDP. This comparison was also made in formulations containing a commercial ultraviolet (UV) absorber (Fig. 13). Those results clearly show the value of a good UV absorber in prolonging the outdoor life of flexible PVC (note the longer time scale on Figure 13 than that on Figure 12); they also show that even with good UV protection the inherently stable linear plasticizer has a significant advantage over the branched-chain material. Antioxidants such as bisphenol A (BPA) are usually added to commercial grades of branched-chain plasticizers to reduce their susceptibility to oxidation and free-radical formation. Without BPA, the performance of the DIDP in Figures 12 and 13 would have been considerably worse.

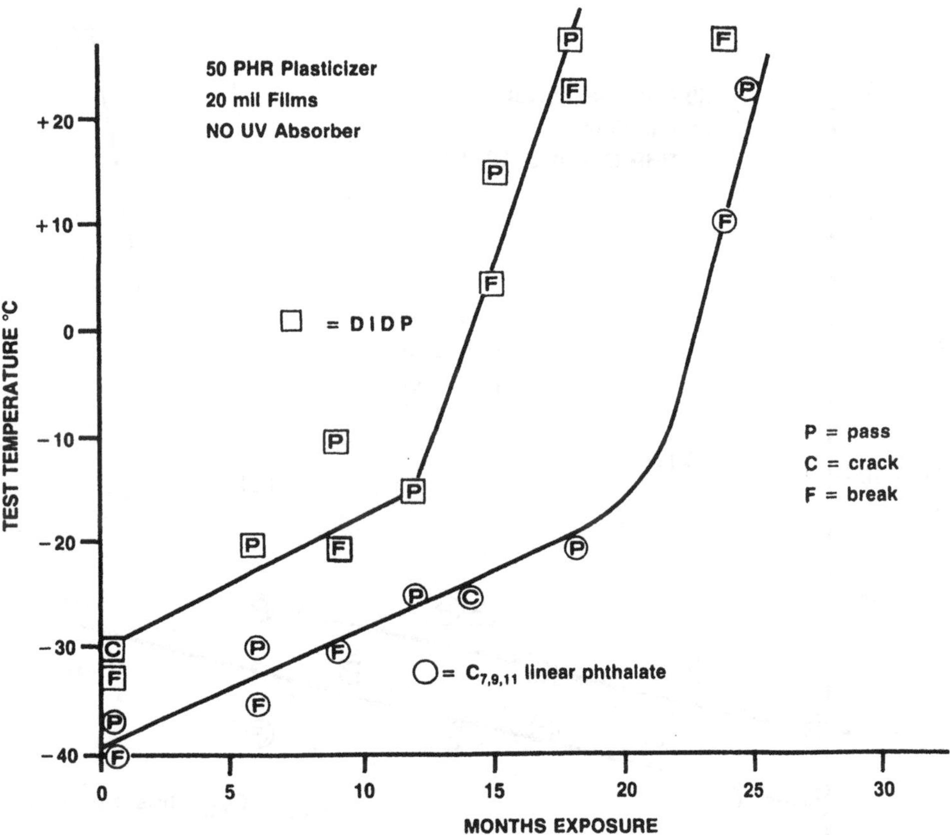

Figure 12. Loss of low-temperature performance in plasticized 20-mil vinyl films aged outdoors in Florida. The film plasticized with the branched-chain phthalate DIDP (squares) degenerates faster than the linear plasticized film (circles). Films contained 50 PHR plasticizer and no UV adsorber.

8.6 Viscosity

Plasticizer viscosity is one of the few per se properties strongly influenced by chain branching; it is also a property of importance to the plastisol segment of the vinyl industry. In most manufacturing operations employing plastisol technology, the viscosity of the plastisol is an important process variable that must be closely controlled. Although almost all formulation ingredients have some effect on plastisol viscosity, the quantity and viscosity of the plasticizer are two of the major determining factors. Table 11

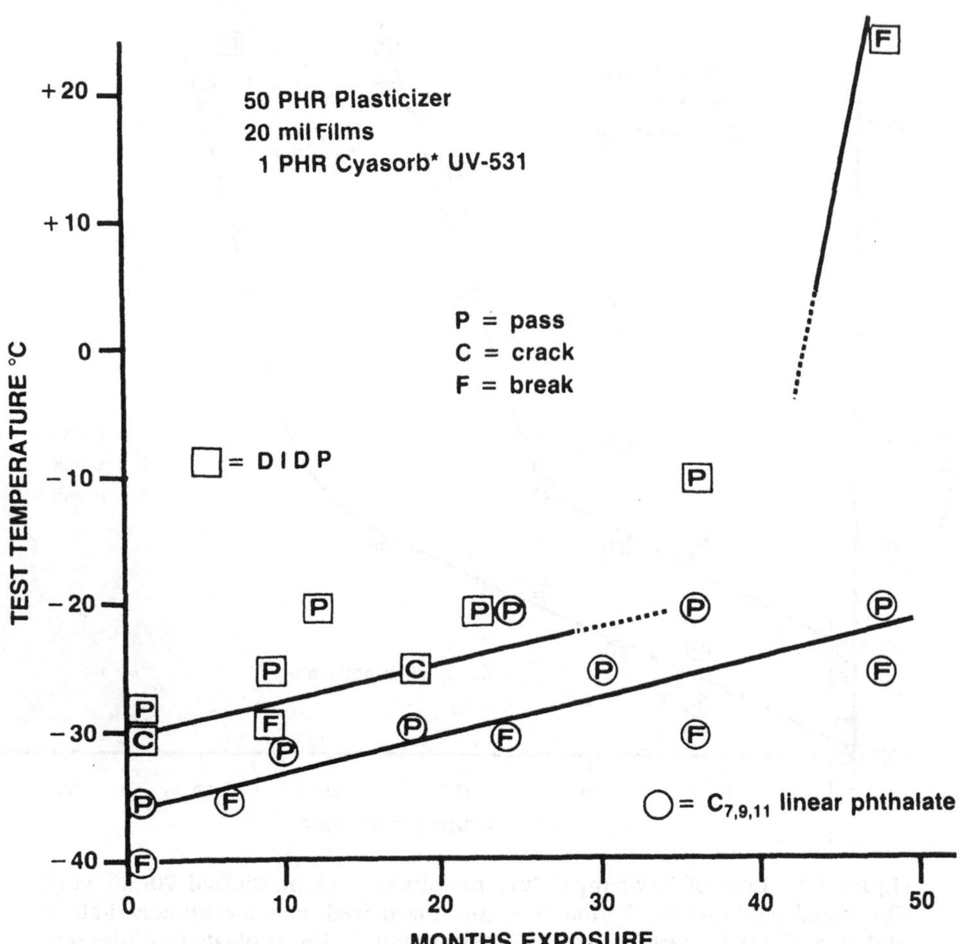

*Trademark American Cyanamid Co.

Figure 13. Degradation of UV-stabilized 20-mil vinyl films aged outdoors in Florida. Films contained 50 phr if plasticizer and 1 phr of a commercial UV absorber. These films lasted longer than the films of Figure 12; still, the film plasticized with the linear ester (circles) was more weather resistant than the film containing the branched material (squares).

Table 11 Relationship Between Structure and Viscosity in Dialkyl Phthalate Plasticizers

Plasticizer	Carbon number of alkyl groups	Structure of alkyl groups	Viscosity (cst) at 25°C
DnOP	8	Linear	31.8
DEHP	8	Branched	58.3
DIOP	8	Branched	55.0
DnDP	10	Linear	34.4
DIDP	10	Branched	90.2

illustrates how branching in the alkyl groups affects the per se viscosity of phthalate esters. In the majority of plastisol uses, low viscosity is desirable to facilitate pumping and application of the plastisol. As shown in Figure 14a, linear phthalates give plastisols of lower viscosity than do branched-chain esters of similar molecular weight. As plastisols age, their viscosity typically rises as the plasticizer slowly begins to solvate and swell the resin particles. This lack of viscosity stability is more pronounced at higher temperatures. Figure 14b and c show the aging characteristics of plastisols at temperatures that simulate storage in a hot warehouse or plant environment. Under these conditions the viscosity stability of DOP plastisols is seriously affected, while the linear phthalate and DINP still show only moderate increases in viscosity.

Many applications for plastisols involve coating the plastisol onto some type of substrate, such as a fabric. For production economy, these coating operations are run at the fastest line speeds consistent with good quality. Frequently, the limiting factor on line speed is the dilatancy of the plastisol, that is, the degree to which its viscosity increases in response to increasing shear rates. If the plastisol is too dilatant, it will thicken so much at high line speeds that it will coat the substrate poorly or not at all. The response of plastisol viscosity to shear rate is influenced by several factors, including the plasticizer; Figure 15 illustrates the effect of plasticizer choice. In the shear range of most coating operations, plastisols made with branched-chain phthalates exhibit marked dilatancy. Under the same high shear conditions, linear phthalates give lower viscosities and practically no dilatancy. Because of this, linear phthalate plasticizers provide more flexibility in formulating for coatings applications and frequently permit greater throughput rates on high-speed coating lines.

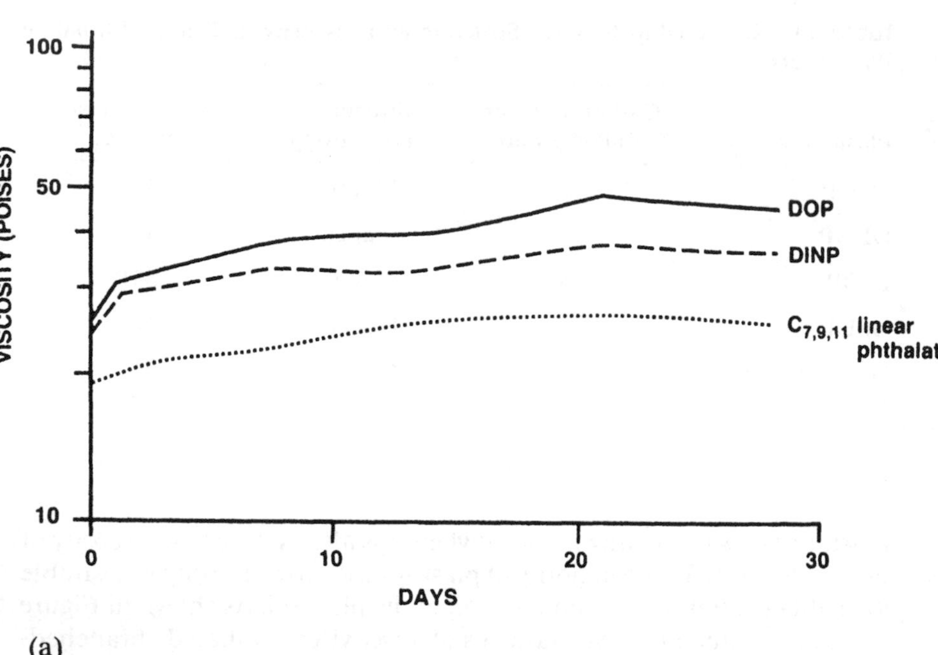

(a)

Figure 14. Viscosity of plastisols made with 65-phr phthalate plasticizer. At room temperature (25°C) linear phthalate gives a plastisol of lower viscosity than do branched-chain esters. (a) In elevated temperature (40 and 50°C, respectively) aging (b and c) viscosity of linear and DINP plastisols is more stable than that of DOP plastisol.

9 COMMERCIAL APPLICATIONS OF LINEAR PLASTICIZERS

Before the utilization of the economical "oxo" route in the late 1960s, linear plasticizers had been largely confined to specialty high-performance applications. The use of linears jumped dramatically with the introduction of products priced close to parity with the commodity branched-chain materials. Phthalates, trimellitates, and adipates have been or are being manufactured from linear alcohols; of these, the phthalates are used in by far the largest quantities.

9.1 Low-Temperature Applications

Probably the biggest single factor in the commercial success of the linear phthalates has been their good low-temperature performance, coupled with low cost. As shown in Table 12, linear phthalates give low-temperature performance equal to 70/30 DOP/DOA blends at a much lower cost. Because of this, linears have almost completely displaced these

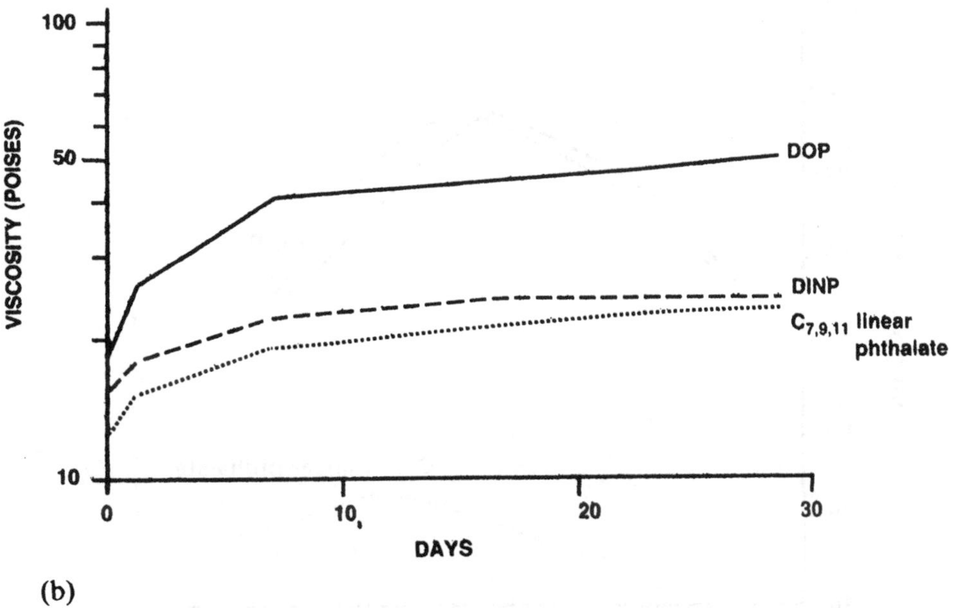

100
50
10
VISCOSITY (POISES)
DOP
DINP
$C_{7,9,11}$ linear phthalate
0
10
20
30
DAYS

(b)

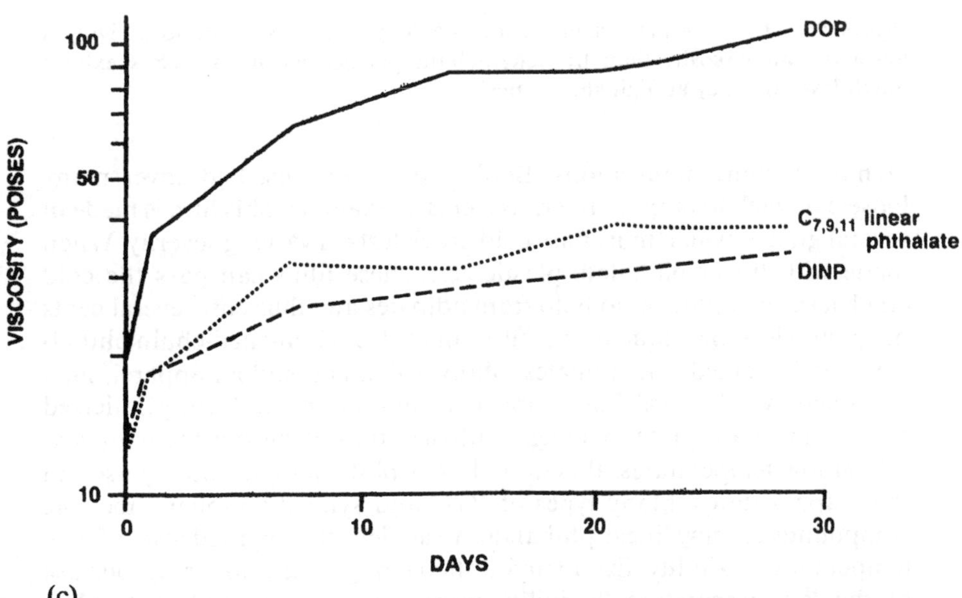

100
50
10
VISCOSITY (POISES)
DOP
$C_{7,9,11}$ linear phthalate
DINP
0
10
20
30
DAYS

(c)

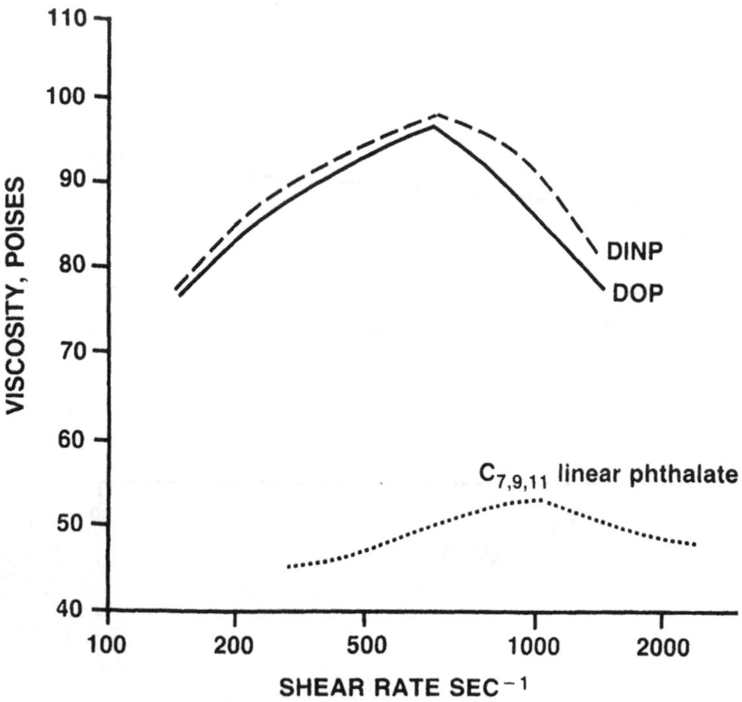

Figure 15. Effect of plasticizer on the rheology of plastisols. Besides giving a lower overall viscosity than branched-chain plasticizers, linear esters exhibit much less dilatency at high shear rates.

blends in many applications. Bookbinding film, used to cover many looseleaf notebooks and binders, is a classic example. This film is made in several grades, which must pass cold crack tests of varying severity. When made with linear phthalate plasticizers, these films can pass the cold crack tests with little or no help from adipates and thus cost several cents per pound less than comparable films made from branched-chain phthalates supplemented with adipates. Many articles of outdoor apparel, such as raincoats, "slickers," boots, and galoshes, are made from plasticized PVC, as are many pieces of luggage. Most of these items must remain flexible at low temperatures; therefore, linear phthalates are widely used in these applications. Many types of PVC and synthetic rubber shoe sole compounds employ linear phthalates to achieve the required level of low-temperature flexibility. Even vinyl wallcovering makes advantageous use of the low-temperature flexibility imparted by linear phthalate plasticizers. The PVC compounds used on wallcovering typically have a low plasticizer level, around 25 PHR, and high loadings of filler and pigments

Table 12 Performance Comparison of Linear Phthalate with DOP/DOA Blend in PVC Compound[a]

	DOP	DOP/DOA blend 70/30	7-, 9-, 11-carbon linear phthalate
43-phr plasticizer			
Shore A hardness	86	86	87
Low-temperature flex, T_f (°C)	−22.3	−26.1	−26.7
Volatility			
1 day	4.2	6.6	1.7
6 days	19.3	28.0	8.2
67-phr plasticizer			
Shore A hardness	70	69	71
Low-temperature flex, T_f (°C)	−40.0	−47.7	−48.1
Volatility			
1 day	4.1	6.7	2.0
6 days	18.3	29.2	7.6
100-phr plasticizer			
Shore A hardness	54	54	54
Low-temperature flex, T_f (°C)	−56.2	−63.1	−62.3
Volatility			
1 day	3.6	6.3	1.7
6 days	16.6	28.5	7.3
Plasticizer cost (cents/lb) 1987	43.0	49.9	45.5

[a]Test formulation: 100 parts PVC resin, medium molecular weight
As shown Plasticizer
0.5 part Dibasic lead stearate (stabilizer)

to achieve opacity. Compounds of this type become brittle at even moderately low temperature; the extra margin of safety contributed by the linear phthalates protects against damage during shipping in the winter months. Similarly, the thin, semirigid clear vinyl film used in "window" packages for many small items is often made with linear phthalates to better withstand rough handling in transit. When severe requirements dictate the use of some adipate, those made from linear alcohols deliver the best performance, thereby minimizing the amount needed.

9.2 Elevated-Temperature Applications

Lower volatility is another important commercial advantage of linear phthalates over branched-chain esters. PVC compounds having lower plasticizer volatility are more durable, being better able to retain their softness and flexibility when subjected to long-term elevated temperatures. Among plasticizers for PVC, linear phthalates are unique in having this favorable combination of low volatility, good low-temperature performance, and low cost. This combination is the primary reason why they are the preferred plasticizers in many PVC wire and cable applications. Most telecommunication and automotive wire insulations employ linear esters in the nine-carbon range, and UL-rated building wire is usually jacketed with vinyl compounds that contain these esters. The primary insulation on building wire must now pass the Underwriters' Laboratory specifications for a 90°C compound and so requires a plasticizer of even lower volatility. Linear phthalates in the 11-carbon range are widely used in this and other 90°C insulation applications. Linear trimellitates are used for the same reasons in higher-temperature applications.

Automotive undersealants are another vinyl application that utilizes higher-molecular-weight linear phthalate plasticizers. These materials are applied on auto rocker panels before painting. Since their purpose is to protect against impacts from road gravel, they must remain flexible at low temperatures. However, they must also be low-volatility materials to retain their flexibility after passing through the body paint bake cycle. Another advantage of linear phthalates in this application is their low affinity for automotive paint resins. This minimizes the chances of any deleterious interactions between the two.

9.3 Sunlight- and Weather-Resistant Applications

There are a large number of uses for flexible PVC which must withstand prolonged exposure to sunlight and weather. For maximum durability, most uses require good low-temperature flexibility and low plasticizer volatility. Because of their superior performance in all three areas, linear phthalate esters are widely recognized as the best plasticizers for these demanding applications. Typical end uses of this type are swimming pool covers, liners for aboveground pools, tent and awning fabrics, outdoor furniture components, and liners for irrigation ditches and landfills. Other applications include traffic control cones, premium vinyl sheet for commercial roofing, and vinyl-coated fabric for automobile "vinyl hardtops" and convertible tops.

9.4 Nonfogging Applications

Vinyl-coated fabric for automobile seats and dashboard skins has all the requirements of outdoor applications plus the requirement for low "fogging." Fogging of auto windshields is caused by the evaporation of volatile components from various parts of a closed car's interior in hot weather, with subsequent condensation on the relatively cooler inner glass surfaces. Only a few percent of a very volatile component will produce an unacceptable level of fogging. Plasticizers in the vinyl upholstery and dashboard padding have been blamed for this, sometimes rightly so, although in many cases other interior systems have been at fault. To eliminate the vinyl components as possible fogging sources, these parts are now usually made with plasticizers such as linear phthalates and trimellitates that have low fogging characteristics. Many of these coated fabric products are manufactured using plastisol technology. In these cases, the superior plastisol rheology obtained with linear plasticizers allows higher line speeds and thus better production economics.

10 SUMMARY

Plasticizers derived from linear alpha olefins are used primarily in the flexible vinyl industry. Phthalates are the most widely used linear esters, although adipates and trimellitates have significant commercial applications.

As a group, linear plasticizers have several performance advantages over their branched-chain counterparts: better low-temperature performance, lower volatility, better oxidative stability, and better plastisol rheology. This unique combination of attributes is itself a significant advantage of linears over other plasticizers; it is why linear plasticizers have long been the materials of choice in most high-performance vinyl applications. Now largely cost competitive with branched-chain esters, linears have displaced these products in a broad range of applications in which economy as well as performance are important.

REFERENCES

1. Sears, J. K., and Darby, J. R. *The Technology of Plasticizers,* Wiley, New York (1982).
2. ASTM D 3291, *Compatibility of Plasticizers in Poly(vinyl Chloride) Plastics Under Compression,* American Society for Testing and Materials, Philadelphia, Pa.

3. ASTM D 2240, *Durometer Hardness,* American Society for Testing and Materials, Philadelphia, Pa.
4. ASTM D 638, *Tensile Properties of Plastics,* American Society for Testing and Materials, Philadelphia, Pa.
5. Clash, R. F., and Berg, R. M., *Ind. Eng. Chem., 34:*1218 (1942).
6. ASTM D 746, *Brittleness Temperature of Plastics and Elastomers by Impact,* American Society for Testing and Materials, Philadelphia, Pa.
7. ASTM D 1790, *Brittleness Temperature of Plastic Sheeting by Impact,* American Society for Testing and Materials, Philadelphia, Pa.
8. ASTM D 1203, *Volatile Loss from Plastics Using Activated Carbon Methods,* American Society for Testing and Materials, Philadelphia, Pa.
9. Leonard, J., and Gaillard, J. F. Make octenes with dimersol X, *Hydrocarbon Process., 60:*99 (1981).
10. Dmuchovsky, B., and Knowles, W. S., A relationship between low temperature flex performance and molecular structure of dialkyl phthalates, *Organic Chemicals Division Progress Report,* Monsanto Chemical Company, St. Louis, Mo., 1967.
11. Sears, J. K., A semimicro brittleness test to monitor weathering of plasticized PVC, *J. Vinyl Technol., 6:*57 (1984).

CHAPTER 6

Surfactants: oxo

CLIVE WILNE Imperial Chemical Industries Plc, Middlebrough, Cleveland, England

1 COMMERCIAL BACKGROUND

Over the last 20 years, oxo-derived nonionic and anionic surfactants have become the most cost-effective and environmentally acceptable surfactants in domestic and industrial detergent formulations.* In the early 1950s, alkylbenzene sulfonate (ABS) synthesized from propylene tetramer achieved rapid growth but was replaced in the mid-1960s by the more biodegradable linear alkylbenzene sulfonate (LABS). Another class of surfactant products—alcohol sulfate (AS), alcohol ethoxylate (AE), and alcohol ethoxysulfate (AES)—was based on higher alcohols in the range C_{12}-C_{18}, which were initially manufactured from natural fats and oils such

* Surfactants may be defined as a class of chemical products whose molecules are able to modify the properties of an interface (e.g. air/liquid, liquid/liquid) by lowering the interfacial tension. This facilitates soil removal, emulsification, wetting, demulsification, solubiliation, and so on. A detergent may be defined as a formulation containing a surfactant with additional ingredients such as builders, foam regulators, stabilizers, antiredeposition agents, perfume, dye, and so on. Builders are molecules that are added to improve the cleaning properties of detergents. They act by increasing soil removal and reducing soil redeposition. The most important builder is sodium tripolyphosphate.

as coconut oil and beef tallow. In terms of performance and biodegradability, the higher-alcohol surfactants derived from natural fats and oils compared favorable with LABS, but their growth was to an extent limited by significant raw material price fluctuation (Fig. 1). In contrast, using petrochemical feedstocks, Zeigler, and oxo chemistry, the production of detergent-range alcohols increased dramatically. During the last 15 years oxo surfactants with carbon chain lengths in the range C_9–C_{18}, predominatly C_{12}–C_{16}, have grown from about 0.2 million metric tons in the late 1960s to about 0.7 million metric tons worldwide in 1985 (COMECON countries not included). This situation, together with the widespread use of synthetic fibers, has stimulated the increased use of these surfactants in detergent formulations in the form of fatty alcohol ethoxylates, alkylphenol ethoxylates, ethoxysulfates, fatty amine derivatives, and so on, based on ethylene, ethylene oxide, linear olefins, and phenol.

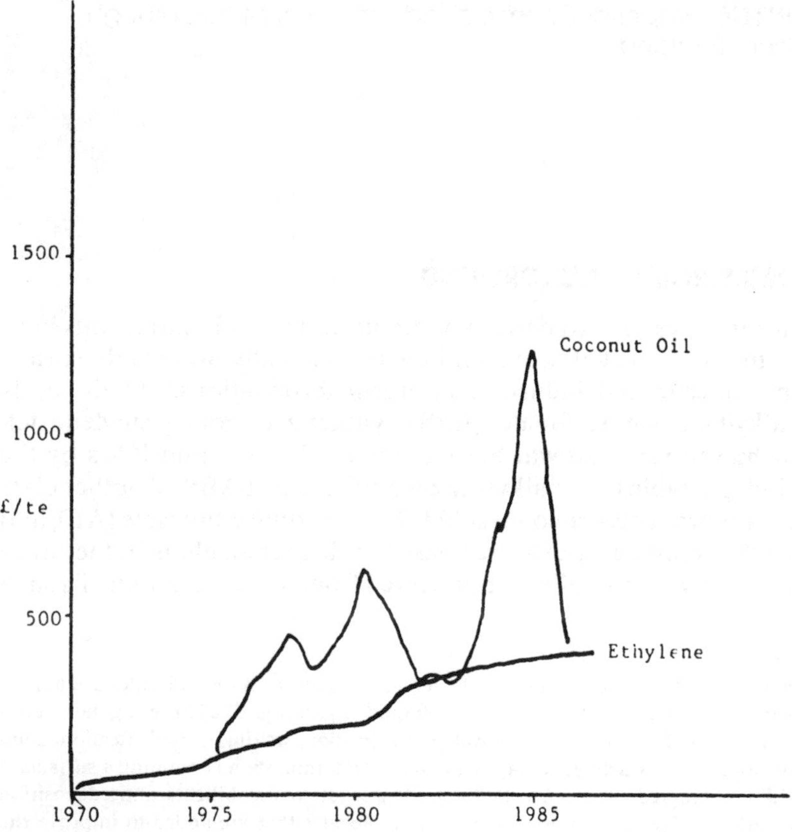

Figure 1 European market prices—ethylene and coconut oil.

1.1 Market Size

Major surfactant actual and forecast consumption for the period 1980–1990 is shown in Table 1. Taken together, in 1985, these surface-active agents represent about 67% of the total surfactant usage worldwide not including COMECON countries. The percentage in the United States is 63% and in Western Europe 72%. Figure 2 shows how the total surfactant market is divided in the United States and Western Europe. For the United States the domestic, household, and personal care sectors account for 45%. Industrial and other uses account for the remaining 55%. In Western Europe the corresponding figures are 61% and 39%.

A potential change of great significance to surfactant consumption and design not considered here is the use of surfactants for tertiary or enhanced oil recovery. Although some geological structures and crude oils are likely to favor enhanced oil recovery by surfactant methods, exploitation on a large scale is unlikely before the mid-1990s. The size of the operation is potentially so large that substantial new investment with long lead times will be required.

Whereas the many uses for surfactants in industrial applications will continue to extend the range and scope of surfactant technology, it is in the household and personal care sectors that changes are most likely to influence surfactant demand and performance significantly over the next five years.

1.2 Surfactant Market Trends and Influences in Household and Personal Care Products and Industrial Processes

Household Products

In 1985 the household products sector accounted for about 71% and 73% of the total demand for alcohol ethoxylates in the United States and Wes-

Table 1 Surfactant Consumption (Thousand Metric Tons)

	United States			Western Europe		
Surfactant type	1980	1985	1990	1980	1985	1990
Alkylphenol ethoxylates	160	185	205	50	66	40
Alcohol (C_{12}–C_{16}) ethoxylates	280	297	334	250	260	285
Linear alkylbenzene sulfonates	750	790	830	450	460	455
Soap	415	350	400	334	298	280

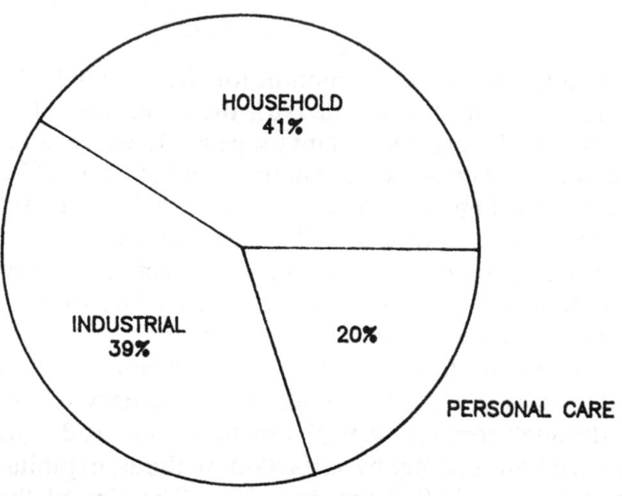

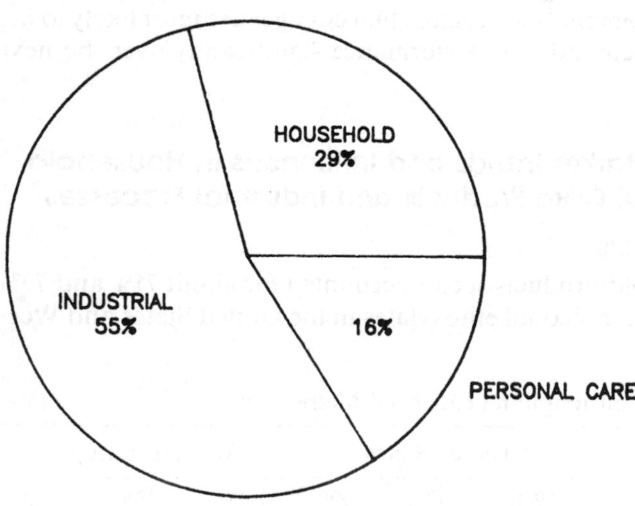

Figure 2 Surfactant end-use markets, 1982.

tern Europe, respectively. Table 2 shows the requirement for alcohol ethoxylates in household products in the period 1980–1990 in the United States and Western Europe. The changes taking place in the household sector are influenced by three factors: increased energy costs leading to reduced fabric washing temperatures, below 40°C; increased use of synthetic fibers requiring wash temperatures below 60°C; and legislative changes. (Figure 3.)

As a consequence of the trend toward lower wash temperatures, the nonionics content of heavy-duty laundry powders is increasing at the expense of anionics, mainly LABS. Alcohol ethoxylates with shorter chain lengths in the range C_{11}–C_{12} are sometimes preferred, as they are considered to be more effective at low temperatures than either LABS or the longer-chain C_{13}–C_{15} alcohol ethoxylates. As the washing machine market in Western Europe moves in favor of front-loading automatic washing machines, low-foaming surfactants such as alcohol ethoxylates together with C_{16}–C_{18} soaps will be used to reduce the sudsing action of LABS. At present, some heavy-duty powders have more nonionic than anionic surfactant in their formulations, which can lead to processing problems in the spray drying towers, where their higher rates of oxidation and volatilization can increase process costs and cause localized atmospheric pollution.

After cost and performance, the most important factor influencing the future use of surfactants relates to the environmental characteristics of detergents: biodegradability, eutrophication potential, and toxicity. In the future there will be an increased requirement for surfactants to be readily oxidized by bacteria in activated sludge or trickling filter treatment processes such that the resultant effluent has a very low biological oxygen demand (BOD). Eutrophication is caused by the condensed phosphate builders present in detergents. Their effect on surfactant usage is discussed in Section 4.2. In general, oxo surfactants exhibit low oral and dermal toxicity to animals and humans (see Section 4.3) and are similar to naturally derived alcohol surfactants.

Alternatives to condensed phosphates (e.g., sodium tripolyphosphate) as the builder in heavy-duty laundry powders are zeolites (calcium

Table 2 Alcohol Ethoxylates in Household Products (Thousand Metric Tons)

United States			Western Europe		
1980	1985	1990	1980	1985	1990
185	210	230	165	190	210

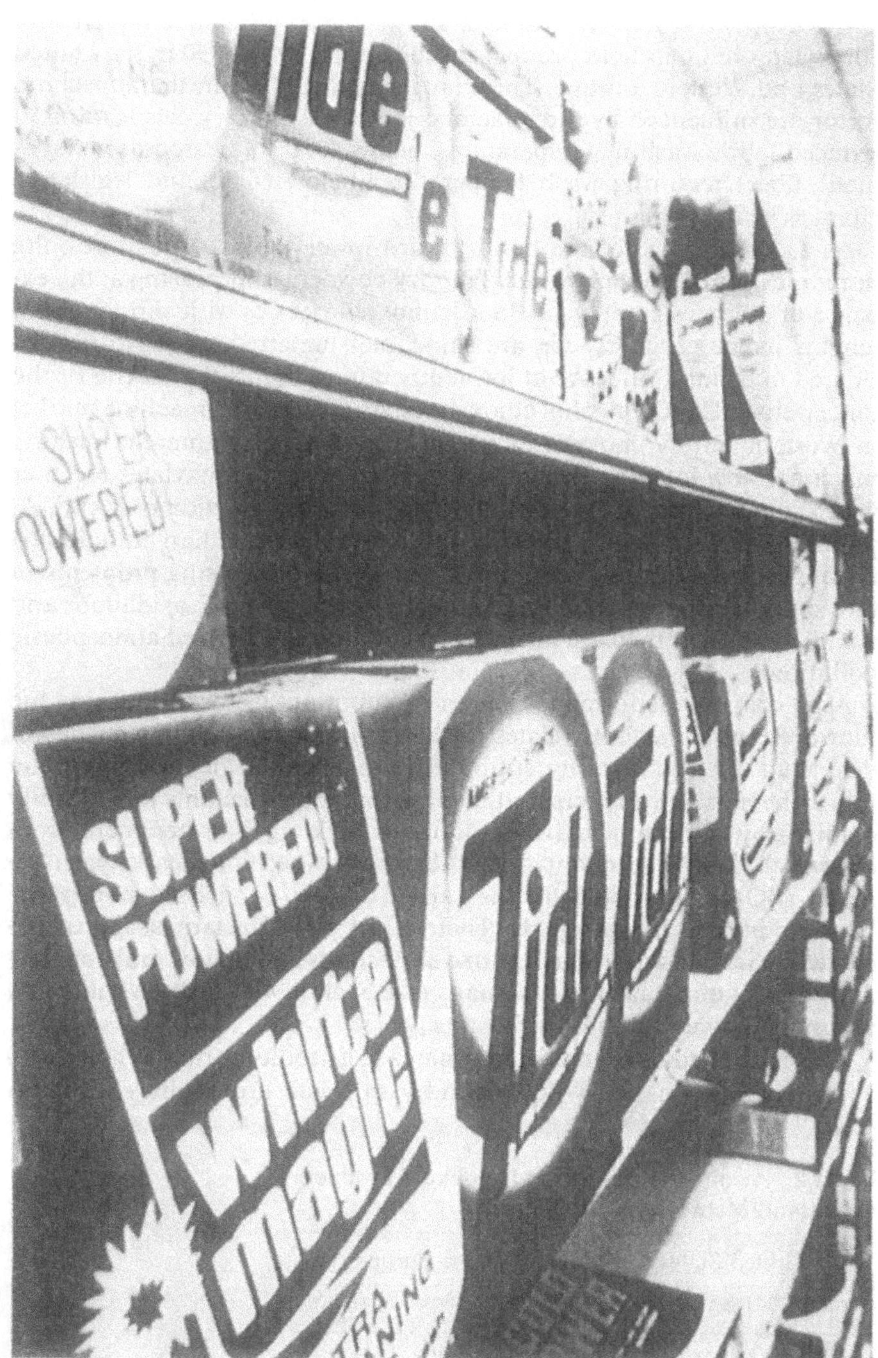

Figure 3 Typical household detergents.

aluminosilicates), trisodium salt of nitrilotriacetic acid (NTA), sodium silicate, and sodium carbonate. However, zeolites are not a total substitute for phosphates but can be used to reduce phosphate levels by up to 50% w/w. Although a minimum of 20% w/w of STPP in the detergent formulation is required for adequate performance at these levels of zeolite addition, the overall builder content of the formulation is usually increased (about 1.5 times phosphate removed). To further improve the sequestration of Ca^{2+} and Mg^{2+} ions, the nonionic surfactant level is also increased. An alternative would be to use alcohol ether sulfates, which are less sensitive to water hardness ions but are high-foaming. In parallel with the reduction in the amount of phosphate used in fabric washing powders, there will be increased use of perborate bleach with bleach activators and low-temperature bleaches (e.g., peracids). In the United States the problem of accommodating reduced washing temperatures and reduced phosphate levels is made somewhat easier by the separate addition of sodium hypochlorite bleach at the end of the wash cycle.

The reduction in the permitted level of phosphates is an important factor favoring the introduction of heavy-duty laundry liquids. At the present time, fabric washing liquids, although significant in the United States with about 25% of the market, have a very small percentage share of the Western European market, about 5%. This is because it is technically difficult to create a stable high-performance liquid that can be used in existing washing machines not normally designed to dispense a fluid. The liquid product must contain surfactants, builders, bleach, and enzymes if it is to compare favorably in terms of wash performance with existing heavy-duty laundry powders. It must also be cost-effective. Both aqueous and nonaqueous systems have been patented, and several aqueous formulations have been commercialized, notable, Wisk (Lever Brothers), Liz (Henkel), and Vizir (Proctor & Gamble) in Europe, and Lever Brothers in the United States. At the present time, none of the laundry liquids can completely match the performance of European heavy-duty powders.

Consumer preference for softer and more fragrant garments will continue to stimulate growth in fabric softeners. Virtually all fabric softeners sold in Western Europe are liquids. Dry paper-impregnated sheets, which are a fast-growing sector of the American market, are not used much in Europe because of the comparatively small number of domestic tumble driers. The use of fabric softener in laundry detergents is also small in Europe with the exception of Procter & Gamble's Bold 3 (United Kingdom) and Dash 3 (France and Germany). Fabric softeners have been in use for nearly 20 years in Western Europe, particularly in areas of hard water, which causes stiffening of the fibers. Overall growth in Western Europe is expected to be about 4% per year and about 3 to 4% in the United States over the next decade. The move toward higher active fabric softeners containing three to four times the level of fabric softener is likely to increase as consumers recognize the cost benefits that can be obtained.

Fabric softeners usually contain a dispersion of cationic surfactant (fatty amine quaternary or imidazoline compunds) in the range 5 to 15%. These cationic surfactants are substantive to fibers and impart softening and antistatic properties to the washed cloth as well as facilitating ironing. Cationic surfactants are normally derived from natural oils and fats. In the last four years ICI PLC has introduced a range of cationic surfactants based on an C_{13}-C_{15} hydrophobe. These synthetic cationic surfactants offer benefits of high purity, liquidity, and ease of processing compared to the naturally derived C_{12}-C_{14} and C_{16}-C_{18} cationics.

Personal Care Products

The current and present demand for alcohol ethoxylates in personal care products is given in Table 3. In this sector, growth is expected to be in proportion to changes in population and per capita income. While liquid soaps are an alternative to the traditional bar soap it is not yet apparent that the consumer prefers them. The incorporation of emollients in the formulation is an attractive idea, as is the ease of application of liquid soaps, but their cost-effectiveness is questionable. Shampoos for hair, on the other hand, are likely to continue to be an attractive growth area with 3 to 4% per year in Europe and 3% in the United States. In future, with increased consumer use, amphoterics, which minimize skin irritancy, are likely to become more important.

Industrial Processing Applications

Industrial surfactants accounted for about 10% of the total demand for alcohol ethoxylates in Western Europe in 1985. The actual and projected demand for alcohol ethoxylates in industrial processing applications in the period 1980–1990 is given in Table 4. Some 50% of the demand is believed to be accounted for by the textile industry. The remainder of the consumption is fragmented. Aside from textile applications, other ap-

Table 3 Alcohol Ethoxylates in Personal Care Products (Thousand Metric Tons)

United States			Western Europe		
1980	1985	1990	1980	1985	1990
15	18	22	36	45	55

Table 4 Alcohol Ethoxylates in Industrial Processing (Thousand Metric Tons)

United States			Western Europe		
1980	1985	1990	1980	1985	1990
45	50	60	25	28	32

plications include agricultural applications, plastics and rubber processing aids, metalworking fluids, paper products, vehicle cleaning, and polishes. Deman is unlikely to increase in the depressed textile sector. Elsewhere, growth is likely to be a modest 2 to 3% annually, in line with increases in gross national product.

The demand in the United States for alcohol ethoxylates in industrial surfactants is given in Table 4. Industrial surfactants accounted for about 16% of the total demand for alcohol ethoxylates in 1985 and are expected to show similar growth to Western Europe in the period to 1990.

2 LINEAR PRIMARY OXO ALCOHOLS FROM LINEAR OLEFINS FOR OXO SURFACTANTS

The development of alcohol-based surfactants is based on the oxo synthesis, discovered by Roelin in 1938. In the following 40 years, several commercial-scale processes were developed using simple rhodium and cobalt hydrocarbonyls as catalysts at high partial pressures of hydrogen and carbon monoxide to give a range of aldehydes and alcohols with chain lengths in the range C_8–C_{16}. More recently, work has concentrated on modifications of the oxo catalysts by varying the metal atom and using other ligands or by changing the reaction phase. Progress has also been made in the control of the hydroformylation reaction and associated technology.

The linear olefin feedstock required for the production of oxo alcohols is obtained by ethylene oligomerization, dehydrogenation of *n*-paraffins, or wax cracking. Chemical routes to oxo surfactants are outlined in Figure 4.

2.1 Ethylene Oligomerization

Ethylene oligomerization is the most important source of alpha olefins and is preferred to *n*-paraffin dehydrogenation or wax cracking because it allows predictable product distributions and higher alpha olefin content—about 95% in the final product. In contrast, *n*-paraffin dehydrogena-

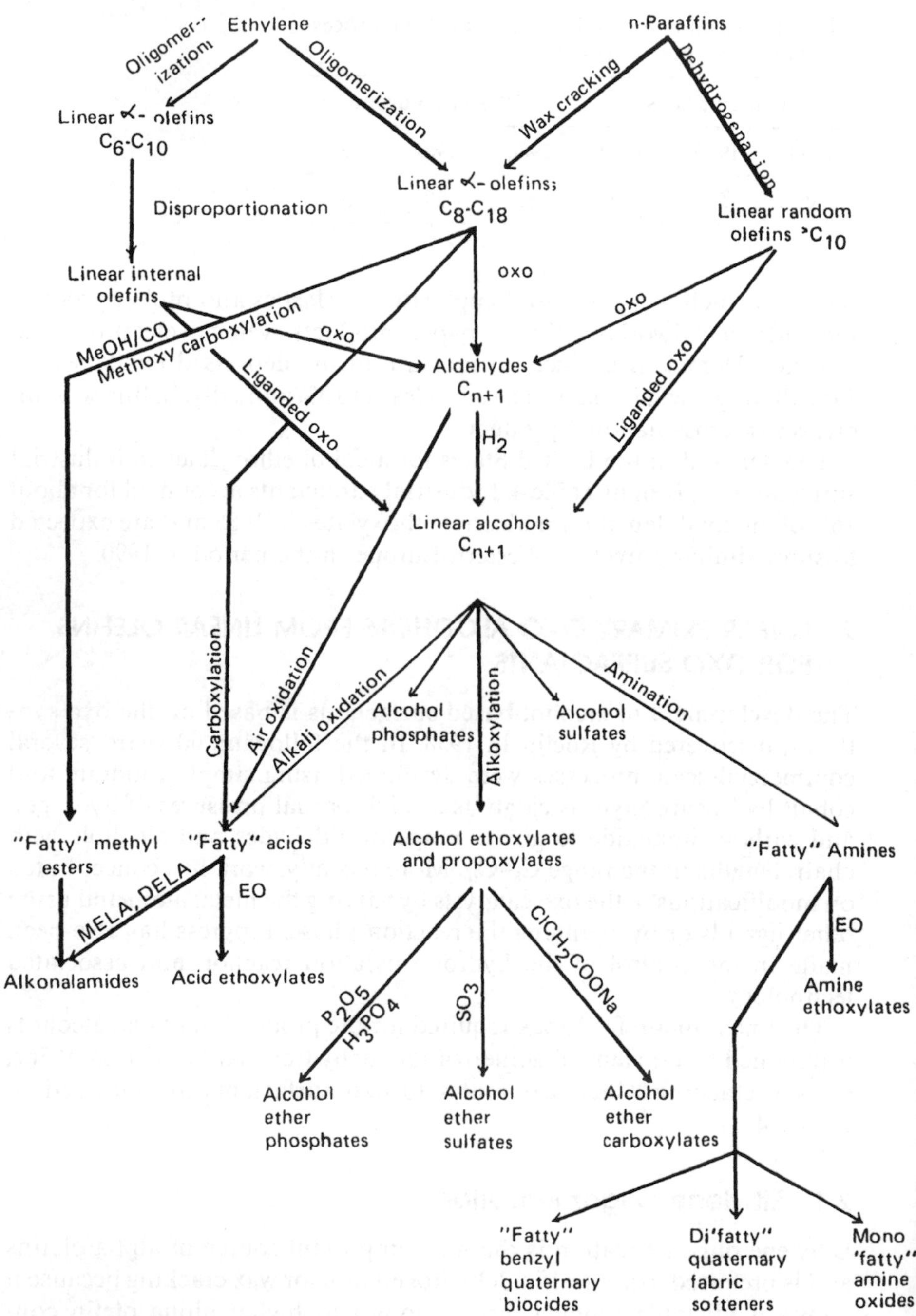

Figure 4 Oxo routes to surfactants.

tion produces linear random internal olefin. In the 1950s, Zeigler first demonstrated the chain growth of ethylene on alkylated metal catalysts, particularly aluminium alkyls. This is the basis of the Gulf process. A disadvantage of this approach is the broad product distribution. In a further development of the process Ethyl Corporation has incorporated a transalkylation step in which excess C_4–C_{10} alpha olefins are used to displace the alkyl chains from the growth stage of the process. This causes the Poisson distribution to peak in the desired carbon number range. The Shell process (SHOP) uses Zeigler-Natta catalysts in a solvent that is immiscible with the product alpha olefins, thus facilitating separation of the product from dissolved catalyst. The Shell process for alpha olefins, which includes oligomerization, isomerization, and disproportionation, together with Shell's own hydroformylation process, can use linear internal olefin feed and avoids the Poisson distribution of alpha olefins, only making alpha olefins of the most economic chain length, C_8–C_{18}.

2.2 Wax Cracking

The raw material for wax cracking is normal paraffins in the range C_{20}–C_{34}, obtained from the waxy distillate of suitable crude oils. The alpha olefins derived from this feedstock have both even and odd numbers of carbon atoms, in contrast to the ethylene oligomerization process using Zeigler catalysts, which produces only even numbers of carbon atoms. By this process it is possible to obtain 70 to 75% of linear alpha olefins, but the product also contains internal and branched olefins, diolefins, aromatics, and paraffins. In general, alpha olefins obtained from wax cracking are significantly lower in purity and quality than those obtained from ethylene growth.

2.3 Dehydrogenation of *n*-Paraffins

In this process, linear internal mono-olefins are produced by dehydrogenating linear paraffins extracted from kerosene via molecular sieve processes over a precious metal catalyst. The internal olefins are extracted from the unreacted paraffins by an absorptive separation process using crystalline metal aluminosilicates. Table 5 shows a comparison of olefin compositions derived from ethylene oligomerization, wax cracking, and *n*-paraffin dehydrogenation.

2.4 Primary Alcohols by Ethylene Oligomerization

In 1962 the first synthetic alochol was prodeuced for surfactant application when Conoco extened its process for ethylene oligomerization to in-

Table 5 Olefin Compositions

	Linearity (%)	Carbon number distribution	Alpha olefin content (%)
Ethylene oligomerization	90–98	Even	87–96
Wax cracking	>80	Odd, even	70–75
n-Paraffin dehydrogenation	90	Odd even	10–15

clude an air oxidation and water hydrolysis stage to release linear alcohols and an alumina slurry. Because the alkyl chain length follows a Poisson distribution, a wide range of alcohols is produced, which is undesirable. The alcohols so produced, however, have high linearity (>98%). In a way exactly analogous to the production of a selective range of carbon chain lengths for alpha olefins, this process can be modified to include intermediate displacement and transalkylation stages. This results in a narrowing of the molecular weight range of the chain, giving a much larger fraction in the range C_{12}–C_{14}.

2.5 Primary Alcohols by Hydroformylation of Linear Olefins

In the hydroformylation process, olefins are reacted with CO and H_2 to form aldehydes that may be hydrogenated to alcohol. Under certain conditions (e.g., high hydrogen partial pressure and hydroformylation catalyst also possessing hydrogenation characteristics), the aldehydes can be converted immediately to primary alcohol. The catalyst used commercially for liquid-phase hydroformylation is homogeneous and is essentially a hydrocarbonyl complex of either cobalt or rhodium. Cobalt-containing catalysts are the principal catalysts used for detergent alcohol processes, but rhodium complex catalysts are gaining in importance because of their potentially higher selectivity and activity. Their use to date has been restricted, mainly because of the need for highly efficient catalyst recovery. The use of cobalt or rhodium catalysts in various forms for oxo alcohol manufacture determines the type and degree of branching present in the final oxo surfactant hydrophobe.

Nonliganded Cobalt and Rhodium Catalysts

The mechanism and kinetics of the oxo process with conventional cobalt catalyst have been described by the Heck-Breslow mechanism [1]. A similar mechanism also describes the oxo process using rhodium catalyst [2]. Increasing the reaction temperature significantly increases the rate of olefin conversion but reduces the yield to aldehyde and also decreases catalyst stability. In practice, it is usual to maximize feedstock efficiency, which often dictates a lower first-pass conversion than is otherwise achievable, with recovery and recycle of unconverted olefin.

Nonliganded cobalt and rhodium catalysts have several shortcomings. The catalyst is unstable at high temperatures above about 170°C, particularly if the CO partial pressure is low. This requires the use of high-synthesis gas pressures, 200 to 300 bar. The mechanism of the oxo reaction with alpha olefin to form primary alcohol is such that both normal aldehyde and 2-methyl aldehyde are formed. This is especially true for hydroformylation involving rhodium catalyst, where the alpha olefin can undergo rapid isomerization before hydroformylation. When linear internal olefins are used as feedstock, the resulting aldehydes are very branched, ⩾90%. In general, aldehydes giving branched alcohols are not desirable as feedstocks for oxo surfactants because of their reduced rate of reaction and biodegradation.

Liganded Cobalt and Rhodium Catalyst

Following work by Lautenschlager [3] and others, Shell, through the work of Slaugh and Mollineaux [4], modified the classical cobalt catalyst by partial substitution of the CO ligands by alkyl and aryl phosphines. The effect of this cobalt modification is to increase hydroformylation selectivity, increase catalyst thermal stability, increase isomerization and hydrogenation activity, and decrease oxo activity. The decrease in oxo reactivity requires that both reaction remperature and reactor volume be increased. The effect of reaction temperature on the ratio of branched to unbranched isomers is a function of the type of ligands used. This ratio is significantly increased by liganded catalyst compared to unmodified catalysts. Because liganded catalyst are much more stable, they can exist at higher temperatures and in the absence of carbon monoxide. This has led to processes for the recovery of catalysts in the active form by in situ removal of aldehyde and unreacted olefin.

For all cobalt- and rhodium-catalyzed oxo processes it is essential to use high-purity alpha olefin feedstock and synthesis gas free from oxygen, sulfur compunds, and dienes, which can act as catalyst poisons for the hydroformylation or subsequent hydrogenation reactions. They may also form unwanted by-products. Commercial processes have been developed for each of the four possible ways of catalyzing the oxo reaction using cobalt and rhodium. A comparison of these processes is given in Tables 6 and 7.

Thus liganded cobalt and rhodium catalysis is preferred for the production of high normal/iso ratio C_{12}–C_{16} alcohols by the oxo process. Currently, only Shell has a commercially feasible process for detergent-range

Table 6 Commercially Available Processes for OXO Alcohols

	Cobalt		Rhodium	
Metal process	Ruhr Chemie/ BASF	Shell	Ruhr Chemie	UCC/JM/ McK
Catalyst	$HCo(CO)_4$	$HCo(CO)_3L$	$HRh(CO)_4$	$HRh(CO)_3L$
Temperature (°C)	110–180	160–250	100–140	60–120
Pressure (bar)	250–300	50–150	200–300	7–50
Metal concentration (% w/w on olefins)	0.1–1	0.6	10^{-5}–0.01	0.01–0.1
Liquid space velocity (hr^{-1})	0.5–2.0	0.1–2.0	0.3–0.6	0.1–0.25
Hydrocarbon formation	Low	High	Low	High
Reaction products	Aldehyde	Alcohols	Aldehydes	Aldehydes
n/iso ratio	4:1	7:1	1:1	>11:1

alpha olefin feedstock that utilizes a liganded cobalt catalyst. The Union Carbide Corporation/Davey McKee/Johnson Matthey process using a liganded rhodium catalyst system has so far been restricted to a low-carbon-number alpha olefin (propylene). In this process the alpha olefin and the corresponding aldehyde are sufficiently volatile to be separated from the homogeneous catalyst system in situ in the reactor. More recent research work using rhodium has aimed at devising a system, often based

Table 7 Comparison of Ligand Modified Cobalt- and Rhodium-Catalyzed Processes

	Cobalt	Rhodium
Total yield	Same	
Heavy end formation	Low	High
Paraffin formation	High	Low
Yield of unbranched products	Lower	Higher
Yield of branched products	High	Low
Temperature	High	Low
Pressure	High	Low
Flexibility	High	Low

on changing the reaction phase, that would allow separation of the homogeneous catalyst phase from the unreacted alpha olefin and product C_{12}–C_{16} aldehyde. Table 7 indicates that success in this endeavor would lead to a process with a lower capital cost, through the use of lower pressure, high normal-to-isomer ratio, high feedstock efficiency, and lower catalyst usage relative to cobalt processes existing at the present time.

3 OXO SURFACTANTS: MANUFACTURE AND PROPERTIES

As described in Section 2, surfactant intermediates are manufactured from a range of mainly linear olefins and by a variety of defferent processes. The key products of these processes (fatty acids, amines, and alcohols) differ from each other and natural fatty chemicals in carbon chain distribution and proportion of normal and branched isomers. These differences can influence the course of their subsequent conversion to surfactants and affect both properties and performance.

3.1 Alcohol Ethoxylates

Reaction Mechanism

Reaction of alcohol with ethylene oxide is catalyzed by bases. Typically, potassium hydroxide or sodium methoxide are used industrially at concentrations of 0.1 to 2% w/w on the hydrophobe. The reaction sequence is

$$ROH + KOH \xrightarrow[\text{vacuum}]{120°C} RO^-K^+ + H_2O \tag{1}$$

$$RO^-K^+ + \underset{\text{O}}{CH_2{-}CH_2} \xrightarrow[\text{1–6 bar}]{120–160°C} ROCH_2CH_2O^-K^+ \tag{2}$$

$$ROCH_2CH_2O^-K^+ + n(\underset{\text{O}}{CH_2{-}CH_2}) \rightarrow RO(CH_2CH_2O)^-_{n+1}K^+ \tag{3}$$

$$RO(CH_2CH_2O)^-_nK^+ + ROH \rightarrow RO(CH_2CH_2O)_nH + RO^-K^+ \tag{4}$$

$$RO(CH_2CH_2O)^-_nK^+ + H^+X^- \rightarrow RO(CH_2CH_2O)_nH + KX \tag{5}$$

As reaction (3) is slightly preferred to reaction (4), alcohol ethoxylates have a wide homolog distribution and contain appreciable amounts of unreacted alcohol until at least 10 mol of ethylene oxide per mole of alcohol has been added. This is shown in Figure 5. Molecular weight distributions for an homologous series of oxo alcohol ethoxylates and a range of 6- to 7-mol ethoxylates of commercially available oxo alcohols are shown in Figures 6 and 7. It is clear that the narrowest molecular weight distributions are obtained with those alcohols with the highest proportion of n-isomers. This is because branched alcohols ethoxylate more

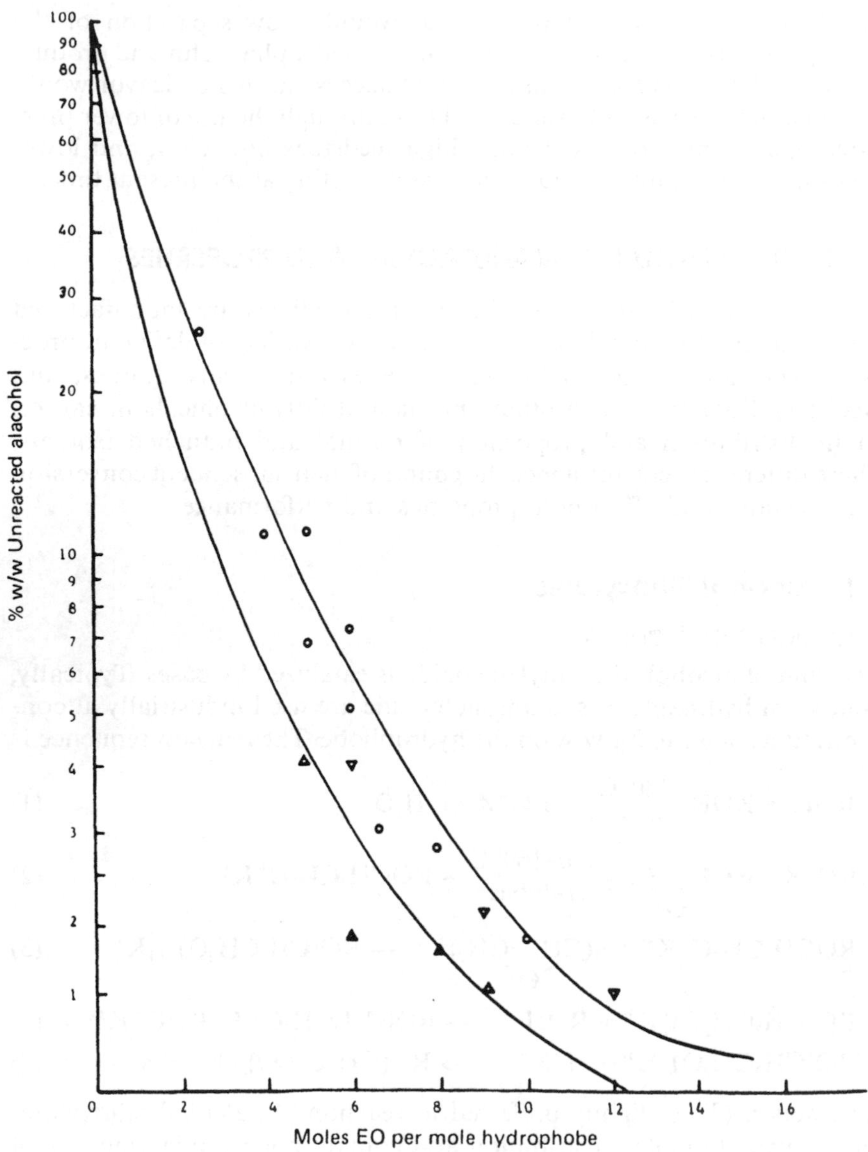

○ oxo $C_9 + C_{11}$ or $C_{13} + C_{15}$ 50% *n*-isomers
▽ oxo $C_9 + C_{10} + C_{11}$ 80% *n*-isomers
△ oxo $C_{12} + C_{13} + C_{14} + C_{15}$ 80% *n*-isomers

Figure 5 Oxo alcohol ethoxylates unreacted alcohol content (50% w/w to 80% w/w n-isomers).

slowly than their linear counterparts. Ethoxylates with a narrow molecular weight distribution also contain less unreacted alcohol, although this is not apparent from Figure 5 because of the scatter of the experimental results. In terms of performance the variations in isomer distribution are only significant for 1- to 3-mol adducts used for the production of alcohol ether sulfate. In a similar way, lower-molecular-weight alcohols react more rapidly than those with a higher molecular weight, but most commercial detergent alcohols have similar molecular weights.

Narrower Homolog Distributions

For comparison the molecular weight distribution using barium nonylphenate as an ethoxylation catalyst is shown in Figure 8. These and similar alkali earth catalysts have been the subject of many recent patents to Conoco/Vista Chemicals, Union Carbide Corporation, and others. [5].

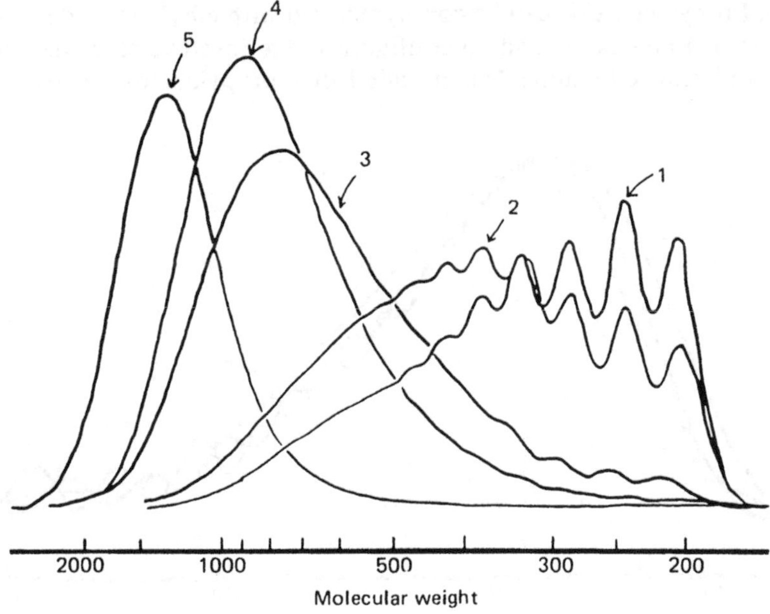

1 2 mole ethoxylate
2 3 mole ethoxylate
3 7 mole ethoxylate KOH catalyst
4 11 mole ethoxylate
5 20 mole ethoxylate

Figure 6 C_{13}–C_{15} oxo alcohol ethoxylates (50% n-isomers) molecular weight distribution (by GPC).

Compared to conventional catalysts, they give significantly narrower molecular weight distributions and less unreacted alcohol, together with similar makes of by-product polyethylene glycols. The colors of the product ethoxylates are good and there are no significant levels of dioxan, dioxolane, or aldehyde, which are normally associated with such Lewis acid catalysts as boron trifluoride or zinc perchlorate and which give ethoxylates of narrow molecular weight distribution. However, as they are active on a molar basis, the barium catalyst residues need to be removed by filtration following neutralization with acid, normally acetic. This adds to the production cost compared to the standard process, in which residual potassium acetate is left in the product.

Alcohol Ethoxylates: Typical Properties

Manufacturers' typical properties for a range of oxo-based alcohol ethoxylates, including the most commonly encountered types, are detailed in Table 8. Propylene oxide can be copolymerized with ethylene oxide onto oxo alcohols in block or random configuration to improve ethoxylate liquidity and reduce foaming. When added first, propylene oxide also in-

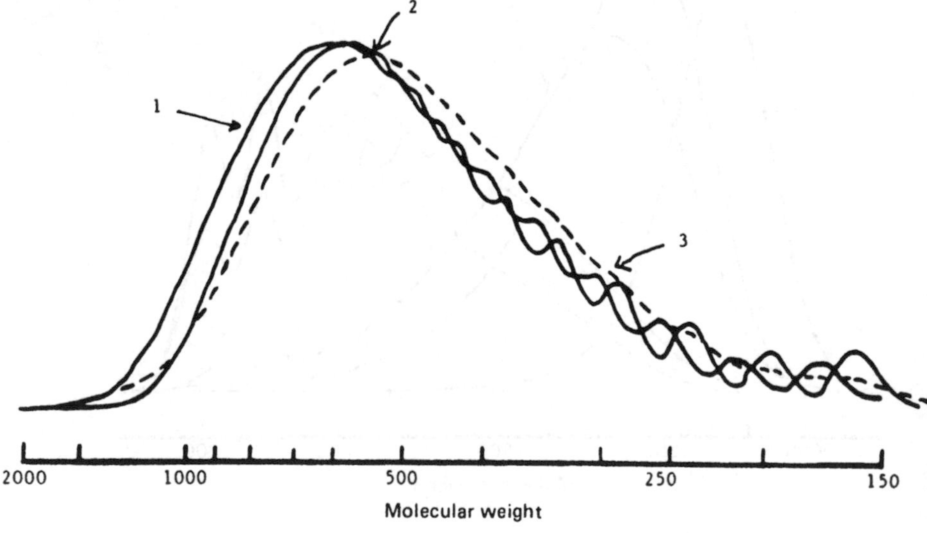

1 oxo $C_9 + C_{11}$ 50% *n*-isomers
2 Zeigler $C_8 + C_{10}$ 100% *n*-isomers
3 oxo $C_9 + C_{10} + C_{11}$ 80% *n*-isomers

Figure 7 $C_8 \rightarrow C_{11}$ alcohol 6 mole ethoxylates molecular weight distribution (by GPC).

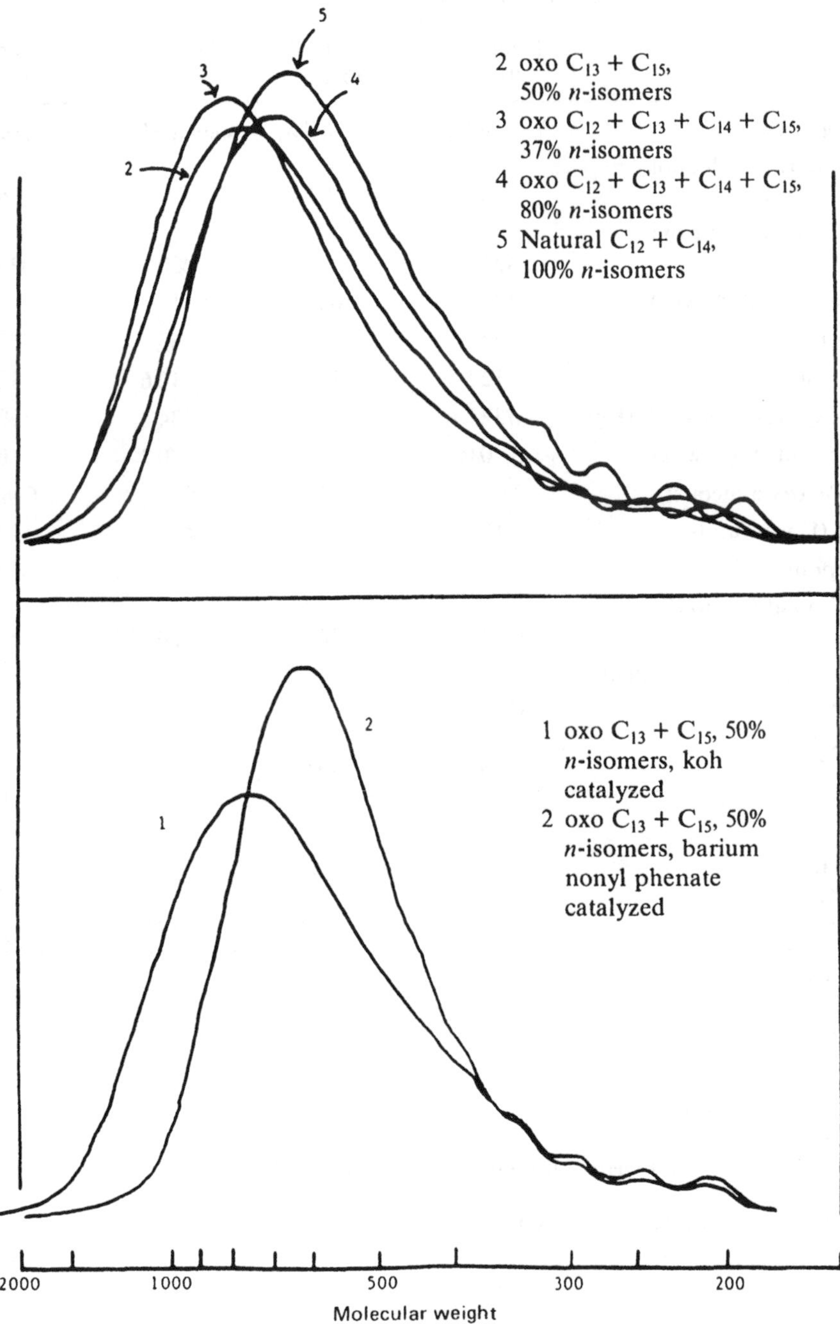

Figure 8 $C_{12} \rightarrow C_{15}$ alcohol 7 mole ethoxylates molecular weight distribution (by GPC).

Table 8 Typical Properties of OXO Alcohol Ethoxylates

	Neodol[a] 91-6	Lutensol[b] ON6	Ukanil[c] 50	Synperonic[d] 91/6
Hydrophobe	C_9-C_{10}-C_{11}	C_9-C_{10}-C_{11}	C_9-C_{10}-C_{11}	C_9-C_{11}
EO groups per alcohol (mol/mol)	6			6
Molecular weight from OH value	425		420	434
Active content (% w/w)	100	100	[f]	[f]
EO content (% w/w)	62	60	63	63
Calculated HLB	12.5	12	12.6	12.5
Hydroxyl value (mg KOH/g)	132		134	129
Water content (% w/w)	0.02		<0.2	0.2
pH (1% w/v aqueous)	6	7	5–7[g]	6–8[h]
Color (Hazen units)	10		<75	20
Pour point (°C)	7	<0		2
Cloud point (1% w/v aqueous)	52	36	50	52
Polyethylene glycol content (% w/w)	(<1)[f]		<2	1
Specific gravity				
25°C	0.991	0.99		
40°C				0.981
50°C				
Viscosity (cSt)				
40°C	23			22
50°C			14	

[a]Shell Chemical Co.
[b]BASF.
[c]Kuhlmann.
[d]ICI PLC.
[e]Chimica Augusta SpA.
[f]Not normally quoted.
[g]5% w/v.
[h]Acid value max 0.2 mg KOH/g; Alkalinity value max. 0.2 mg KOH/g.
[i]Insoluble.
[j]10% w/w in 25% butyldiglycol/75% water.

Neodol[a] 91-8	Ukanil[c] 85	Synperonic[d] 91/8	Neodol[a] 25-3	Synperonic[d] A3
C_9-C_{10}-C_{11}	C_9-C_{10}-C_{11}	C_9-C_{11}	C_{12}-C_{13}-C_{14}-C_{15}	C_{13}-C_{15}
8.4	9	8	3	3
529	545	533	336	348
100	[f]	[f]	100	>99
70	72	70	40	39
14	14.4	14	7.9	7.8
106	102	105	167	161
0.02	<0.3	0.2	0.02	<0.2
6	5–7[g]	6–8[h]	[i]	6–8[h]
10	<75	20	10	<50
16		14	4	5
80	85	80	[i]	57[i,j]
	<2	2	(<1)[f]	(<1)[f]
1.002			0.925	
		1.006		
				0.900
30		32.5	19	
	20.4			23

(continued)

Table 8 *Continued*

	Lialet[e] 125L/3	Lialet[e] 125H/3	Neodol[a] 25-7
Hydrophobe	C_{12}-C_{13}-C_{14}-C_{15}	C_{12}-C_{13}-C_{14}-C_{15}	C_{12}-C_{13}-C_{14}-C_{15}
EO groups per alcohol (mol/mol)	3	3	7
Molecular weight from OH value	337	342	519
Active content (% w/w)	100	100	100
EO content (% w/w)	39	39	61
Calculated HLB	7.8	7.8	12.2
Hydroxyl value (mg KOH/g)	166	164	108
Water content (% w/w)	<0.1	<0.1	0.02
pH (1% w/v aqueous)	[f]	[f]	6.0
Color (Hazen units)	<50	<50	10
Pour point (°C)	5–11	8–14	21
Cloud point (1% w/v aqueous)	55[j]	56[j]	50
Polyethylene glycol content (% w/w)	<2	<2	[f]
Specific gravity			
25°C			
40°C			
50°C	0.923	0.921	0.967
Viscosity (cSt)			
40°C			34
50°C	11.5	13.3	

Synperonic[d] A7	Lialet[e] 125L/7	Neodol[a] 25-9	Synperonic[d] A9	Lialet[e] 125L/9
C_{13}-C_{15}	C_{12}-C_{13}-C_{14}-C_{15}	C_{12}-C_{13}-C_{14}-C_{15}	C_{13}-C_{15}	
7	7.5	9	9	8.9
533	537	610	602	583
>99	100	100	>99	100
61	62	67	63	65
12.2	12.4	13.3	12.5	13
105	104	92	93	96
<0.3	<0.5	0.02	<0.3	<0.5
6–8[h]	5.5–7.5	6.0	6–8[h]	5.5–7.5
<50	<50	10	<70	<50
21	14	24	24	22
47	77[j]	74	65	60
(<1)[f]	<3	(2)	(<2)[f]	<3
0.958	0.97	0.982	0.976	0.98
21	25	41	25	27

creases effective hydrophobe chain length and can improve the surfactant properties of oxo alcohols containing 10 or fewer carbon atoms. However, it also reduces the rate of surfactant biodegradation to an unacceptable level if more than about 5 mol of propylene oxide per mole of alcohol is added.

Aqueous Cloud Point and HLBs

The relationship between aqueous cloud points and HLBs (hydrophobe/lyophobe balance) quoted by major manufacturers of detergent-range oxo alcohol ethoxylates in shown in Figure 9. This indicates that for the same polarity/HLB, the cloud point increases with the proportion of normal to branched isomers in the alcohol. This is consistent with the lower content of unethoxylated alcohol and water-insoluble lower homologs, as discussed above.

Surface tension lowering as a function of HLB is shown in Figure 9.

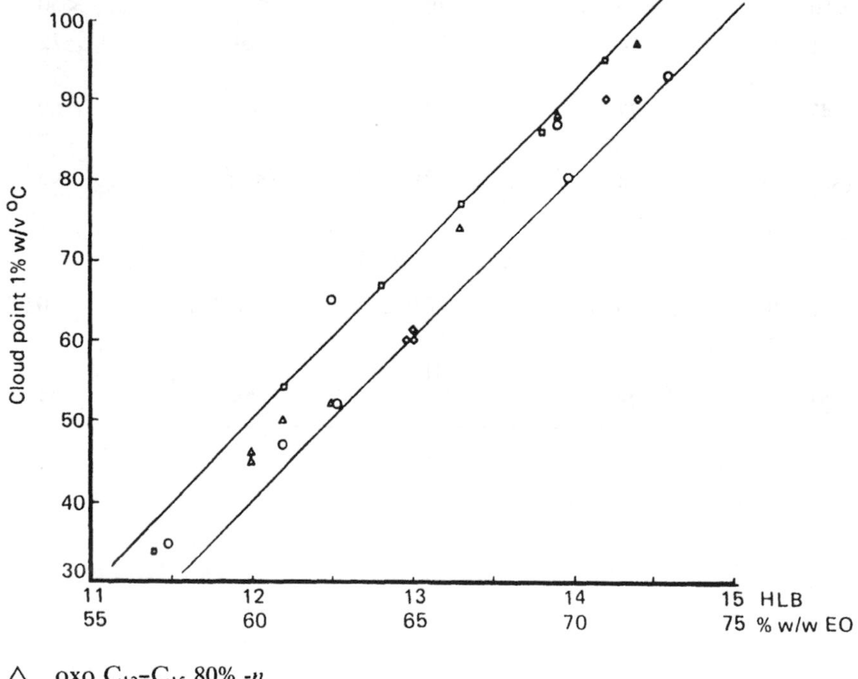

Figure 9 Aqueous cloud point vs. HLB.

Above the critical micelle concentration (CMC)* oxo C_9–C_{10}–C_{11} alcohols give significantly lower surface tensions for each specific HLB value. However, as the CMCs with carbon numbers >12 alcohol ethoxylates are below those of C_9–C_{10}–C_{11} ethoxylates, they are more effective at low concentrations (0.001% w/v). In each family of surfactants shown in Figure 10, the most active surfactants are those whose cloud point is closest to the temperature of the solutions tested (25°C). At higher temperatures the minimum surface tension shifts to products with higher cloud points/HLB.

Wetting Performance

Wetting performance follows a pattern similar to that shown in Figure 11. At 0.1% w/w, C_9–C_{11} oxo alcohol ethoxylates above their CMCs are better wetting agents than the derivatives with carbon numbers ⩾12. Optimum wetting is achieved by the surfactants in each series with cloud points closest to the test temperature (25°C).

Static Foam Heights

Ross Miles (static) foam heights for a range of oxo alcohol ethoxylates with 50 to 80% *n*-isomer content are shown in Figure 12 at 25°C. Within experimental error, isomer distribution does not affect foaming. The C_9–C_{10}–C_{11} alcohol ethoxylates are higher foaming than the ethoxylates of carbon chain lengths ⩾C_{12} but their foam is more open and less stable. Foam performance reaches a peak at HLB 12.5–13, but foam stability declines as the temperature is increased, due to breakdown of the hydrogen bonding between interstitial water and polyether oxygen atoms which stabilize the bubble surfaces (Fig. 13).

Surface Properties and Detergency Performance

Most oxo alcohols are converted to either 1- to 3-mol ethylene oxide adducts. The former are sulfated (see Section 3.2) for use in hand, dish, and fabric wash liquids and shampoo, bubble bath, shower gel, and liquid soap formulations, and the latter are used in heavy-duty detergent powders for domestic and industrial laundries. With the continuing trend toward energy-saving wash programs, cool (25 to 40°C) water, and increasing ownership of polyester cotton texiles and crease/soil resistant cottons, the choice of the nonionic surfactant component for the formulation is critical for premium wash performance. Most detergent powder

---

* The critical micelle concentration may be defined as the surfactant concentration at which molecular aggregation begins.

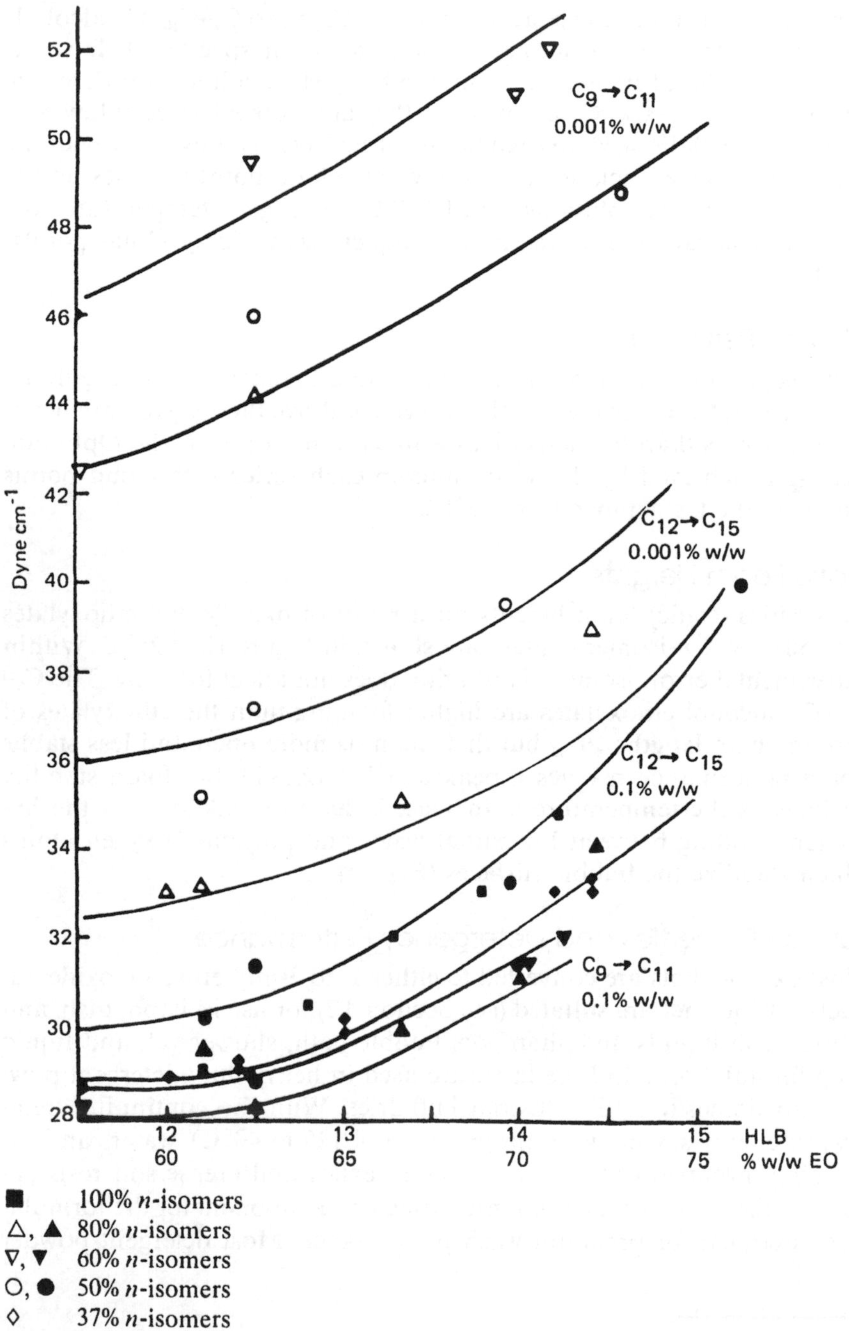

■ 100% *n*-isomers
△, ▲ 80% *n*-isomers
▽, ▼ 60% *n*-isomers
○, ● 50% *n*-isomers
◇ 37% *n*-isomers

Figure 10 Oxo alcohol ethoxylates, surface tensions, 25°C in deionized water.

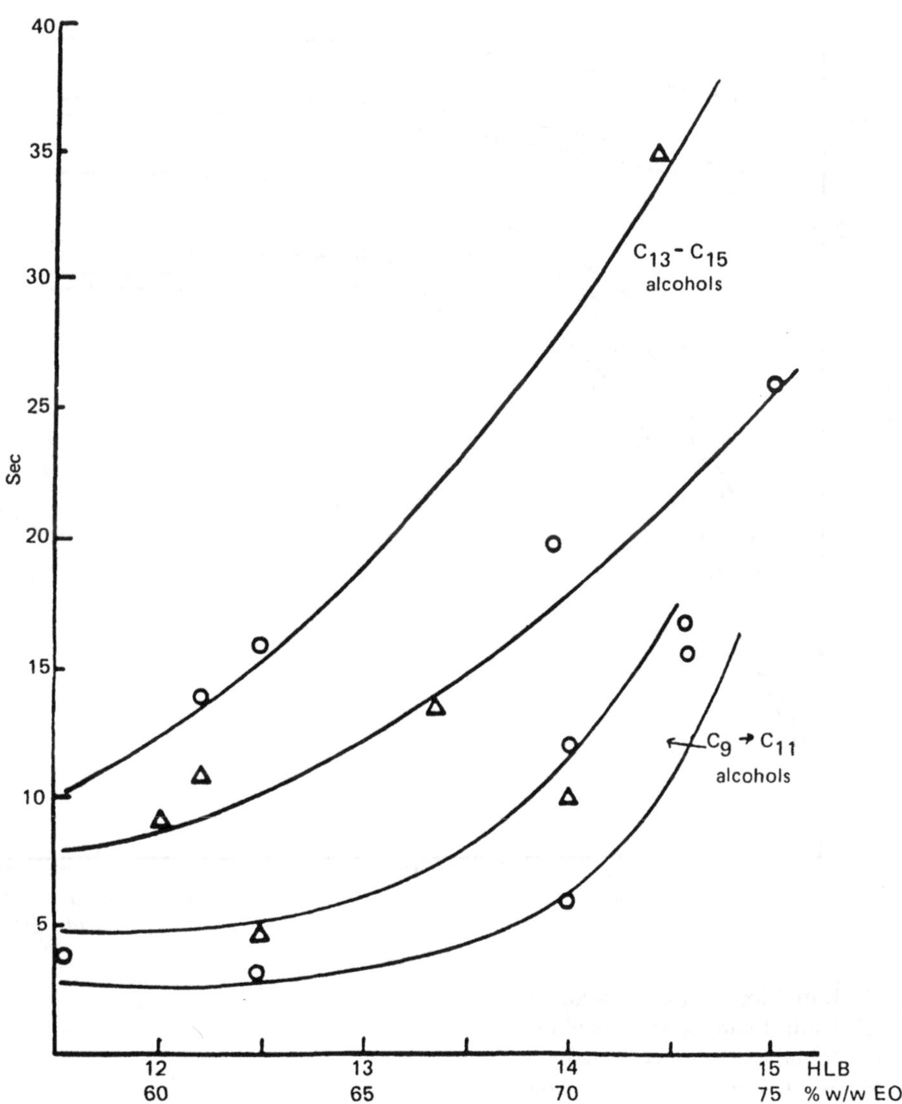

△ 80% *n*-isomers
○ 50% *n*-isomers

Figure 11 Oxo alcohol ethoxylates, Draves wetting times (cotton), 25°C, 0.1% w/w concentration in deionized water.

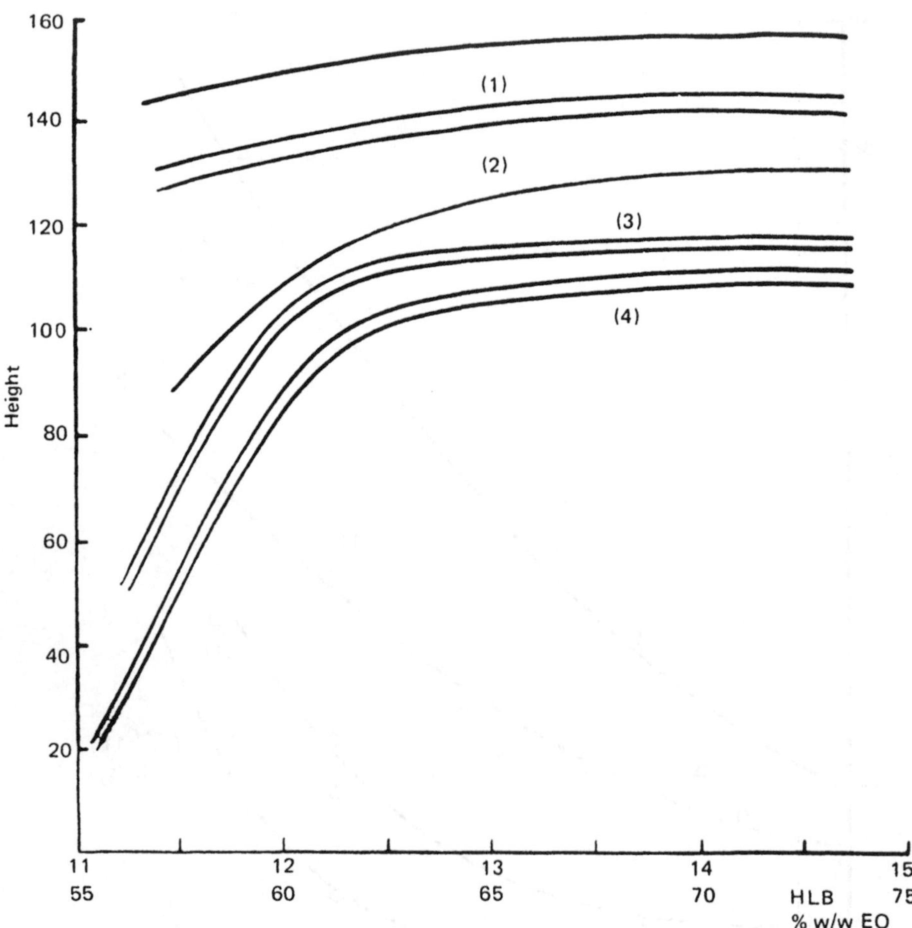

(1) Initial foam $C_9 \rightarrow C_{11}$ alcohols
(2) 5-min foam $C_9 \rightarrow C_{11}$ alcohols
(3) Initial foam $C_{12} \rightarrow C_{15}$ alcohols
(4) 5-min foam $C_{12} \rightarrow C_{15}$ alcohols

Figure 12 Oxo alcohol ethoxylates, Ross-Miles foam heights 25°C, 0.1% w/w concentration, 50 and 300 ppm water hardness, $Ca^{++}Mg^{++}$ ratio 4:1, n-isomer range 50–80%.

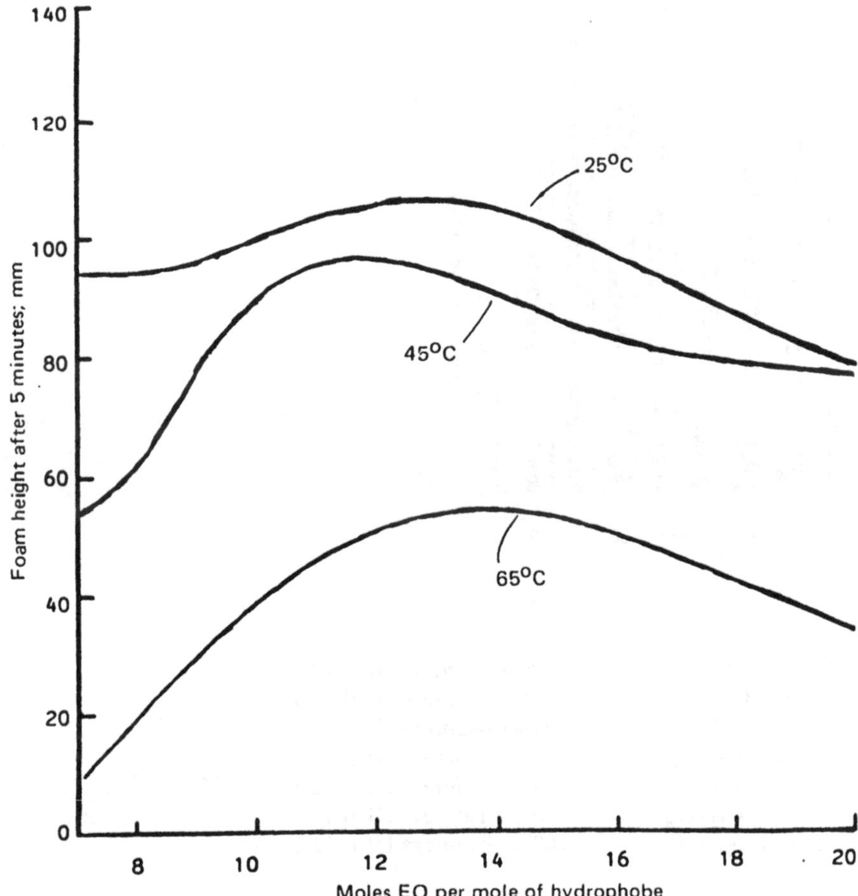

Figure 13 Foaming properties, C_{13}–C_{15} oxo ethoxylates, 50% n-isomers (Ross-Miles method, static).

producers currently use C_{12}–C_{15} hydrophobe mixtures with an average carbon chain length of $C_{13.5}$, ethoxylated to a cloud point of 45 to 55°C. Surface properties and detergency performance of a range of oxo alcohol ethoxylates of similar HLBs (12.2 to 12.5) are shown in Figures 14 to 17. From the figures it is apparent that:

1. Wetting time increases with hydrophobe linearity and molecular weight.
2. Surface tension lowering and foaming are unaffected by normal-to-branched isomer ratio.

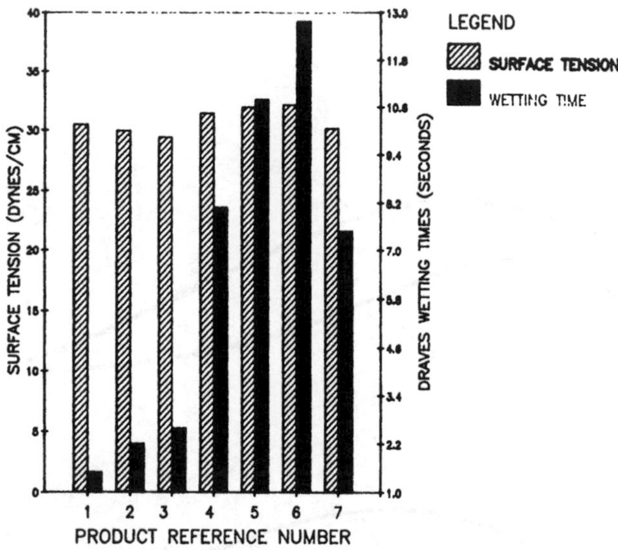

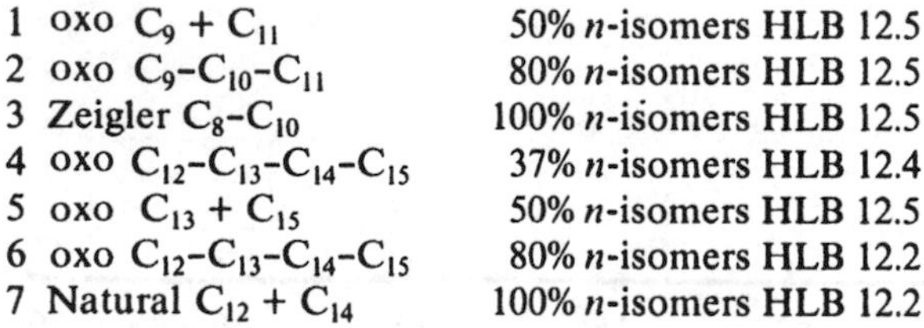

1 oxo $C_9 + C_{11}$ 50% *n*-isomers HLB 12.5
2 oxo C_9–C_{10}–C_{11} 80% *n*-isomers HLB 12.5
3 Zeigler C_8–C_{10} 100% *n*-isomers HLB 12.5
4 oxo C_{12}–C_{13}–C_{14}–C_{15} 37% *n*-isomers HLB 12.4
5 oxo $C_{13} + C_{15}$ 50% *n*-isomers HLB 12.5
6 oxo C_{12}–C_{13}–C_{14}–C_{15} 80% *n*-isomers HLB 12.2
7 Natural $C_{12} + C_{14}$ 100% *n*-isomers HLB 12.2

Figure 14 Alcohol ethoxylates surface active properties, 25°C, 0.1% w/w concentration, 300 ppm water hardness, Ca^{++}/Mg^{++} ratio 4:1.

3. C_9–C_{11} ethoxylates give markedly lower viscosities when in aqueous solution compared with alcohol ethoxylates with chain lengths of >12, but variation in *n*-isomer content from 37 to 100% has no significant effect.
4. Wash performance decreases slightly as linearity increases but is unaffected by carbon number distribution within the range tested. These results could be interpreted in terms of the linear alcohol ethoxylates forming crystalline micelles at lower temperatures, thereby reducing their wash performance. However, in fully formulated built detergent powders, it is unlikely that changing the structure of the alcohol ethoxylate in this way will have any significant effect on performance.

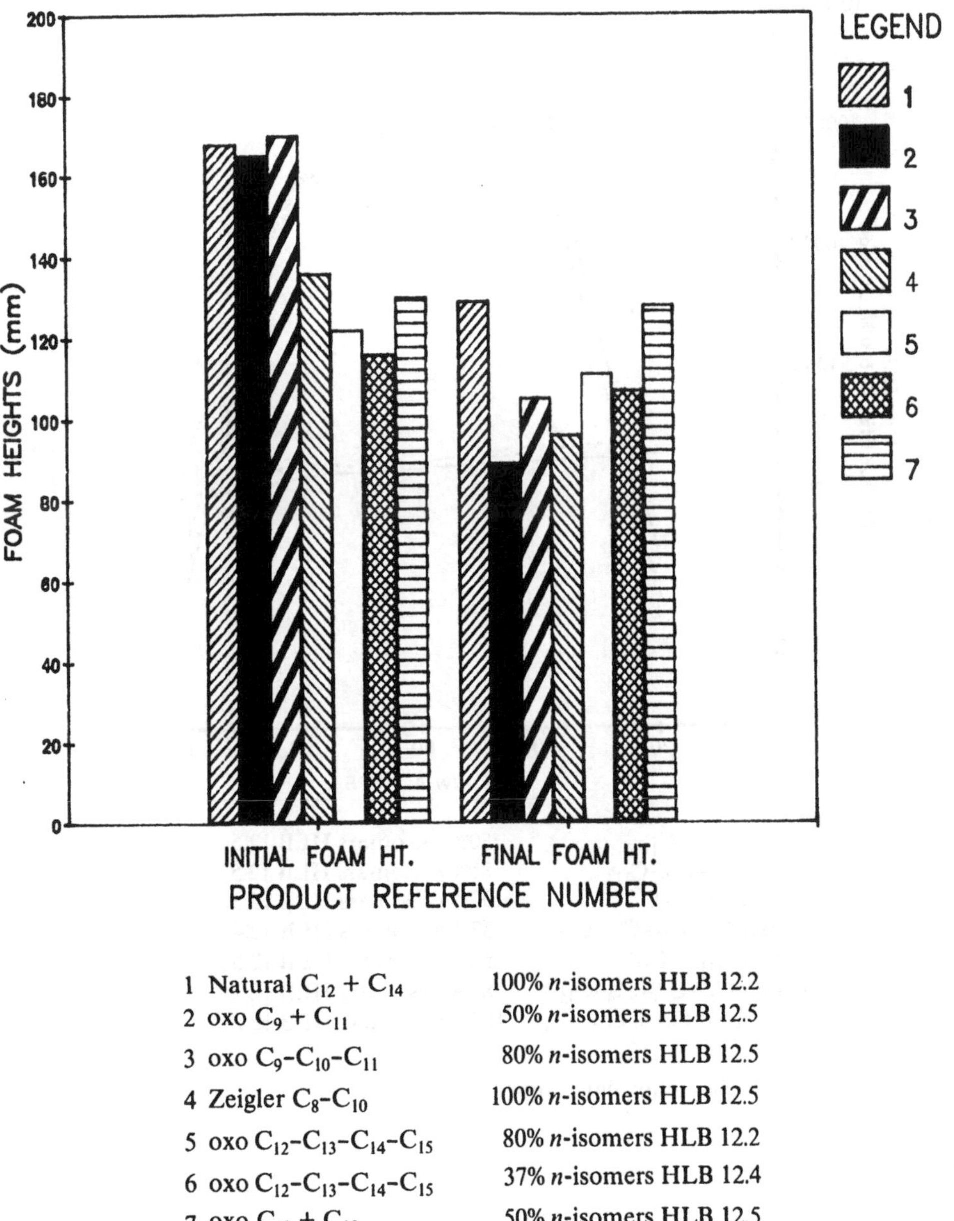

Figure 15 Alcohol ethoxylates Ross-Miles foam heights, 25°C. 0.1% w/w concentration, 300 ppm water hardness, Ca^{++}/Mg^{++} ratio 4:1.

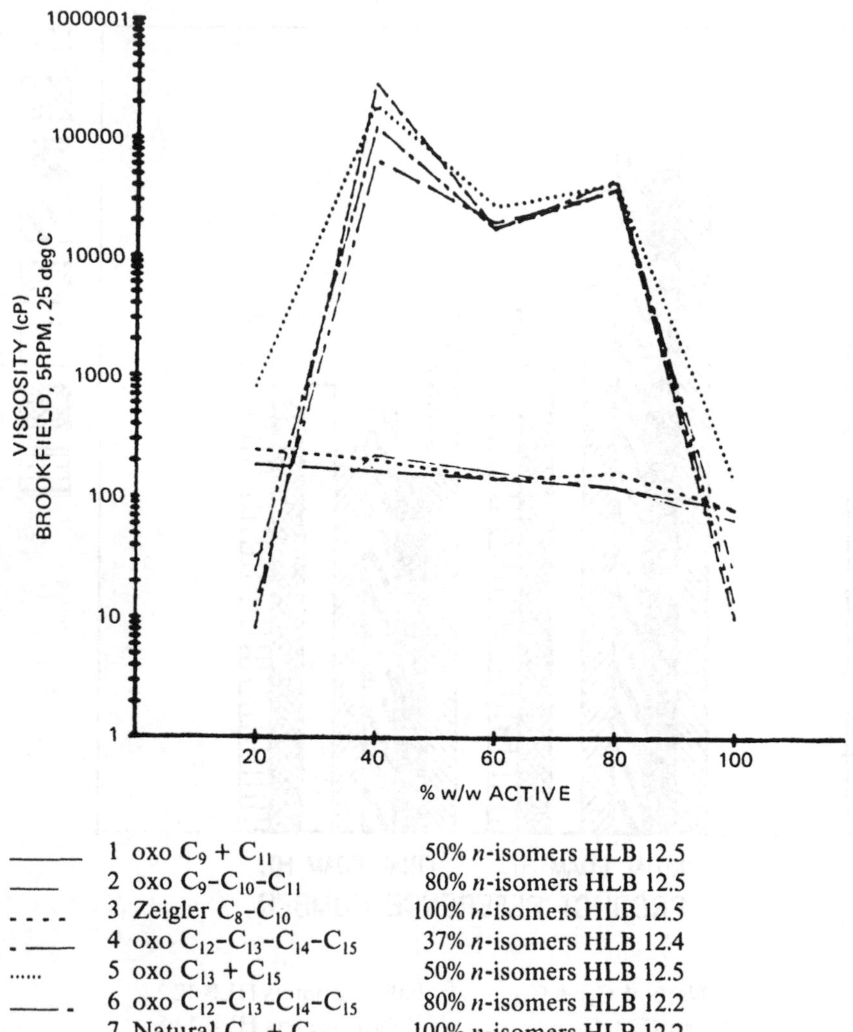

Figure 16 Alcohol ethoxylates aqueous viscosities 25°C.

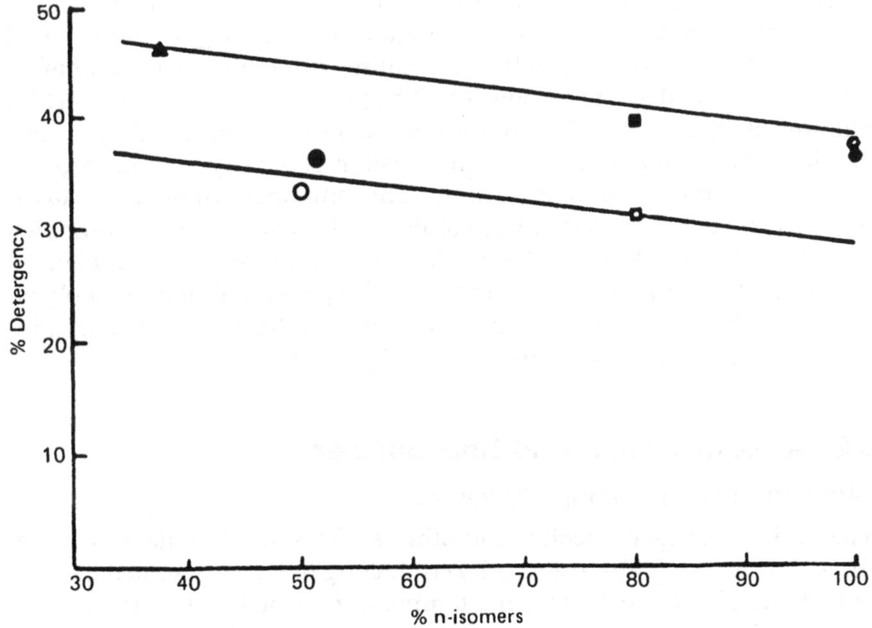

Tergotometer conditions: 10-min wash
1 g/L ethoxylate concentration
300 ppm Ca^{++}/Mg^{++} 4:1
100 rpm

▲ oxo $C_{12} + C_{13} + C_{14} + C_{15}$ + 7EO, 37% *n*-isomers
● oxo $C_9 + C_{11}$ + 6EO, 50% *n*-isomers
○ oxo $C_{13} + C_{15}$ + 7EO, 50% *n*-isomers
■ oxo $C_9 + C_{10} + C_{11}$ + 7EO, 80% *n*-isomers
□ oxo $C_{12} + C_{13} + C_{14} + C_{15}$ + 7EO, 80% *n*-isomers
⬭ Zeigler $C_8 + C_{10}$ + 6EO, 100% *n*-isomers
Natural $C_{12} + C_{14}$ + 7EO, 100% *n*-isomers

Figure 17 Alcohol ethoxylates: HLB 12.2–12.5 detergency: Krefeld soil, cotton, and polyester cotton, 40–60°C.

Historically, the change from using tallow C_{16}–C_{18} alcohol ethoxylates in detergent powders at levels of 2 to 4% w/w to oxo C_{12}–C_{15} with 7-mol ethylene oxide at levels of 3 to 10% w/w caused difficulties in the manufacture of spray-dried detergent powders. The problems encountered were (1) pluming of organic volatiles from spray tower exit, (2) blackening of powder deposits in the tower, and (3) combustion of the dry powder to give a low-pressure dust/aerosol explosion. The pluming problem is caused primarily by a higher level of free alcohol in 7-mol ethoxylates compared with the 11-mol adducts. The oxidation and combustion problem is characteristic of all ethoxylates and is due to peroxidation of the polyoxyethylene chain. This peroxidation decreases as the ethoxylate chain increases. These effects are illustrated in Figure 18.

3.2 Alcohol Sulfates and Ether Sulfates

Manufacture of Anionic Surfactants

The bulk of detergent alcohol and ether sulfates are manufactured continuously in falling film reactors by contacting them with diluted gaseous SO_3, in air, about 3 to 4% v/v, at a temperature of 30 to 40°C. Reaction is complete in a few seconds (95 to 98% conversion), and the products have good color and low levels of impurities such as UOM, (unsulfated organic matter), alkenes and ethers, sulfones, and dioxan. On leaving the reactor, the sulfonic acids are immediately neutralized with base, usually NaOH, NH_3, or triethanolamine. Ethanol, ployethylene glycols, or electrolytes are also added to modify the viscosity and elasticity properties. Typical properties of some commercial products are shown in Table 9.

Changes in alcohol carbon chain length, branching, and minor impurities require fine tuning of the sulfation process to maintain quality and composition. Impurities such as paraffins in the alcohol increase UOM; olefins form olefin sulfonates and small amounts of sulfones. High residual aldehyde levels (e.g., 0.1% w/w) give poorly colored products. These problems occur with both natural and oxo detergent alcohols.

Branching and carbon chain distribution alter the melting points and viscosities of alcohols and their lower ethoxylates. On completion of the sulfation reaction, poorly colored products may be bleached using hydrogen peroxide, but not sodium hypochlorite. A slight stoichiometric excess of neutralizing alkali is normally used, ensuring a 5 to 10% w/v aqueous pH of 7 to 9, which prevents sulfate hydrolysis on storage.

Small-scale and specialty sulfates are also made batchwise using chlorosulfonic acid as the sulfating agent. It is more expensive than SO_3 and contaminates the product with inorganic chlorides as well as sulfates from residual SO_3 and neutralizing base, which have a marked effect on solution viscosities (see below).

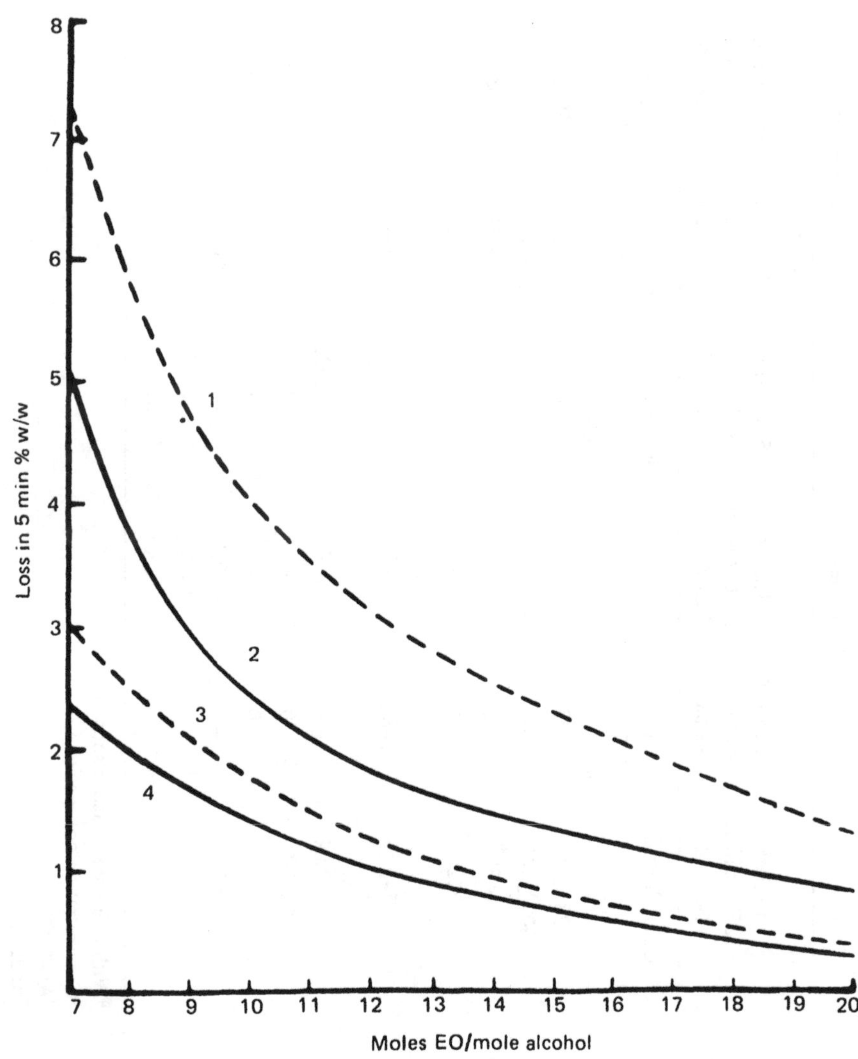

1 200°C in air
2 200°C in N_2
3 170°C in air
4 170°C in N_2

Figure 18 Thermal and oxidative stabilities of C_{13}-C_{15} oxo alcohol ethoxylates (50% n-isomers).

Table 9 Alcohol Sulfates and Ether Sulates

	Alcohol sulfates				2-mol ether sulfates	
	Lialet[a] 125L sulfate Na salt	Synprol[b] sulfate Na salt	Lialet[a] 125L sulfate TELA[c] salt	Synprol[b] sulfate TELA[c] salt	Lialet[a] 125L 2 EO sulfate Na salt	Synperonic[b] 2570 sulfate Na salt
Appearance at 20°C	Clear liquid	Clear pale yellow liquid	—	Clear pale yellow liquid	Clear liquid	Pale yellow viscous liquid
Anionic active (% w/w)	28 ± 1	28	—	40	27 ± 1	70 ± 2
Mean-molecular weight	307 ± 3[g]	—	434 ± 3	435	293 ± 3[g]	400
Unsulfated organic matter (% w/w)	<3[h]	2.0	<3[h]	1.0	<3[h]	<3
Inorganic content (% w/w)	<2	1.5	—	—	<2	<1.5
pH						
5% w/v aqueous	7–9	7–8.5	7–8[i]		7–9	7–9
10% w/v aqueous				8		
Density 20°C (g/mL)	1.04	—	—	1.025	—	1.130
Viscosity 20°C (poise)	20–30	70	—	1	—	60[l]
Pour point (°C)		0	—	—	—	—
Color (Klett units, 4-cm cell)	<100[n]	—	<100[n]	—	<100[n]	<50[n]
TELA[c], sulfate[h] (% w/w)			<2.5	—		
Free TELA[c] (% w/w)			<5	—		

[a]Chemica Augusta SpA.
[b]ICI PLC.
[c]Triethanolamine.
[d]Menro Products Ltd.
[e]Shell Chemical Co.
[f]Contains 13–15% w/w ethanol.
[h]With reference to active substance.
[i]Concentration not specified.
[j]At 40°C.
[k]At 25°C.
[l]Oscillating can rheometer.
[m]At 100°F.
[n]5% w/v AM.

	3-mol ether sulfates						
	Lialet[a] 125L 3EO sulfate Na salt	Synperonic[b] 3627 sulfate Na salt	Manro NEC[d] lauryl 3EO sulfate Na salt	Synperonic 3660 sulfate Na salt	Neodol[e] 2570 sulfate Na salt	Synperonic A3 S70 sulfate Na salt	Manro NEC 70 lauryl 3EO sulfate Na salt
Appearance at 20°C	Clear liquid	Colorless/ pale yellow mobile liquid	Clear white liquid	Pale yellow mobile liquid		Pale yellow viscous liquid	Pale yellow mobile gel
Anionic active (% w/w)	—	27 ± 0.5	>27	60 ± 1[f]	59[f]	70 ± 2	>68
Mean molecular weight	337 ± 3[g]	—	—	—	437	450	—
Unsulfated organic matter (% w/w)	<3	<2	<1	<3	1.4	<3	<2.5
Inorganic content (% w/w)	<2	<1.5	<1	<2	0.4	<1.5	<1.5
pH							
5% w/v aqueous	7–9	7–8.5	6–7	7–8.5	7–7	7–9	7–8
10% w/v aqueous							
Density 20°C (g/mL)	1.04[j]	—	1.05	—	1.05[k]	1.130	1.05
Viscosity 20°C (poise)	2–3	—	30	—	47[m]	60[l]	—
Pour point (°C)	—	—	—	—	—	—	—
Color (Klett units, 4-cm cell)		<50	—	<100	35	<50	—
TELA[c], sulfate[h] (% w/w)							
Free TELA[c] (% w/w)							

Static Foam Heights and Viscosity Characteristics

As the majority of alcohol and alcohol ether sulfates are used in high-foaming liquid detergent and personal care formulations and are produced as aqueous solutions, their most important properties are (1) foam stability and texture, especially in the presence of oily and greasy soils in hard water, and (2) viscosity characteristics related to storage, handling, and transportation, and their ease of adjustment in formulations. Apart from impurities such as UOM, these parameters are directly dependent on hydrophobe chain length and branching, ethylene oxide/hydrophobe ratio, and neutralizing cation.

The formation of foam may be regarded as an indication of the effectiveness of a detergent or shampoo. Comparative foam performance for a range of oxo ether sulfates is shown in Figure 19. Ether sulfate foaming is effectively insensitive to calcium ions and improves with the addition of magnesium ions, probably by forming more interfacially active magnesium ether sulfates. Foam performance is also insensitive to changes in monovalent cations, although triethanolamine sulfates and ether sulfates have reduced viscosities compared with sodium and ammonium sulfates and ether sulfates. Fabric and hard surface detergency of sulfates and ether sulfates are both adversely affected by hydrophobe branching. Loss of performance is typically 5 to 10% with 50% normal/50% 2-methyl C_{13}–C_{15} oxo alcohol anionics.

Maximum foam stability is found with predominantly (>75% *n*-isomers) linear C_{15} alkyl sulfate and C_{13} 3-mol ether sulfates. Foam volume and stability decline with increasing ethylene oxide/hydrophobe ratio (i.e., $EO_0 > EO_1 > EO_2$) and to a lesser extent by decreasing the ratio of *n*/2-methyl isomers in oxo alcohol sulfates and ether sulfates.

The viscosity of aqueous alcohol ether sulfate solutions reach a maximum in the range of 30 to 60% w/w anionic active content, forming stiff gels. They are usually supplied with anionic activities of 27 to 28% or 69 to 71% w/v or 60% w/v containing about 15% ethanol. This is shown in Figure 20. Subsequent formulation/dilution of high active ether sulfates is achieved by adding the sufactant to warm water with good mixing, thereby avoiding the gel region. Conversely, the addition of electrolytes, such as NaCl, has a marked effect on the aqueous viscosities of ether sulfates both alone and formulated. This is a property that enhances consumer appeal by providing a formulated product of optimum texture, especially for shower gels, shampoos, and washing-up liquid. The effects of ethylene oxide/hydrophobe ratio for an homologous series of oxo ether sulfates and of hydrophobe molecular weight and branching are shown in Figures 21 and 22. It is apparent that:

1. The viscosity at any specific level of sodium chloride content is in-

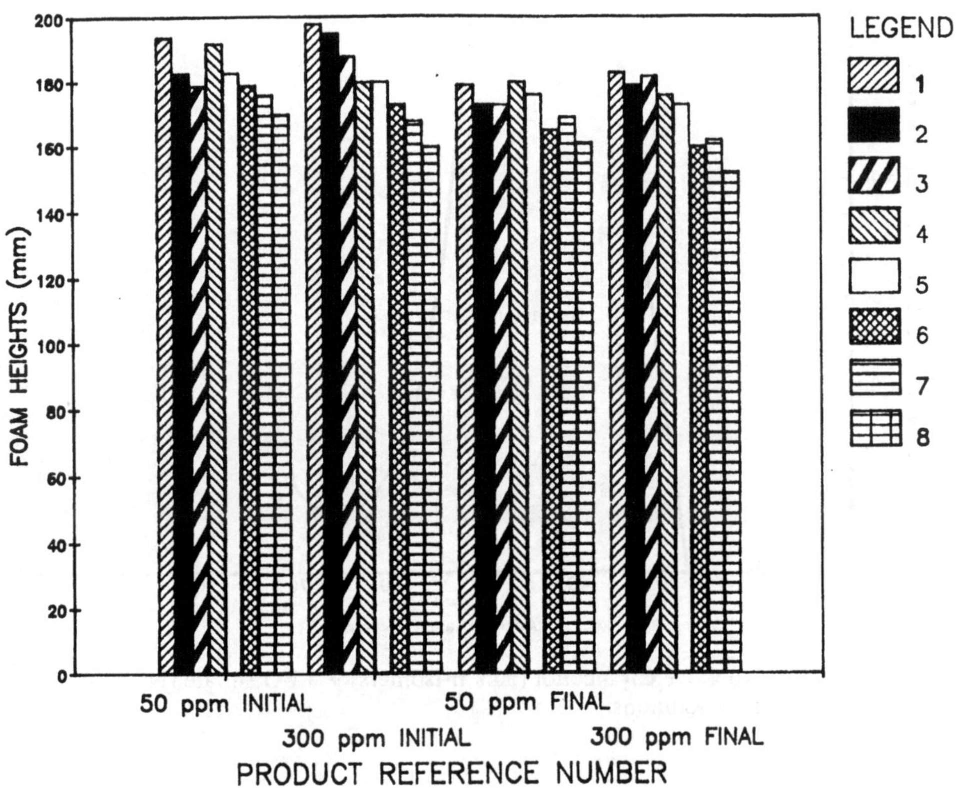

1 oxo $C_{12} + C_{13}$	2EO	80% *n*-isomers
2 Natural $C_{12} + C_{14}$	2EO	100% *n*-isomers
3 oxo $C_{13} + C_{15}$	2EO	50% *n*-isomers (1)
4 oxo $C_{13} + C_{15}$	2EO	50% *n*-isomers (2)
5 Natural $C_{12} + C_{14}$	3EO	100% *n*-isomers
6 oxo $C_{12} + C_{13} + C_{14} + C_{15}$	3EO	80% *n*-isomers
7 oxo $C_{13} + C_{15}$	3EO	50% *n*-isomers (1)
8 oxo $C_{13} + C_{15}$	3EO	50% *n*-isomers (2)

Figure 19 Ether sulfates Ross-Miles foam heights, 25°C; 0.1% w/w concentration, 50 and 300 ppm water hardness, Ca^{++}/Mg^{++} ratio 4:1.

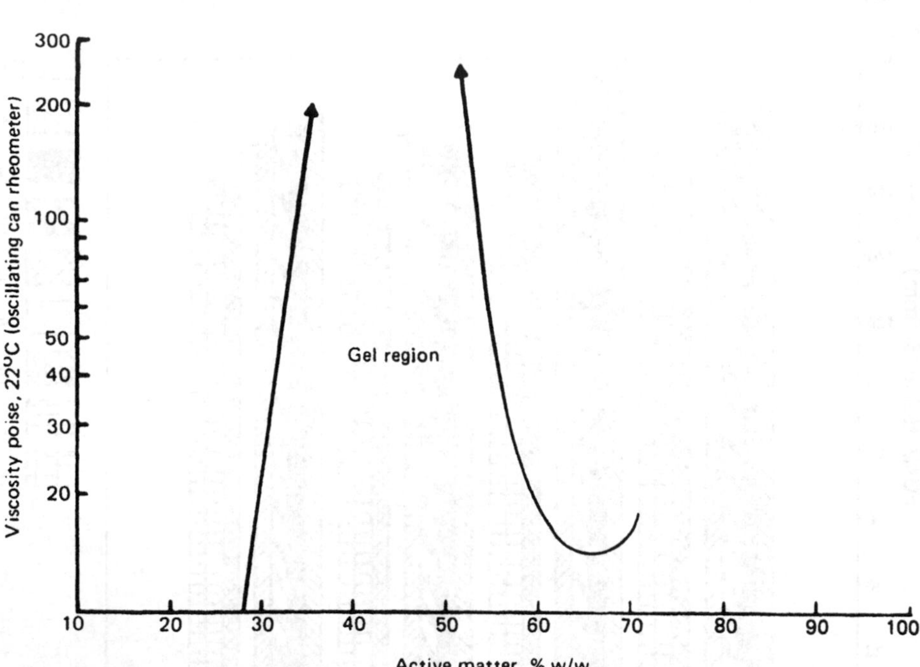

Figure 20 Oxo C_{13} + C_{15} alcohol (50% n-isomers) + 3 EO sulfate (Na salt) viscosity of aqueous solutions.

creased by increasing hydrophobe molecular weight or by reducing the *n*-isomer content.

2. Increasing the ethylene oxide/hydrophobe ratio causes progressively lower increases in viscosity per unit of NaCl. This parallels the trend in foam stability, suggesting that it is related to the viscosity in the surface of the bubble.

The judicious use of linear and oxo alcohol ether sulfates thus helps the detergent formulator achieve the desired texture and performance without a cost penalty.

3.3 Cationics: Fatty Amines and Derivatives

Properties

Most fatty amines are based on natural fats and oils, mainly beef tallow and coconut oil. These fatty amines and their derivatives have a linear

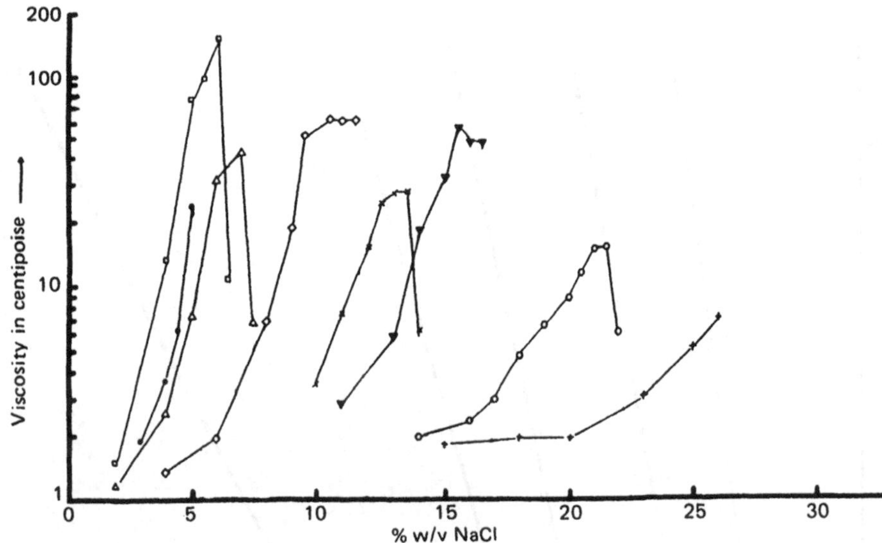

● Alcohol sulfate
□ 1 Mole ether sulfate
△ 2 Mole ether sulfate
◇ 3 Mole ether sulfate
× 4 Mole ether sulfate
▼ 5 Mole ether sulfate
○ 6 Mole ether sulfate
\+ 7 Mole ether sulfate

Brookfield viscometer, 25°C, 7.3 rev sec^{-1}
0.5% w/v concentration

Figure 21 C_{13} + C_{15} oxo alcohol ether sulfates (50% n-isomers): effect of NaCl.

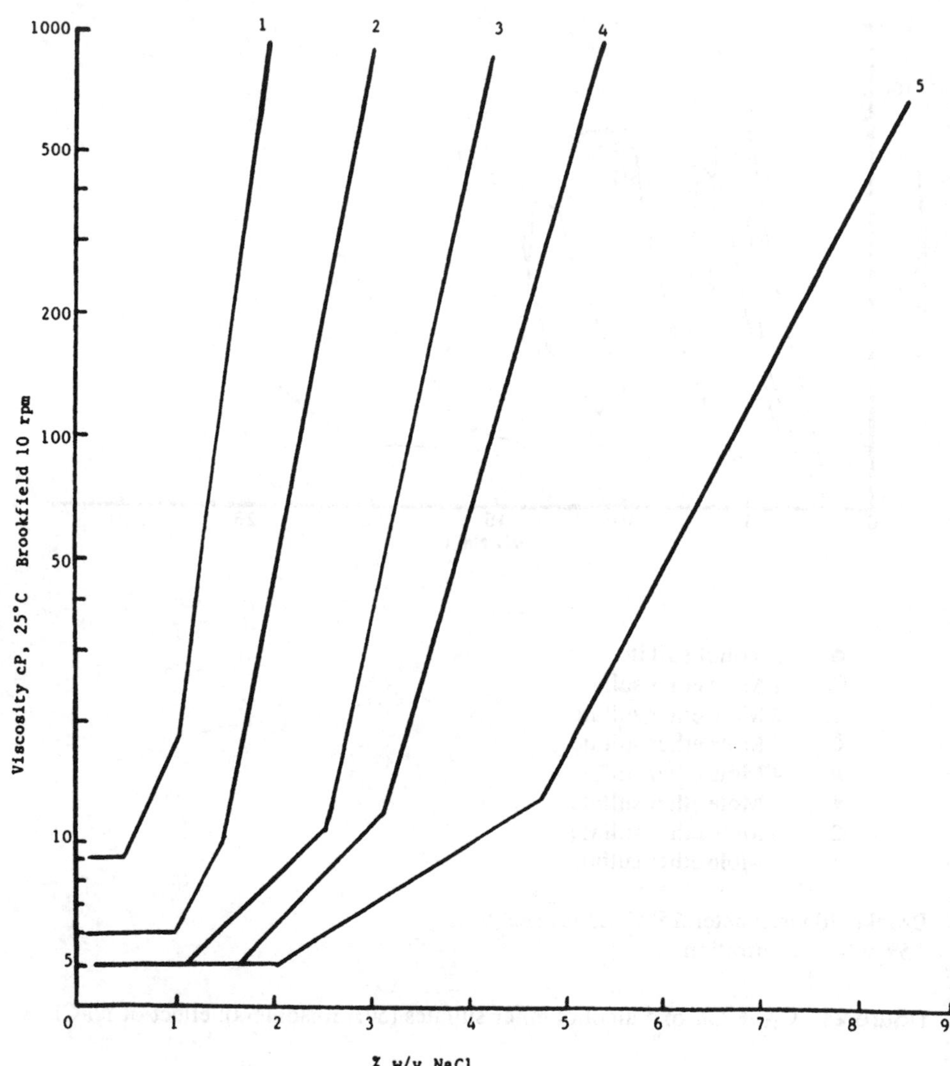

1 15% AM oxo $C_{13} + C_{15}$, 50% *n*-isomers
2 10% AM oxo $C_{13} + C_{15}$, 50% *n*-isomers
3 5% AM oxo $C_{13} + C_{15}$, 50% *n*-isomers
4 5% AM oxo $C_{12} + C_{13}$, 80% *n*-isomers
5 5% AM oxo Natural $C_{12} + C_{14}$, 100% *n*-isomers

Figure 22 Alcohol 2 mole ether sulfates (Na salts) effect of NaCl on solution viscosity.

hydrophobe with carbon numbers generally in the range C_{12}–C_{14} and C_{16}–C_{18}. Recently, ICI PLC has developed a range of fatty amines and derivatives based on an oxo C_{13}–C_{15} hydrophobe that contains about 50% of the normal isomer and 50% of the 2-methyl isomer. The isomer distribution influences the properties and reactivity of the amines and associated derivatives in the following ways. The branched isomers enhance amine liquidity (Table 10), reduce pour point and viscosity, and im-

Table 10 Effect on Liquidity of Amine Derivatives

Primary amines $R-NH_2$ (R = fatty alkyl)

Trade name	Alkyl chain (predominantly)	Freezing point (°C)
Synprolam 35	13/15	12
Synprolam 35T	13/15	9
Farmin 20	Lauryl	20–26
Farmin 20D	Lauryl	22–28
Farmin C	12/14	17–23
Farmin CD	12/14	18–24
Armeen C	12/14	11–15
Armeen CD	12/14	13–17
Armeen 12D	12	24–29
Amine 12SP	12	23
Amine 12D	12	23
Amine 14D	14	32

Tertiary amines $R-N(CH_3)_2$ (R = fatty alkyl)

Trade name	Alkyl chain (predominantly)	Freezing point (°C)
Synprolam 35DM	13/15	−21
Synprolam 35DMT	13/15	−15
Amine 2M12	12	Below −10
Amine 2M14	14	−7
Farmin DMC	12/14	−14 → −8
Farmin DM20	Lauryl	−20 → −14
Farmin DM40	Myristyl	−8 → −2

Table 11 Comparison of Aqueous Solubility of Amine Derivatives

Supplier	Product	Alkyl chain	Activity (%)	IPA (%)	Flash point (°C)
ICI PLC	Synprolam 35 DMBQC (80)	13/15	80	2	>65
Millmaster onyx	BTC-E-8358	12/14 (10% 16)	80	5–10	40
Kenogard	Querton 246	12/14 (10% 16)	80	15	30

prove aqueous solubility of derivatives (Table 11). The fatty amines derived from an oxo hydrophobe contain no residual unsaturation and are less colored than their natural counterparts. Table 12 compares the colors of a range of amine ethoxylates. All oxo-based cationic derivatives are substantially free of odor because they contain no residual nitriles. In addition, the oxo chemistry route allows production of compounds not accessible from natural feedstocks. The structure below shows an amine oxide of a (precise) monoethoxylate. However, a branched hydrophobe sterically hinders access to the nitrogen atom, thereby inhibiting quaternization and oxidation. In practice, this is overcome by extending the reaction time and increasing reaction temperature, giving excellent color and high conversion despite the more forcing conditions. Typical properties for a range of oxo amine derivatives are given in Table 13.

Table 12 Comparison of Color of Amine Ethoxylates

Amine ethoxylates $R{-}N\begin{matrix}\diagup (CH_2CH_2O)_nH \\ \diagdown (CH_2CH_2O)_pH\end{matrix}$ $(n + p = x)$, R = fatty alkyl

Feedstock		Color of ethoxylates[a] where x = 2	5	10	15
Oxo	C_{13}/C_{15}	50H	5G	6G	6G
Hardened tallow	C_{16}/C_{18}	6G	10G	—	10G
Coconut	C_{12}/C_{14}	6G	8G	11G	13G

[a]H, Hazen; G, Gardner.

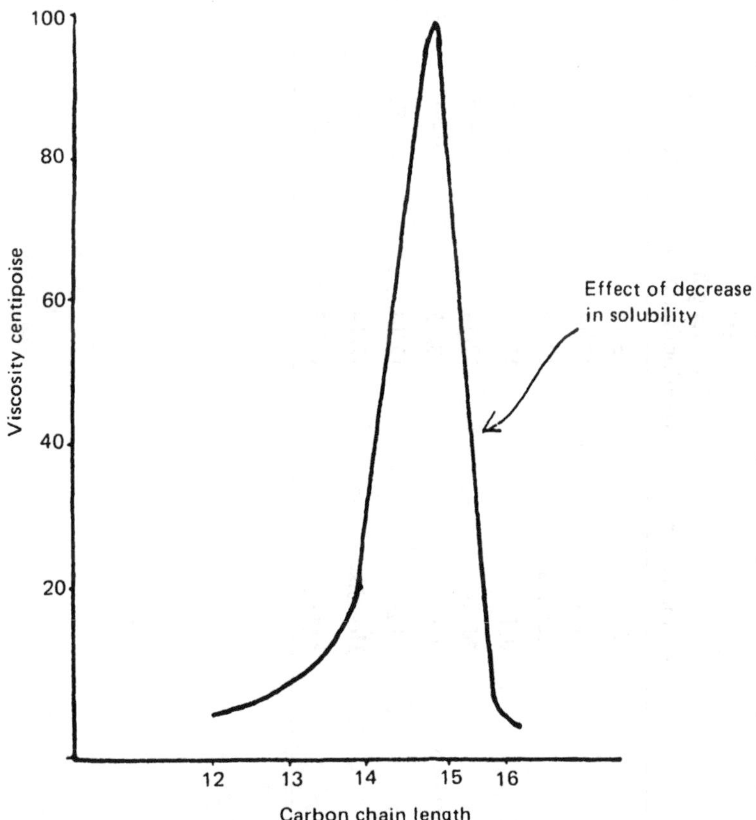

Figure 23 Effect of amine oxide alkyl chain length on the viscosity of an oxidizing electrolyte, sodium hypochlorite.

Table 13 Properties of Fatty Amine Derivatives

Type of derivative[a]	Trade name	Manufacturer	Alkyl chain length (predominantly)	Activity (% w/w)	Color[b]	pH
(i)	Synprolam 35DMO	ICI	13/15	29–31	<50H	6.7–7.0
	Aromox DMMCDW	Akzo	12/14	>29	<1G	7.0–8.0
	Lilaminox M24	Kenogard	12/14	29–31	<1G	6.5–7.5
(ii)[c]	Synprolam 35TMQC	ICI	13/15	34–36	<50H	6.0–8.0
	Arquad C33-W	Akzo	12/14	32–35	<2G	6.0–9.0
(ii)[e]	Synprolam 35DMBQC (50)	ICI	13/15	49–51	<100H	6.0–8.0
	Arquad DMMCB-50	Akzo	12/14	49–52	<2G	6.0–9.0
	Dodigen 5462	Hoechst	12/14	50–52	—	6.0–7.0
	Querton KKBCL50	Kenogard	12/14	50	<3G	6.0–7.0

[a](i) $R{-}N(CH_3)_2{-}O$, (ii) $R{-}N^{+}(R')(CH_3){-}CH_3\ X^{-}$.

[b]H, Hazen; G, Gardner.
[c]$R' = CH_3$, $X^- = Cl^-$.
[d]Amine plus amine salt.
[e]$R' = C_6H_5CH_2$, $X^- = Cl^-$.

Fabric Conditioning

The largest consumption of cationic surfactants is in fabric-softening formulations estimated at 92 MM pounds in Western Europe and 118 MM pounds in the United States in 1985. The di-C_{13}-C_{15} dimethyl amine quaternary chloride has inadequate softening properties, but a compound based on the C_{13}-C_{15} oxo hydrophobe with the structure shown below

$$(C_{13}C_{15})-\overset{\displaystyle CH_3}{\underset{\displaystyle CH_3}{\overset{|}{\underset{|}{N^+}}}}-CH_2CH_2O\overset{\displaystyle O}{\overset{\|}{C}}-R' \qquad CH_3SO_4^-$$

where $R' = C_{15}$-C_{17}. This molecule exhibits a good combination of fabric softener properties: substantive to the fiber, liquid at ambient temperatures (pour point about 15°C), and readily formulated to active levels in the range 10 to 20% w/w compared to about 5% w/w for most products available commercially. Table 14 shows a comparison between a 15% w/w Synprolam FS formulation and three commercially available fabric conditioner formulations, about 15% w/w, based on quaternized difatty hydrogenated tallow amine. Synprolam FS is shown to give comparable softening performance and improved rewettability relative to fabric conditioners based on a natural amine derivative.

Disinfectants

Cationic surfactants are also used widely in disinfectant formulations, where fatty dimethyl benzalkonium chloride has been shown to have good biocidal properties. Using *Escherichia coli* as the test organism, measurements of the concentration of biocides required to achieve a reduction in population concentration in 10 minutes of 4 log 10 gave the result shown in Table 15.

Stability in Alkaline Media

Certain cationics exhibit stability in alkaline and oxidizing media. Fatty amine ethoxylate quaternary compounds containing 12 to 15 mol of ethylene oxide can be used to solubilize nonionics in aqueous alkaline solutions. Fatty amine oxides may be used, alone or in combination with

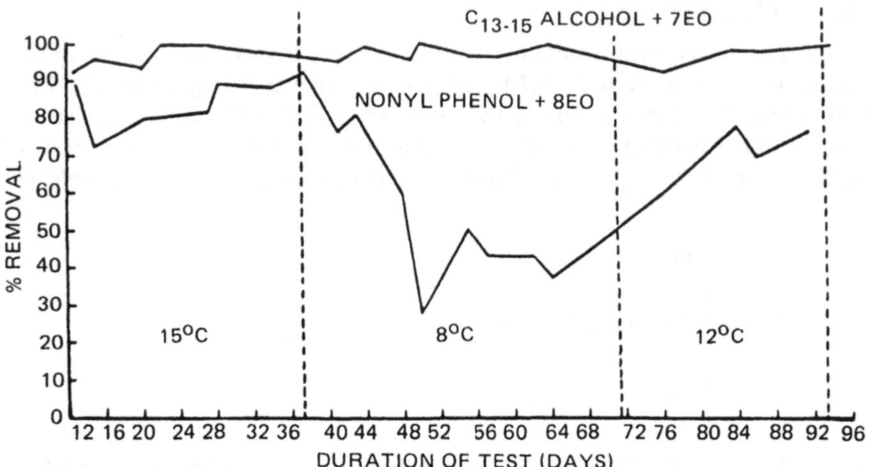

Figure 24 Biodegradability of alcohol and nonyl phenol ethoxylate by continuous activated sludge digestion, showing the effect of temperature (ICI porous pot method; Torbay settled sewage; 10 mg/l surfactant; mean residence time 6 hr).

metal soaps or alcohol sulfates, to thicken sodium hypochlorite. In this application there is a marked effect depending on the length of the hydrophobe. This is shown in Figure 25 for a system containing about 8% available chlorine. The viscosity peaks strongly at an alkyl chain length of 15 carbon atoms. For alkyl chain lengths greater than 16 carbon atoms, the solubility of the surfactant is too low for significant viscosity building. The C_{15} hydrophobe therefore represents the optimum combination of increased thickening and reduced solubility with increasing alkyl chain length. The linearity of the oxo hydrophobe is also significant, those with high linearity being preferred.

Other Applications

Oxo cationic derivatives are also used in textile processing, as lubricants; in agrochemicals, as wetters; in plastic processing, as antistatics; in

Table 14 Comparison Between Synprolam FS and Commercially Available Fabric Softeners

	Activity (% w/w)	Softening[a,b]		Rewettability expressed as percent of untreated desized cloth[a]	
		Treatment 1	Treatment 2	Treatment 1	Treatment 2
Synprolam FS formulation	15	69	72	49	63
Product A	⩽15	71	72	26	35
Product B	⩽15	68	66	26	21
Product C	⩽15	64	68	35	32

[a]Treatment 1, 8 washes with softener added at the end of the eighth wash; treatment 2, 11 washes with softener added at the end of the ninth, tenth, and eleventh washes.

[b]Softening was determined by the energy required to pull the test material through a metal cylinder. The lowest the number, the greater the softening effect.

Table 15 British Standard Test (BS6471): Test Organism *Escherichia Coli*

Biocide[a]	Test time (min)	Population reduction	Biocide concentration[b] (ppm)
$(C_{13}C_{15})-N^+(CH_3)_2(R')\ Cl^-$	10	4 log 10	100–150
$(C_{12}C_{14})-N^+(CH_3)_2(R')\ Cl^-$	10	4 log 10	150–200

[a]$R' = C_6H_5CH_2$.
[b]The statistical variation in tests of this type would suggest that the two results are very similar.

bitumen additives as emulsifiers and wetters; and for fertilizer anticaking, where their performance compares very favorably with natural products.

4 ENVIRONMENTAL FACTORS

Oxo surfactants are produced at a rate of about 0.4 million metric tons annually in the United States and Western Europe. The areas of application are extensive and include detergency, agriculture, civil engineering, cosmetics, toiletries, foodstuffs, metalworking, leather processing, mining, paints, paper, oil recovery, pharmaceuticals, plastics, textiles, and fibers. It is therefore essential to assess their environmental acceptability and effect on human safety.

4.1 Biodegradation

Surfactant biodegradability may be measured in terms of primary breakdown, which has to date been the basis for the establishment of government standards, or ultimate breakdown of the surfactant into CO_2, H_2O, and mineral salts. Ultimate biodegradability is under consideration but is not yet part of any government regulations.

Of the four types of surfactants—anionics, nonionics, cationics, and amphoterics—anionics and nonionics are subject to most investigation because they are the most widely used. The biodegradability of the surfac-

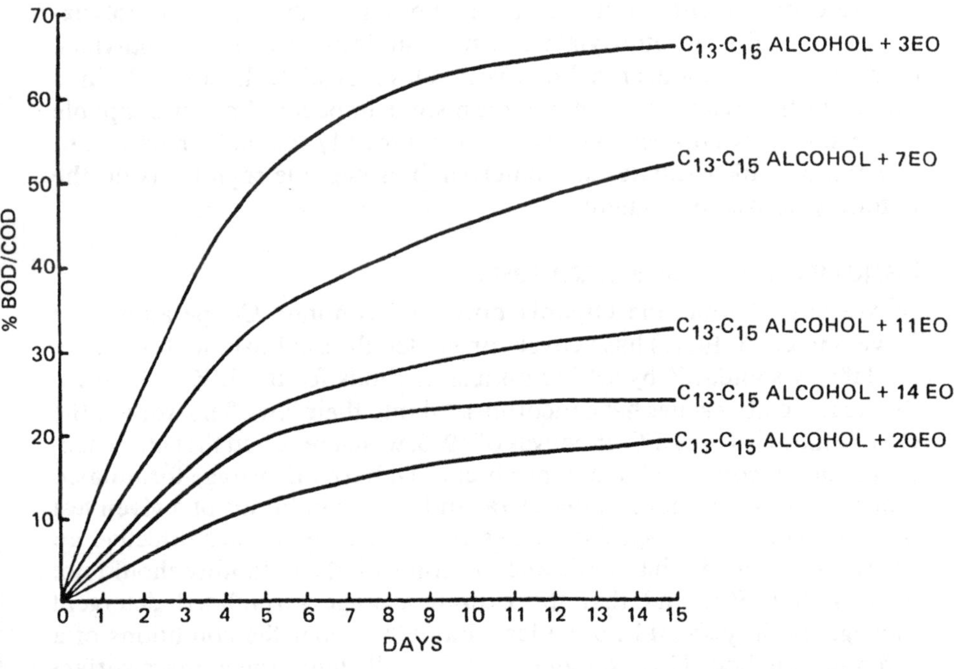

C_{13}–C_{15} ALCOHOL

Figure 25 Biodegradability oxo alcohol (50% n-isomers) etholxylates BOD/COD ratio.

tant molecule is determined by its structure. Most anionics are formed from benzene by alkylation and sulfonation. Alternatively, an olefin, alcohol, or ethoxylated alcohol may be sulfated. The majority of nonionics are alcohol and alkylphenol ethoxylates. Commercial surfactants normally contain hydrocarbon chains between 9 and 18 carbon atoms. For surface activity to be destroyed, the hydrocarbon chain must be reduced to <8 carbon atoms, or the nonionic chain shortened until the molecule becomes water insoluble. The extent of biodegradability is therefore dependent on the ease of shortening the alkyl or polyoxyethylene chains. In ethoxylate chains containing >11 mol of ethylene oxide, the rate of biodegradability is much reduced. If the alkyl chain is branched, the attack by the microorganism is impeded. Enzymes capable of breaking down branched isomers are not readily available in nature. In general, oxo alkyl sulfates and ether sulfates degrade rapidly, as do the naturally derived equivalents.

Regulations and Standard Tests

In Western Europe, the Organisation for Economic Cooperation and Development (OECD) has actively promoted the establishment of biodegradability standards by setting up test methods for the EEC, which individual member states have incorporated into their laws. The work of the OECD followed the EEC directives of 1973, which required that all surfactant material contained in any product must have an average formulated biodegradability of not less than 90% under the conditions of a given test method. The current legislation has broadened the earlier directive an now requires that both anionic and nonionic biodegradability should not be less than 80% and that the overall average formulated detergent biodegradability should not be less than 90% under the conditions of a given test method. This excludes branched alkylbenzene anionic surfactants, but linear alkylbenzenes are acceptable. The OECD "screening test" for biodegradability is the die-away test, in which surfactant is held in dilute aqueous solution in the presence of a bacterial inoculum and any degradation is assessed over a period of 19 days. Alternatively, a continous-flow activited sludge test (the OECD "confirmatory test") may be used. In this test, the surfactant is put through a laboratory model of a sewage treatment activated sludge plant in admixture with a synthetic sewage and the biodegradation is assessed over a period of several weeks.

Primary Biodegradation

Figure 26 shows the biodegradability of an alcohol ethoxylate and an alkylphenol ethoxylate over a range of temperatures as determined by a

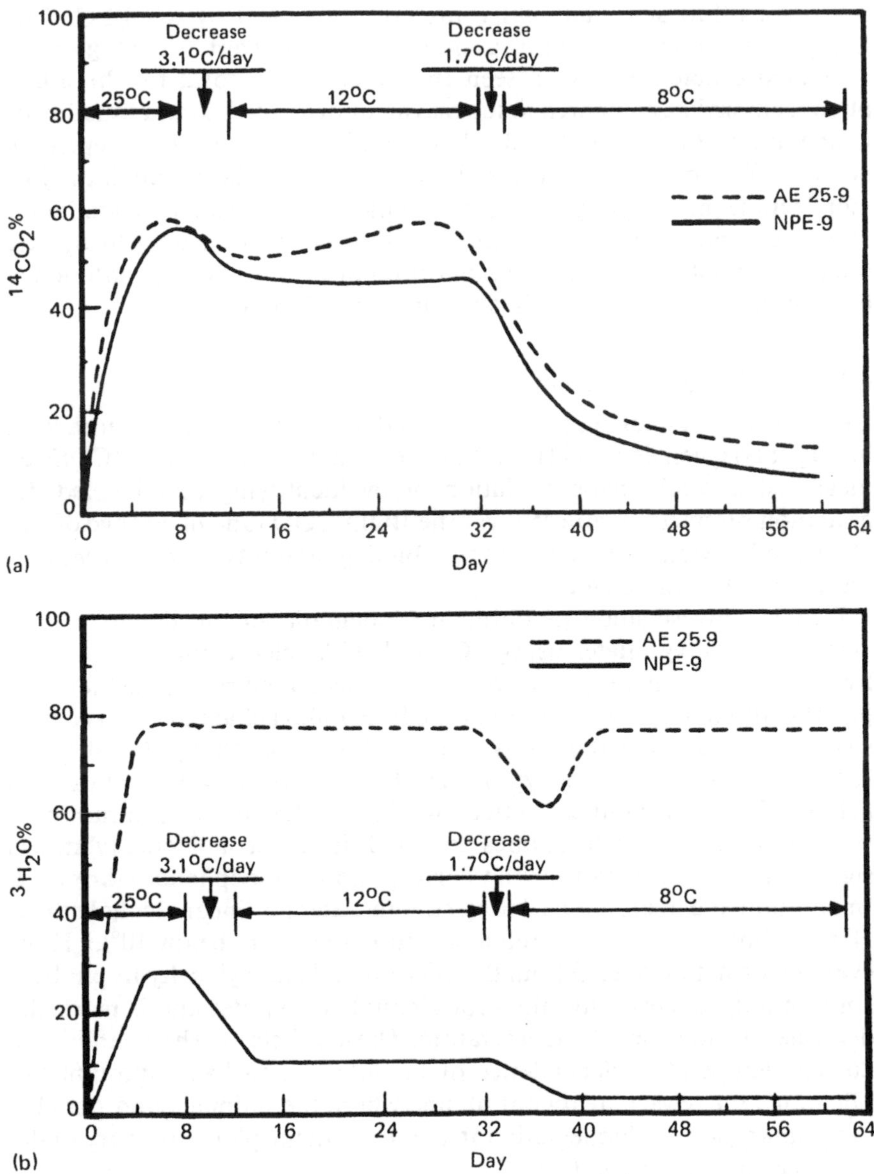

Figure 26 (a) Biodegradability of oxo [C_{12} + C_{13} + C_{14} + C_{15}] 9 EO (80% n-isomer) (AE25-9) and nonyl phenol 9 EO (NPE-9) to $^{14}CO_2$. (b) Biodegradability of ^3H-hydrophobe groups of oxo [C_{12} + C_{13} + C_{14} + C_{15}] 9 EO (80% n-isomer) (AE25-9) and nonyl phenol 9 EO (NPE-9) to ^3H_2O. (L. Kravetz et al., Primary and Ultimate Biodegradation of an Alcohol Ethoxylate and a Nonyl Phenol Etholxylate under Average Winter Conditions in the United States.)

continuous-flow activated sludge method developed by ICI PLC. This tests primary degradation only, as does the die-away test. From Figure 25, a marked difference may be seen in percentage removal for the alkylphenol ethoxylate compared to the oxo alcohol ethoxylate as a function of temperature. Both alcohol and alkylphenol ethoxylates are accepted as being >80% biodegradable using the OECD protocols, although the former degrade more rapidly and more completely. The primary degradation of most commonly used surfactants effectively removes toxicity to aquatic life and the potential to cause foam on waterways. Such degradation has been termed "environmentally acceptable biodegradation."

Ultimate Biodegradation

Ultimate degradation may be estimated using the ratio of surfactant biological oxygen demand (BOD) to chemical oxygen demand (COD) as measured by dichromate oxidation or by identifying radiolabeled degradation products. Figure 25 gives the BOD/COD ratio of a range of oxo alcohol ethoxylates, which shows that biodegradability decreases with increasing EO chain length.

Detailed investigation of ultimate biodegradation of alcohol ethoxylates and ethoxysulfates using ^{14}C- and ^{3}H-labeled compounds on the hydrophobic and hydrophilic groups have been reported in the literature [6]. The ultimate biodegradation of the linear alkyl chain to CO_2 is extensive for both alkylphenol ethoxylates and alcohol ethoxylates at about 25°C. For both surfactant types, the effect of reducing temperature to below 10°C has a significant effect on CO_2 production. In contrast, conversion of the ^{3}H hydrophobes to $^{3}H_2O$ from alcohol ethoxylates is significantly greater than for the corresponding alkylphenol ethoxylate. In addition, the hydrophobe conversion to water for linear alcohol ethoxylates is little influenced by reduction in temperature below 10°C. However the formation of H_2O from the alkylphenol ethoxylate hydrophobe is considerably less extensive than for alcohol ethoxylates and is markedly decreased by lowering the temperature. These effects are shown in Figure 26. The temperature dependence of biodegradation has important implications for sewage treatment plants, where the ultimate, and possibly the primary, level of biodegradation achieved in the plant may vary under summer and winter conditions.

In general, linear fatty alcohols degrade faster than oxo alcohols, which contain methyl isomers, and this trend continues as the number and length of alkyl side chains increases. Insertion of propylene oxide into the ethoxylate chain also reduces biodegradability, although terminal blocks

Table 16 Primary Biodegradation Off End-Capped and Propoxylated Oxo C_{13}–C_{15} Ethoxylates (50% *n*-Isomers)

Alkoxylate	Test method	Percent degraded
C_{13}-C_{15} (PO), $(EO)_5$[a]	Screening[b]	98
C_{13}-C_{15} (PO), $(EO)_3$[a]	Screening	97
C_{13}-C_{15} (PO), $(EO)_9$[c]	Confirmatory[d]	98
C_{13}-C_{15} $(EO)_6$OMe[e]	Screening	99

[a]Block copolymer.
[b]OECD screening test.
[c]Random copolymer.
[d]OECD confirmatory test.
[e]Methyl-end-capped ethoxylate.

of up to 4 to 5 mol will still biodegrade to >80% under favorable conditions. Methyl-end-capped ethoxylates improve alkaline stability while remaining fully biodegradable (Table 16). High-molecular-weight ethylene oxide/propylene oxide copolymers are not biodegradable.

In summary, the major degradation pathway of oxo alcohol ethoxylates appears to be hydrolysis of the ether linkage and subsequent oxidation of the alkyl chain. The polyoxyethylene moiety of the alkyl ethoxylate molecule readily degrades to form lower-molecular-weight polyglycols and ultimately CO_2 and H_2O.

4.2 Eutrophication

Euthrophication is caused by condensed phosphate compounds that are present in heavy-duty laundry products to improve soil removal and suspension. At the present time, there is a wide divergence of view on this ecological problem among American states and member states of the EEC related to their geographical positions and access to the sea. The likely effect on surfactants is an increase in the use of nonionic surfatants relative to anionic surfactants as manufacturers seek to maintain wash performance following reductions in allowable levels of condensed phosphate.

In future, there will undoubtably be legislation attempting to harmonize the levels of phosphates in detergents among the member states of the EEC. Since West Germany has successfully legislated for a reduction

Table 17 Physiological Properties

	Fish toxicity LC_{50} (trout) (mg AM/L)	Acute oral toxicity LD_{50} (rats) (g/kg)	Acute dermal toxicity (rabbit) LD_{50} (g/kg)	Eye irritation draize test (rabbit)	Skin irritation (rabbit)
Ethoxylates					
Coconut alcohol 9EO	0.9	1.6	4.6	Extreme	Moderate
Neodol 9–11 (7EO)		1.1	>4.0	Severe	Severe
Synperonic 9–11 (7EO)	20.0	>2.0		Mild	Moderate
Lial 12–15 (7EO)	—	1.0–2.0		Irritant	Moderate
Neodol 12–15 (7EO)		2.7	2.3	Moderate	Mild
Synperonic 13–15 (7EO)	1.8	2.5–5.0	>5.0	Mild	Mild
Lial 12–13 (7EO)	—	1.0–2.0		Irritant	Moderate
Neodol 9–11 (9EO)		1.0	>4.0	Severe	Severe
Synperonic 9–11 (9EO)	20.0	>2.0		Moderate	Mild
Lial 12–13 (9EO)	—	1.0–2.0		Irritant	Moderate
Neodol 12–15 (11EO)		1.6	2.5	Extreme	Severe
Synperonic 13–15 (11EO)	—	2.5–5.0	>5	Moderate	Mild
Lial 12–13 (11EO)	—	1.0–2.0		Irritant	Moderate

Ethoxy sulfates					
Coconut alcohol sulfate (3EO)	0.9	1–2	>10.2	Severe	Moderate
Neodol 12–13 (3S)	0.8	7.0	7.6	Severe	Moderate
Synperonic 13–15 (3S) 25%	1.0	6–16	>5	Moderate	Moderate
Lial 12–13 (3S) Na		>2.0		None (1% AM)	Low (1% AM 24 hr)
Alcohol sulfates					
Lauryl 12–14 sulfate Na	0.9				
Synprol 13–15 sulfate Na	2.3[a]				
Dobanol 12–15 sulfate Na	0.8				
Lial 12–13 sulfate Na		1.0–2.0		Mild	None
Lial 12–13 sulfate Na		1.0–2.0		—	—
Lial 14–15 sulfate Na		1.0–2.0		Mild	None

[a]Static test.

Table 18 Acute Toxicity of Partially Degraded Surfactants[a]

Surfactant	Test fish	Test conc. (mg/L)	Day 0	
			Percent degraded	Time of death
Oxo C_{13}–C_{15} 7-mol ethoxylate (50% *n*-isomers)	Rainbow trout	10	0	1 hr
Oxo C_{13}–C_{15} sulfate (50% *n*-isomers)	Brown trout	12	0	4 hr
Oxo C_{13}–C_{15} 3-mol ethoxysulfate (50% *n*-isomers)	Brown trout	8	0	24 hr
Lauryl sulfate	Brown trout	12	0	24 hr

[a] General conditions:
1. *Biodegradation stage:* Surfactant in 10 L of tap water containing 2 mL/L treated sewage effluent + BOD nutrient salts 15 ±ξ 0.5°C; Gentle aeration.
2. *Fish toxicity stage:* 7 small trout (rainbow or brown) to each test vessel; temperature 15 ± 0.5°C; aeration to maintain $O_2 > 90\%$ saturation.

of 50%, it is possible that this will form a basis for the drafting of other European standards. This reduction will require reformulation of phosphate-containing products with alternatives, the most likely being zeolites (sodium aluminosilicate) NTA, Na_2CO_3, Na_2SiO_3, and soap.

4.3 Toxicity

Table 17 shows that there are no significant differences in the mammalian and fish toxicity for natural and oxo-derived alcohol ethoxylates.

Table 18 *Continued*

Day 2		Day 4		Day 7	
Percent degraded	Time of death	Percent degraded	Time of death	Percent degraded	Time of death
41	>48 hr	52	>48 hr	96	>48 hr
20	6 hr	69	>48 hr	88	>48 hr
56	24 hr	69	>48 hr	>94	>48 hr
17	24 hr	>96	>48 hr	>96	>48 hr

Alcohol Ethoxylates

Mammalian Toxicity. Although alcohol exthoxylates have adverse effects on plant growth, they show low acute toxicity in experimental animals. Oral LD_{50} values range from 1000 to 5000 mg/kg. Oral mammalian toxicity increases rapidly as the length of the ethoxylate chain increases up to a maximum of about 10 ethylene oxide units per sufactant molecule. The length of the alkyl chain exerts a negligible toxic effect. Rats have been shown to be unaffected by ingestion of about 1% alcohol ethoxylate for 1 month. Acute skin irritation studies with neat alcohol ethoxylate samples gave slight to extreme irritation in rabbits, but these exposures are much greater than exposure through normal use. Eye irritation studies according to the Draize criteria have shown transient irritation at a 1% concentration, with varying degrees of reversible irritation at concentrations of 10 to 100% of alcohol ethoxylate.

Fish Toxicity. Alcohol ethoxylates are toxic to rainbow trout with LC_{50} values generally between 0.9 and 24 mg/L, whereas sublethal effects have been observed at concentrations as low as 0.26 mg/L. The toxicity of alcohol ethoxylates generally increases as the ethoxylate chain increases in length from 2 to 20 ethylene oxide units per mole of hydrophobe. Toxicity also increases as the hydrocarbon chain length is increased at a constant level of ethoxylation. The products of biodegradation are less toxic to fish than the original surfactant, as shown in Table 18.

Alcohol Ether Sulfates

Mammalian Toxicity. The conclusions above also apply to alcohol ether sulfates, which exhibit a low order of acute and chronic toxicity in experimental animals. Oral LD_{50} values range from 1000 to 15,000 mg/kg in the rat. Ingestion at low levels (about 0.5% alcohol ethoxysulfate) produces no adverse effects. Dermal LD_{50} values in rabbits range from 5000 to 10,200 mg/kg. Eye and skin irritation studies in rabbits suggest that care should be taken to avoid direct eye contact or excessive dermal exposure to concentrations of 1 to 2% w/v.

Fish Toxicity. Alcohol ether sulfates are toxic to fish with LC_{50} values in the range 1 to 450 mg/L, depending on the species. Sublethal effects have been noted at a concentration of 0.22 mg/L. The toxicity of alcohol ether sulfates appears to increase with increasing alkyl chain length, with a maximum at C_{16} beyond which toxicity decreases. Toxicity also appears to decrease with increases in ethylene oxide chain length from 2 to 6 ethylene oxide units for alkyl chains of <16 carbon atoms.

Carcinogens

Recently, trace levels of 1,4-dioxan, an animal carcinogen, have been detected in alcohol ether sulfate concentrates. There are, however, no data to indicate that commercial alcohol ether sulfates are mutagenic, teratogenic, or carcinogenic.

REFERENCES

1. *J Am. Chem. Soc., 83:* 4023 (1961).
2. *J Organomet. Chem., 66* (1974).
3. Lautenschlager, F.H., German patent 1046030 (1959).
4. *J. Organomet. Chem., 13:* 469 (1968).
5. Yang, K., Nield, G.L., and Washecheck, P.H., U.S. patent 4,223,164: McGain, J.H., and Foster, D.J., European patent 0026546 (1980, 1985): Edwards, C.L., U.S. patent 4,375,564 (1983).
6. Kravetz, L., et al., *Primary and Ultimate Biodegradation of an Alcohol Ethoxylate and a Nonyl Phenol Ethoxylate under Average Winter Conditions in the United States,* American Oil Chemists Society, Chicago, 1983.

CHAPTER 7
Surfactants: AOS

IZUMI YAMANE and OSAMU OKUMURA Lion Corporation, Tokyo, Japan

1 INTRODUCTION

In this chapter we describe the use of alpha olefin sulfonates (AOS) in the fields of household cleaning and personal care products in the context of its growing application to household commodities. While it was known since the 1930s that, by sulfonating alpha olefins (AO), superior surfactants containing alpha olefin sulfonates (AOS) could be obtained, this could not be realized industrially because the raw materials were not available and because of the difficulty encountered in the process of sulfonation [1]. Industrial AO production by wax cracking, which started in the United States in 1965, permitted AO supply on a commercial basis.

In Japan, Lion Corporation succeeded in 1967 in industrially producing AOS and began marketing the first AOS-based laundry detergent in the world [2]. Later, high-quality AOS requiring no bleaching treatment became available as the result of the development of new sulfonators [3–5] and high-quality Ziegler olefins. These were used in dishwashing detergents and shampoos. These pioneering efforts by Lion Corp. led to the annual consumption of 42 million pounds of AO in Japan in 1985. As shown in Table 1, Japan represents the biggest AOS market in the world. Continued growth based on the development of new formulations is expected.

Table 1 AO Use in AOS for Household Products (Millions of Pounds per Year)

	1965	1970	1975	1980	1985	1990	1995
Japan	0	4	7	38	42	44	46
United States	0	2	14	15	20	29	37
Europe	0	1	1	5	6	8	9
Others	0	0	0	0	6	20	40
Total	0	7	22	58	74	101	132

In the United States, AOS was first introduced to the marketplace in the late 1960s. By 1985 the annual consumption of AO had grown to the level of approximately 20 million pounds. AOS is used for many household products, including liquid soaps, light-duty detergents, and shampoos.

The use of AOS in Europe has been limited to small amounts as light-duty detergents, primarily in France and Germany. A slow but steady increase is anticipated.

In Korea, AOS is consumed in laundry detergents. India and China are now planning the use of AOS. Thus it is expected that in 1995, the annual world consumption will reach 132 million pounds.

2 HOUSEHOLD CLEANING PRODUCTS

2.1 Laundry Detergents

Alkylbenzene sulfonate (ABS) was the first surfactant developed for use in laundry detergents. In the late 1950s and early 1960s, ABS was found to cause foaming problems in rivers and sewage treatment facilities [6] because of the poor biodegradation [7] resulting from the use of the branched propylene oligomers for the alkyl chain. As part of measures against this, more biodegradable types of surfactants such as linear alkylbenzene sulfonate (LAS) replaced ABS during the 1960s in Europe, the United States, and Japan successively.

The problem of water pollution by eutrophication gained public attention in the late 1960s around the Great Lakes in the United States and in about 1970 in Lake Biwa and Seto Inland Sea in Japan. Nurtrients such as nitrogen and phosphorus were mentioned as the causes for such eutrophication. As one of the measures to reduce the phosphorus content of surface waters, phosphate-free formulations were demanded by communities and phosphates were either eliminated from or greatly reduced in the laundry detergent formulations. Countries around the world started adopting governmental and voluntary regulations on phosphates as

Table 2 Phosphorus Regulation for Detergent in the World

Japan	Voluntary regulation by JSDA 1975: P_2O_5 less than 15%; 1976: P_2O_5 less than 12%; 1979: P_2O_5 less than 10% 1980, July: Lake Biwa Eutrophication Prevention Act; nobody can sell phosphate-detergent 1982, Sept.: Lake Kasumigaura Eutrophication Prevention Act; nobody can sell phosphage-detergent
United States	Regulation control by states and cities Indiana, New York, Minnesota, Michigan, Wisconsin, Vermont; Chicago; Dade and Lake counties: $P_2O_5$0% Florida, Maine, Connecticut: P_2O_5 less than 20%
Canada	National regulation: P_2O_5 less than 5%
West Germany, Switzerland	National regulation corresponding to water hardness; NTA was permitted as a substitute for phosphate
Netherland	Voluntary regulation corresponding to water hardness
Italy	Local government regulation: P_2O_5 less than 9%
Norway	Voluntary regulation: P_2O_5 less than 13%
Finland	Voluntary regulation: P_2O_5 less than 15%
Sweden	Voluntary regulation: P_2O_5 less than 17%
France, Belgium, Great Britain, Denmark	No regulation

Source: Ref. 8.

shown in Table 2. A recent trend in European and American countries has been the use of lower temperatures in household washing for the purpose of energy savings, with 30% of U.S. households washing in cold water [9].

Basic Properties

Detergency. Püshel [10] investigated not merely the critical micelle concentration (CMC) and surface tension of AOS solutions but also the detergency, one of the basic functional properties required of laundry detergents. He used 2-alkene sulfonates, the main component of AOS, and found that those having an alkyl chain of length C_{14}–C_{18} exhibit superior physicochemical properties and excellent performance. Additionally, Yamane et al. [11] investigated the influences of the alkyl chain length on the removability of trioleins, paraffins, and natural soils containing inorganic soils. They found that the optimum chain length of AOS lies around C_{16}, while that for LAS and alcohol sulfate (AS) lie around C_{12} and C_{14}, respectively, as shown in Figure 1. Of these optimal surfactants, AOS exhibited the highest oil soil removability and detergency.

Since an AOS contains unsaturated and hydroxyl groups, it, unlike LAS, does not readily form insoluble salts with calcium ions [12]. As illustrated in Figure 2, AOS detergency is not reduced as much as LAS at higher water hardness [3,13]. AOS exhibits higher detergency at low washing temperatures and low surfactant concentration levels than LAS or AS [14,15], as shown in Figures 3 and 4.

Hunter and Roga [16] investigated the yellowing effects exhibited in various fibers when AOS, LAS, and nonionic surfactants were used. As shown in Figure 5, AOS exhibited a better result than LAS with a permanent-press finished polyester/cotton fabric.

Tripolyphosphate, representative of phosphates used as builders, not only chelates such metallic ions as calcium or magnesium [18,19], thus dispersing inorganic soils [20] and providing better detergency, but it also prevents caking when incorporated in the powder degergents. Since simple reduction of phosphates in formulations leads to remarkably inferior performances as shown in Figure 6, detergent and raw material manufacturers have examined numerous compounds as candidates for alternative builders to replace phosphates. From the viewpoints of performance, safety, and economy, zeolite is now considered to be the most promising and is formulated in most of the presently marketed phosphate-free detergent products.

Yamane et al.[22] investigated the effect of simultaneous use of zeolite with AOS. As shown in Figure 7, they found that the addition of type A zeolite, which has a high calcium-ion exchanging capacity, improves the detergency and gives the same degree of detergency as that possessed by phosphate-built formulation. As for the soil redeposition, Okumura [15] clarified that the anti-redeposition effect exhibited by AOS alone was superior to other surfactants, and that when zeolite was added, a good anti-redeposition effect was observed even at a higher water hardness, as shown in Figure 8. He explained these phenomena theoretically using the potential energy of the AOS/zeolite system, which is expected to result in electrostatic repulsion between the fabric and the soil.

Soils adhering to apparel consist generally of oil, inorganic, and proteinaceous solids, the last of which acts as an adhesive [23]. In order to release other soils from this adhesion, the molecular weight of the protein should be reduced to about 10^3 to 10^4 by enzymes, thus permitting the solubilization of soils by coexisting surfactants [24]. In laundry detergents, alkaline proteases, which are produced by *Bacillus subtilis,* are widely used. The activities of alkaline proteases are often decreased by surfactants, depending on their chemical and physical nature; hence the selection of

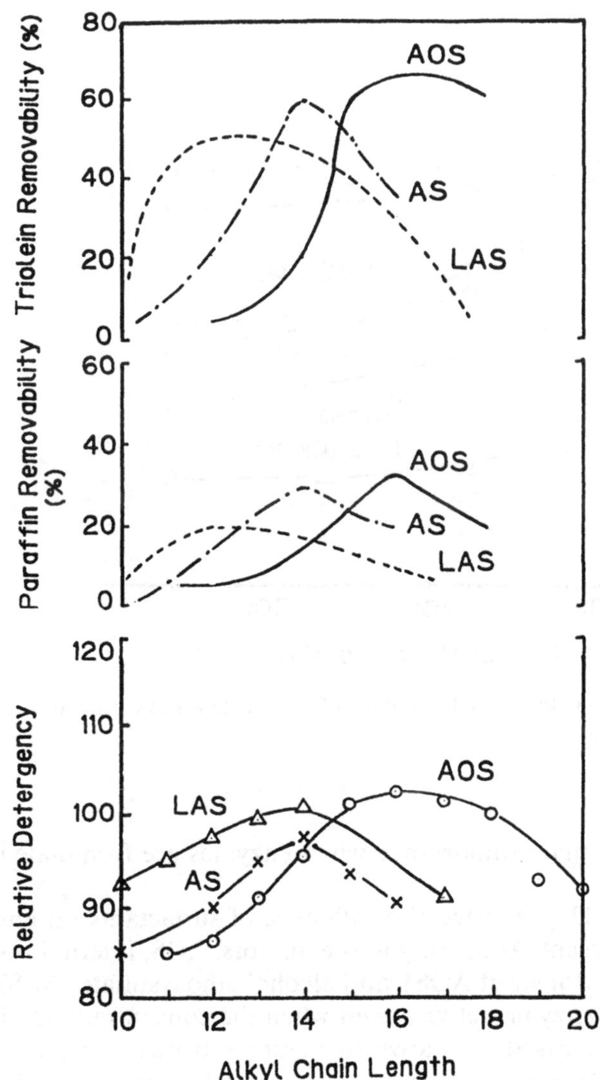

Figure 1 Effect of alkyl chain length of AOS, LAS, and AS on soil removability and detergency.

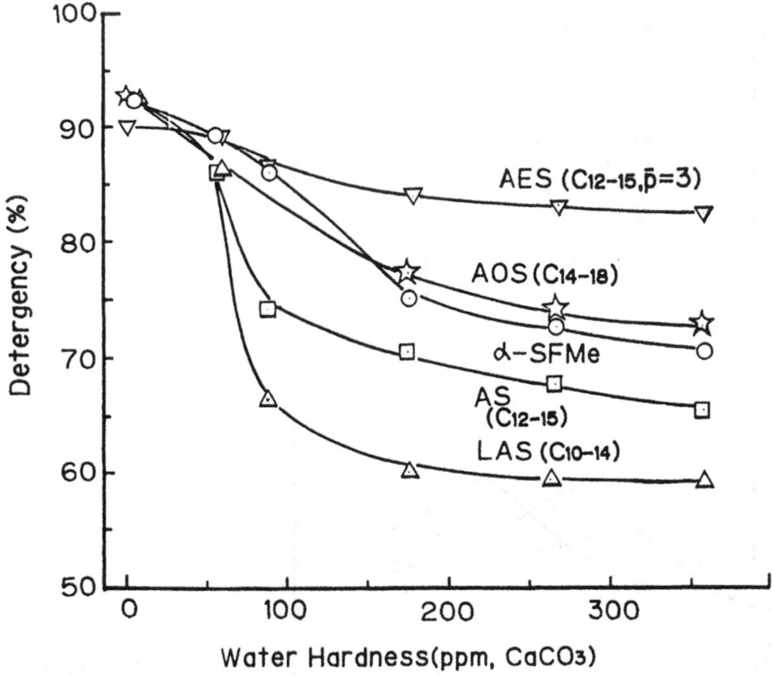

Figure 2 Effect of water hardness on detergency of various anionic surfactants in phosphate-built formulation.

the surfactants is of primary importance when enzymes are formulated [25,26].

Mori and Okumura [27] reported the influence of surfactants on enzyme activity and detergent. According to the authors, AOS, internal olefin sulfonate (IOS, an isomer of AOS), and alcohol ethoxysulfate (AES) had little effect on the enzyme activity even when the concentrations of the surfactants were increased, as shown in Figure 9. It was found that simultaneous use of AOS and enzymes exhibited good detergency even in the absence of presoaking treatment, and that, when presoaking was conducted, the detergency was remarkably improved, as shown in Figure 10.

Foamability. There are observed differences in the demand for the foamability of laundry detergents among Japanese, American, and European consumers. In Japan, consumers prefer rich foams. The impeller-type washing machines commonly used are equipped with pulsators constructed such that foaming does not affect their mechanical power. As

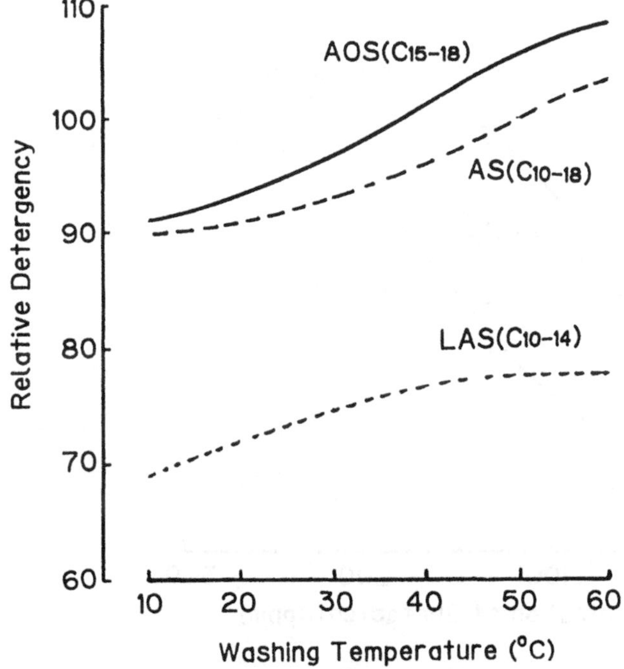

Figure 3 Effect of washing temperature on detergency of AOS, LAS, and AS in phosphate-built formulation.

illustrated in Figure 11, the foamability of detergents differs depending on the chemical type of the surfactants; AOS is recognized in Japan as a suitable high-suds detergent [3].

On the other hand, front-loading washing machines are used in European countries and, to a much lower degree, in the United States. In these, excess foaming weakens the mechanical power and reduces the detergency. AOS, when used in combination with soap and nonionic surfactant, can provide excellent formulations giving good detergency with low suds suitable for use in front-loading washing machines, as shown in Figure 12. Moreover, it was reported that this formulation exhibited better performance than that of LAS-based detergent formulation [27], as illustrated in Figure 13.

Safety. AOS is far more biodegradable than other surfactants, such as LAS or AES, as shown in Figure 14, and as Ohba et al. [29] reported, it exhibits less influence on aquatic organisms.

As for the safety evaluation of AOS using experimental animals, many

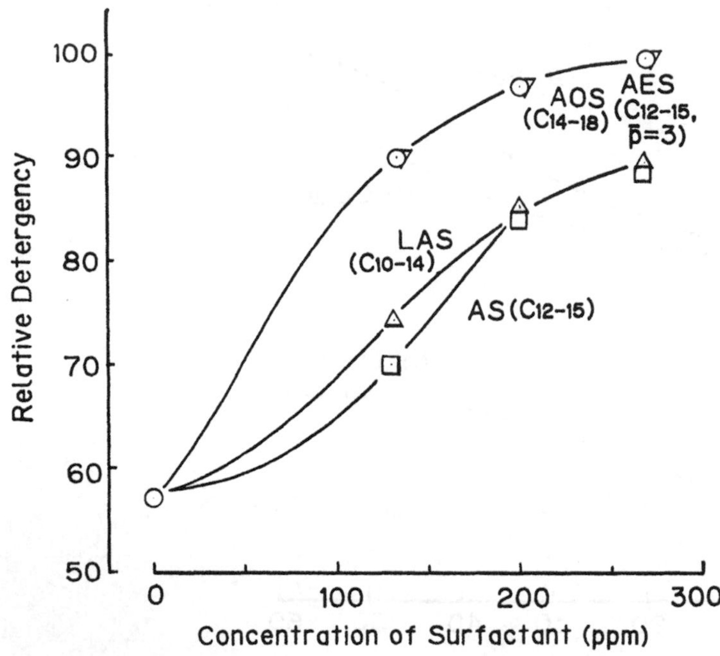

Washing conditions
Soil: artificial soil
Cloth: cotton swatches
Washer: Terg-O-Tometer
Water hardness: 54 ppm (as $CaCO_3$)
Temp.: 25°C
Time: 10 min

Washing liquor
Anionics: 0–270 ppm
Sodium silicate
Sodium carbonate :340 ppm
Sodium sulfate: 600 ppm

Figure 4 Effect of concentration on detergency of various anionic surfactants in phosphate-free formulation.

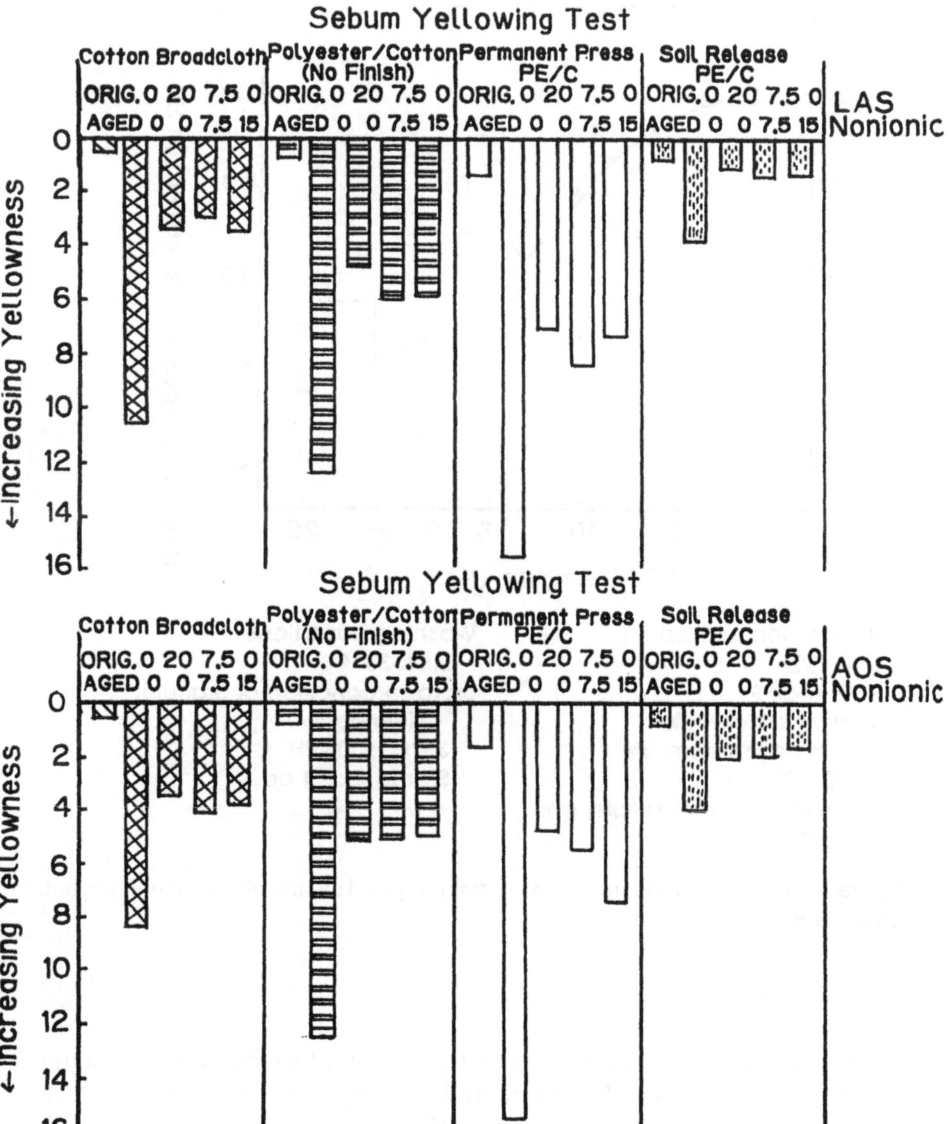

Figure 5 Effect of active ingredient variation on fabric performance (LAS-nonionic, AOS-nonionic), sebum yellowing test. (From Ref. 17.)

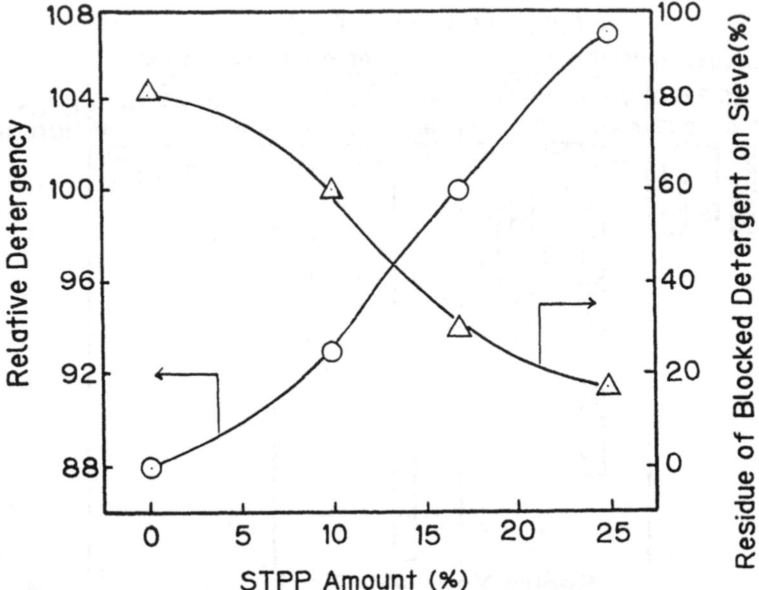

Detergent formulation
LAS: 15%
STPP: 0–25%
Sodium silicate: 5%
Sodium carbonate: 3%
CMC: 1%
Water and sodium sulfate: q.s. to 100

Washing conditions
JIS-K-3371
Storage condition of granular detergent
35°C, 100%RH
Storage for 4 days

Figure 6 Effect of STPP amount on detergency and caking of granular detergent. (From Ref. 21.)

institutions conducted detailed investigatons, and its safety has been appropriately established. The acute oral toxicity to mice and rats of AOS alone is confirmed to be 1 to 3 g/kg, as shown in Table 3, suggesting moderate toxicity [30–33]. Additionally, as summarized in Tables 4 and 5, no abnormalities were induced in subchronic and chronic toxicity tests, and no teratogenicity or adverse reproductive effects were observed [34–48]. According to absorption, distribution, metabolism, and excretion studies, it was found that most of the AOS was excreted in urine, leaving virtually no accumulation in body tissues [49–54].

Since laundry detergents are sometimes used with bleaching agents,

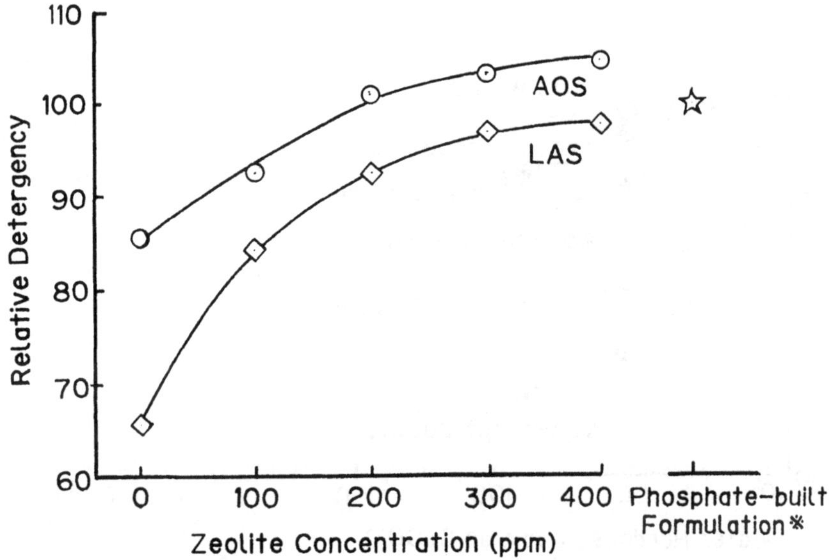

Washing conditions
Soil: artificial soil
Cloth: cotton swatches
Washer: Terg-O-Tometer
Water hardness: 54 ppm (as $CaCO_3$)
Temp.: 25°C
Time: 10 min

Washing liquor
Anionics: 320 ppm
Zeolite: 0–400 ppm
Alkaline builder: 270 ppm

Figure 7 Effect of zeolite concentration on detergency of the binary system, AOS/zeolite and LAS/zeolite in phosphate-free formulation. LAS, 320 ppm; STPP, 230 ppm; alkaline builder, 270 ppm.

Ohba et al. [55] investigated the propensity of AOS when used in combination with a hypochlorite bleaching agent to cause contact dermatitis or skin sensitization. Using maximization tests the authors confirmed that the AOS formulation induced no abnormal reactions. As seen above, AOS in normal use exerts no influence on the environment or on human health, and thus is to be considered a very safe surfactant.

Application of AOS in Laundry Detergents

Heavy-Duty Detergents. A heavy-duty powder detergent should flow freely and not cake. Phosphate-free powders based on various anionic surfactants were prepared and a comparative flowability study was conducted [27]. As shown in Figure 15, AOS, among other surfactants, exhibited su-

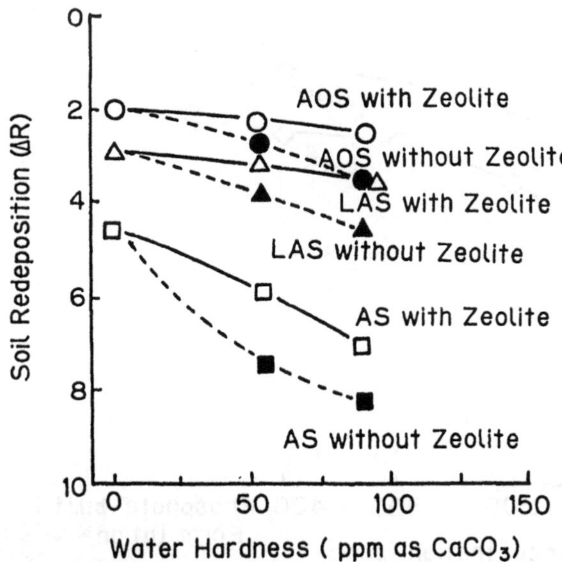

Washing conditions

Soil: artifical soil
Cloth: polyester/cotton (65/35) swatches
Washer: Terg-O-Tometer
Temp.: 25°C
Time: 10 min

Washing liquor

Anionics: 270 ppm
Zeolite: 0, 270 ppm
Sodium silicate :270 ppm
Sodium carbonate
Sodium sulfate: 400 ppm

Figure 8 Effect of water hardness on the soil redeposition of the binary system AOS, LAS, and AS with zeolite.

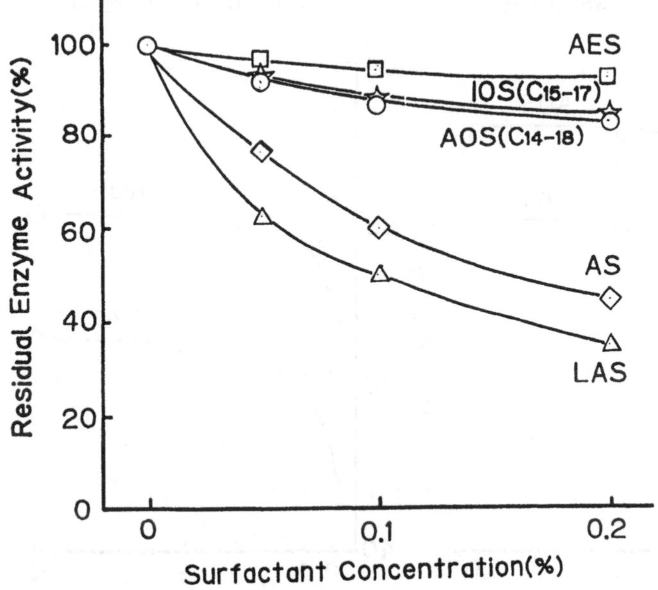

Residual enzyme activity measured by dimethyl casein—TNBS method

Test conditions

Temp.: 25°C

Time: 1 hr

Test liquor

Anionics: 0-2000 ppm

Enzyme (bac. sub. BPN): 10 AU/L

Sodium sulfate: 600 ppm

Calcium ion: free

Figure 9 Effect of surfactant concentration on residual enzyme activity in phosphate-free formulation.

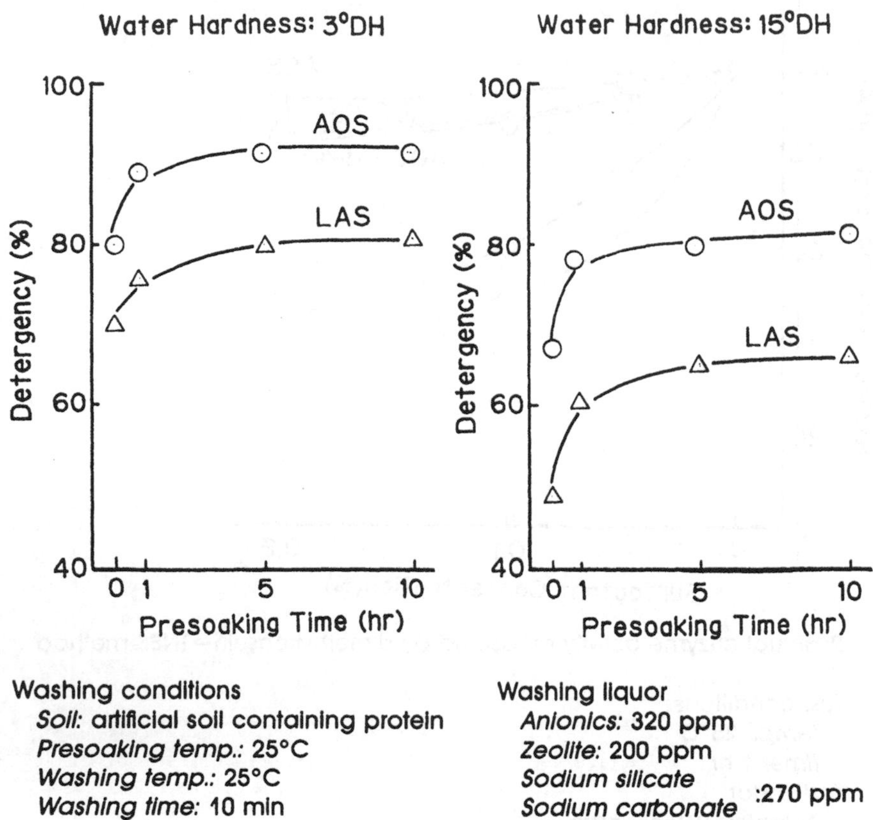

Figure 10 Effect of soaking time on detergency of AOS and LAS in phosphate-free formulation.

perior flow properties. AOS-based powder detergents are less prone to induce caking than are LAS-based detergents, and dry-blending addition of fine zeolite leads to an anti-caking effect equivalent to phosphate-built detergents as shwon in Figure 16.

The patents covering AOS use in laundry detergents describe features such as high detergency against sebum and inorganic soils [57], high foamability [58], ready rinsability [59], free flowability, and anti-caking properties [60]. In addition to these features, numerous technological developments have been observed, such as reduction in the increase in aggregated particles, increasing productivity, and solution of problematical spray drying caused by the elimination of phosphates from powder

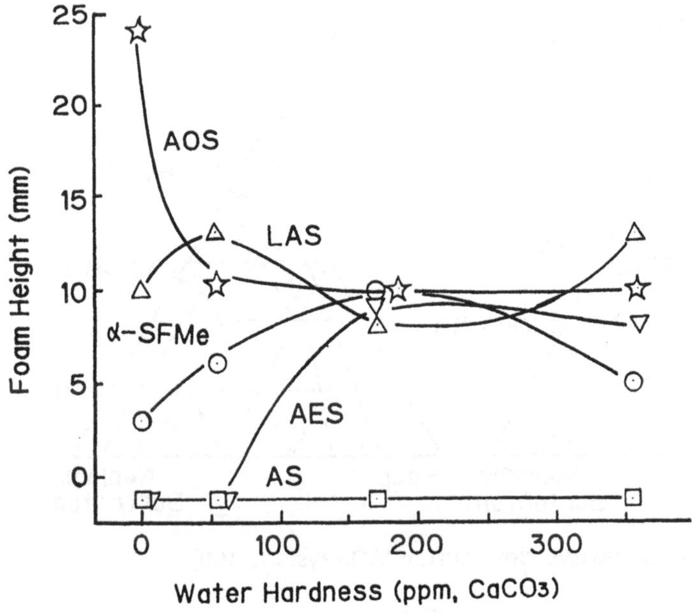

Figure 11 Effect of water hardness on foam height of various anionic surfactants in phosphate-free formulation.

Table 3 Acute Toxicity of AOS

Test material	Species	Route	LD_{50} (mg/kg)		Author
C_{14}–C_{18} AOS	Mouse	IV	♂ 72	♀ 68	Kitazato Univ. (1968)
		SC	♂ 209	♀ 213	
		Oral	♂1120	♀1110	
	Rat	IV	♂ 165	♀ 182	Kitazato Univ. (1968)
		SC	♂ 418	♀ 380	
		Oral	♂1540	♀1480	
C_{15}–C_{18} AOS	Mouse	Oral	♂3000		Ohba and Tamura (1967)
		IV	♂ 90		
		IP	♂ 200		Ogura and Tamura (1967)
		SC	♂1660		
C_{15}–C_{18} AOS (nonbleached)	Mouse	Oral	2500		Webb (1966)

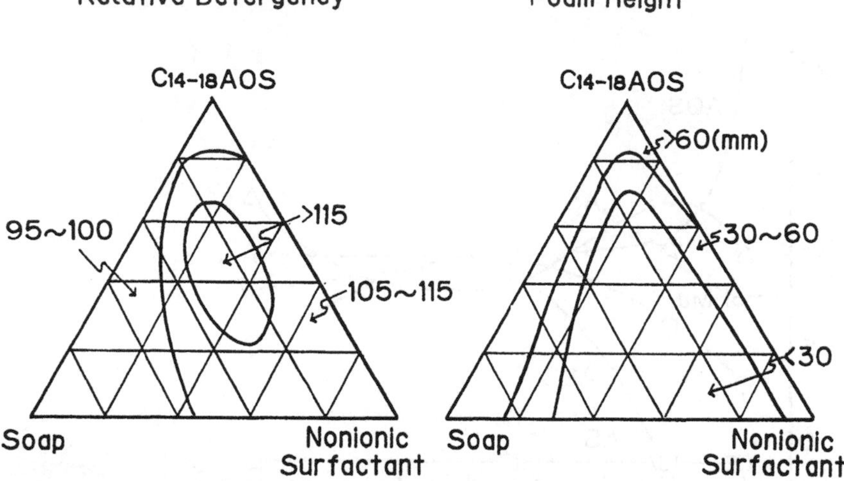

Detergency shown as relative detergency (AOS system: 100)

Washing conditions
- *Soil:* artificial soil
- *Washer:* front-loading washing machine (Westinghouse)
- *Temp.:* 30°C
- *Water hardness:* 3°DH
- *Time:* 20 min

Washing liquor
- *Surfactants:* 700 ppm
- *Zeolite:* 300 ppm
- *Sodium silicate* / *Sodium carbonate* :500 ppm
- *Enzyme:* 0.02 AU/L
- *Sodium sulfate:* 1200 ppm

Figure 12 Detergency and foam height of the ternary system AOS, nonionic surfactant, and soap in phosphate-free formulation.

detergents. These effects have made it possible to produce on an industrial scale highly functional, AOS-based, phosphate-free, granular detergents for laundry use [22,61].

Fine Fabric Washes. Slightly alkaline detergents typically cause shrinkage or fluffy texture when wool or acrylic fibers are washed. For these articles, fine fabric wash calls for a function that can provide softness to apparel in addition to detergency. AOS, in the absence of alkaline builders, is reported to provide more softness to wool, acrylic, or cotton fabrics than other anionic surfactants, including LAS [15].

Others. Soaps form insoluble calcium salts as a result of reaction with calcium ion in water, and the resultant scum is accumulated on the fabric, which causes such undesirable phenomena as yellowing. Nobble and

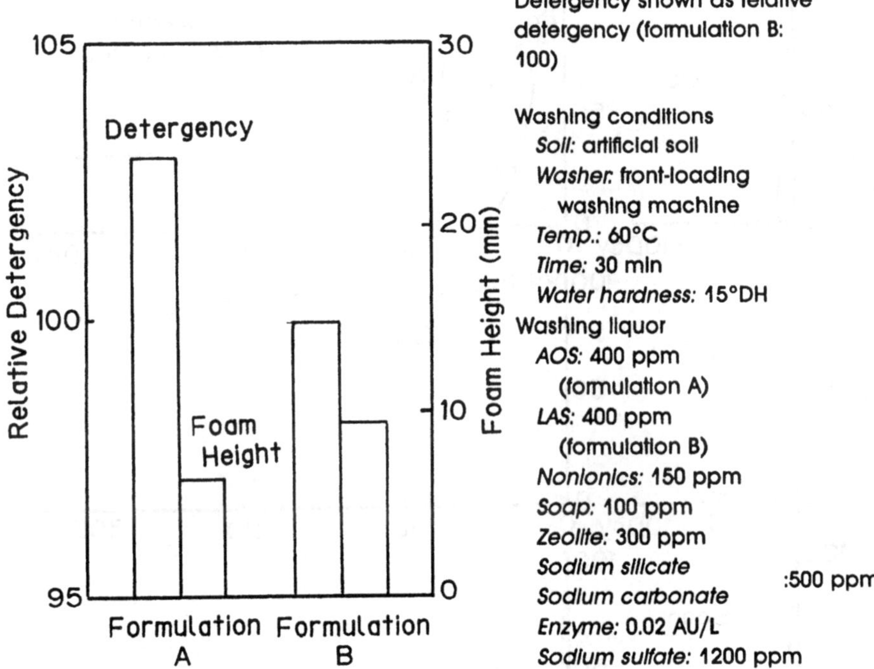

Figure 13 Detergency and foam height of the ternary system anionic surfactant, soap, and nonionic surfactant in phosphate-free formulation under European washing condition.

Linfield [62] investigated lime soap dispersant efficiency and detergency when various surfactants were used in soap-based formulation, and found that AOS exhibits superior effectiveness (Table 6). The satisfactory lime soap dispersing property of AOS was applied to powder soap formulations as claimed in patents [63]. AOS is characterized by its high detergency and mildness to skin, and these advantages were applied to detergent bars as patented [64].

2.2 Dishwashing Detergents

Hand dishwashing detergents are widely used as detergents for dishes and other tablewares. Also, detergents for automatic dishwashers intended solely for dishes are sold in European countries and the United States. High detergency, rich foaming, and mildness to the skin of hands are the basic functional requirements for hand washing detergents.

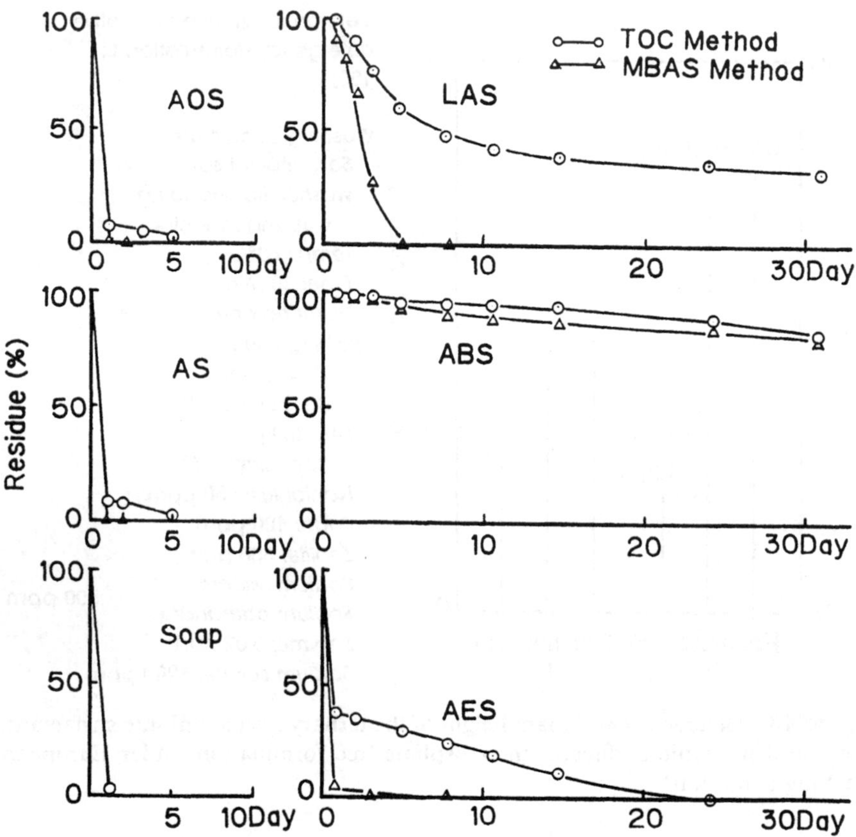

Figure 14 Biodegradability according to TOC method and MBAS method. (From Ref. 28.)

Basic Properties

Soil Removability and Foamability. Effective soil removal and generation of rich, stable foams are required for hand dishwashing detergents. Tuvell and co-workers [65] examined the interfacial tension between surfactant and Crisco, a component of dish soils, as a basic physicochemical property. As shown in Table 7, they reported that interfacial-tension-reducing ability was superior with AOS then with LAS. On the other hand, Mori and Okumura [27] investigated foam-maintaining properties of various anionic surfactants and, as shown in Figures 17 and 18, reported that C_{14} AOS, among AOS of various alkyl chain lengths, exhibited the most satisfactory behavior. These workers also found that the foam-maintaining properties of this particular AOS, vinylidene olefin sulfonate (VOS, an

Table 4 Subchronic and Chronic Toxicity Studies of AOS

Test material (active ingr.)	Species (number/group)	Route	Dose	Duration	Author
C_{14} AOS	Rat (10)	Oral (gavage)	0 (water), 100, 250, 500, 750, 1000 mg/kg/day	1 month	Kitazato Univ. (1968)
C_{14} AOS	Rat (10)	Oral (gavage)	0 (water), 100, 250, 500 mg/kg/day	6 months	Kitazato Univ. (1968)
C_{14}–C_{18} AOS (97.34%)	Rat (20)	Oral (gavage)	0 (water), 125, 250, 500, 1000 mg/kg/day	1 month	Yokosuka Inst. for *Appl. Pharmacol.* (1980)
C_{14}–C_{18} AOS (92.93%)	Rat (♂50 ♀ 50)	Oral (in diet)	0, 1000, 2500, 5000 ppm	2 years	Hunter and Benson (1976)
C_{14}–C_{18} AOS–Mg (92.90%)	Rat (20)	Oral (gavage)	0 (water), 125, 250, 500, 1000 mg/kg/day	1 month	Yokosuka Inst. for *Appl. Pharmacol.* (1980)
C_{14}–C_{18} AOS–Mg (95.54%)	Rat (20)	Oral (gavage)	0 (water), 125, 250, 500 mg/kg/day	6 months	Yokosuka Inst. for *Appl. Pharmacol.* (1980)
AOS	Mouse (5)	Dermal (skin painting)	0 (water), 20%, 0.25 mL/body/day	7 days	Uchida and Sasaki (1980)
C_{14}-C_{19} AOS	Rat (♂10 ♀ 10)	Dermal (skin painting)	0 (water), 1, 10, 30%, 0.5 mL/body/every other day	70 weeks	Tomizawa (1978)
AOS-based detergent (22%)	Rat (♂10 ♀ 10)	Dermal (skin painting)	0 (water), 50%	70 weeks	Tomizawa (1978)
C_{14}–C_{18} AOS–Mg (95.54%)	Rat (20)	Dermal (skin painting)	0 (water), 3, 5, 7%, 0.1 mL/body/day	6 months	Yokosuka Inst. for *Appl. Pharmacol.* (1980)

Table 5 Teratogenicity and Reproductive Studies of AOS

Test material (active ingr.)	Species (number/ group)	Route	Dose	Dosing period (days of gestation)	Sacrificed days of gestation	Author
C_{14}–C_{18} AOS	Mouse (20)	Oral (gavage)	0 (water), 0.2, 2.0, 300, 600 mg/kg/day	6–15	17	Palmer et al. (1975)
C_{14}–C_{18} AOS	Rat (20)	Oral (gavage)	0 (water), 0.2, 2.0, 300, 600 mg/kg/day	6–15	20	Palmer et al. (1975)
C_{14}–C_{18} AOS	Rat (32)	Oral (gavage)	0 (water), 150, 300, 600 mg/kg/day	7–17	20, F1	Life Science Research (1979)
C_{14}–C_{18} AOS (94.88%)	Rat (♂12 ♀ 24)	Oral (in diet)	0, 1, 250, 2500, 5000 ppm	F2		Life Science Research (1980)
C_{14}–C_{18} AOS	Rabbit (13)	Oral (gavage)	0 (water), 0.2, 2.0, 300, 600 mg/kg/day	6–18	29	Palmer et al. (1975)
C_{14}–C_{18} AOS	Rabbit (14–18)	Oral (gavage)	0 (water), 75, 150, 300 mg/kg/day	6–18	29	Life Science Research (1979)

C_{14}–C_{18} AOS (41.0%)	Mouse (31)	Dermal (skin painting 0.25 mL/ 2 × 4 cm)	0 (water); 1, 5, 20% (exposure: 24 hr/day); 20% (exposure: 30 min/day)	1–16	17	Huntington Research Centre (1977)
C_{14}–C_{19} AOS	Mouse (1–4)	Dermal (skin painting)	0 (water), 0.1, 1, 5%, 0.5 mL/body	0–14	18	Sawano (1978)
AOS-based detergent	Mouse (1–3)	Dermal (skin painting)	0.5, 5, 25%, 0.5 mL/body	0–14	18	Sawano (1978)
C_{14}–C_{18} AOS	Rat (20)	Dermal (skin painting 0.25 mL/ 4 × 6 cm)	0 (water), 1, 5; 20% (exposure: 24 hr/day); 20% (exposure: 30 min/day)	1–19	20	Huntington Research Centre (1976)
C_{14}–C_{18} AOS (94.4%)	Rat (32)	Dermal (skin painting)	0 (water), 1.75, 3.5, 7.0%, 0.5 mL/body	7–17	20, F1	Life Science Research (1980)
C_{14}–C_{18} AOS (94.4%)	Rabbit (14–16)	Dermal (skin painting)	0 (water), 0.75, 1.5, 3.0%, 0.5 mL/body	6–18	29	Life Science Research (1979)

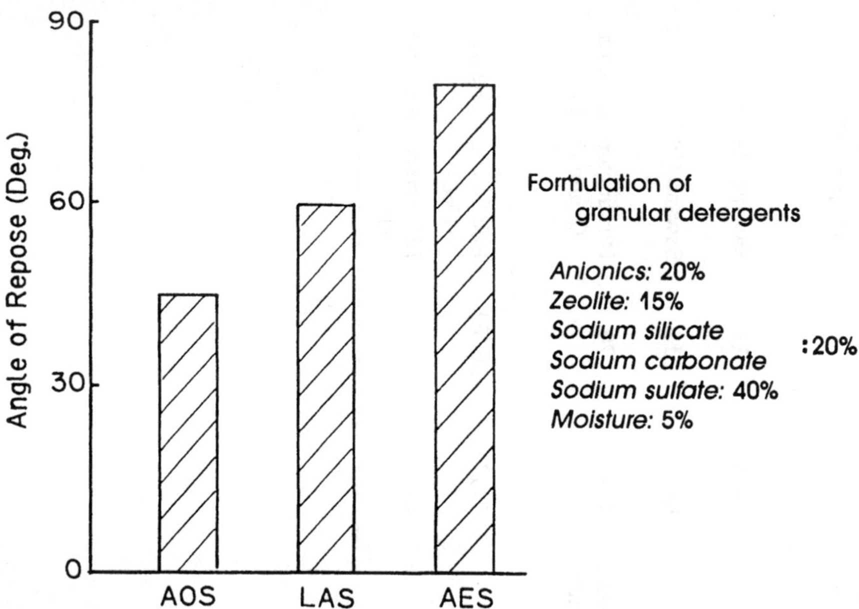

Figure 15 Smoothness of the flow of granular detergents in phosphate-free formulation.

Table 6 Effectiveness of Various Surfactants in a Soap-Based Formulation[a]

Surfactant	LSDR	Detergency (ΔR)[b]		
		TF	EMPA	UST
LAS	40	1	20	3
C_{14} AOS[c]	38	4	26	5
C_{16} AOS[c]	27	11	37	7
C_{18} AOS[c]	25	11	38	7
C_{16}–C_{18} AOS[d]	27	—	—	—
TMS	9	17	33	10
TAM	5	20	38	11
TSB	3	23	37	13
Control detergent[e]		25	40	11

[a] LSDR, lime soap dispersant requirement; TF, test fabrics polyester-cotton (65:35) cloth with permanent-press finish; EMPA, EMPA 101 cotton cloth; UST, U.S. Testing Co. cotton cloth; LAS, sodium linear alkylbenzene sulfonate; AOS, alpha olefin sulfonate; TMS, sodium methyl α-sulfotallowate; TAM, sulfated *N*-(2-hydroxypropyl) tallowamide (sodium salt); TSB, 3-sulfopropyldimethyltallowylammonium hydroxide inner salt (tallow sulfobetain).

[b] Detergency of a formulation consisting of 64% soap, 21% surfactant, 15% sodium silicate ($Na_2O:SiO_2$ = 1:1.6)

[c] Gulf Research and Development Co.

[d] Witco Chemical Co.

[e] Commercial detergent containing about 50% sodium tripolyphosphate.

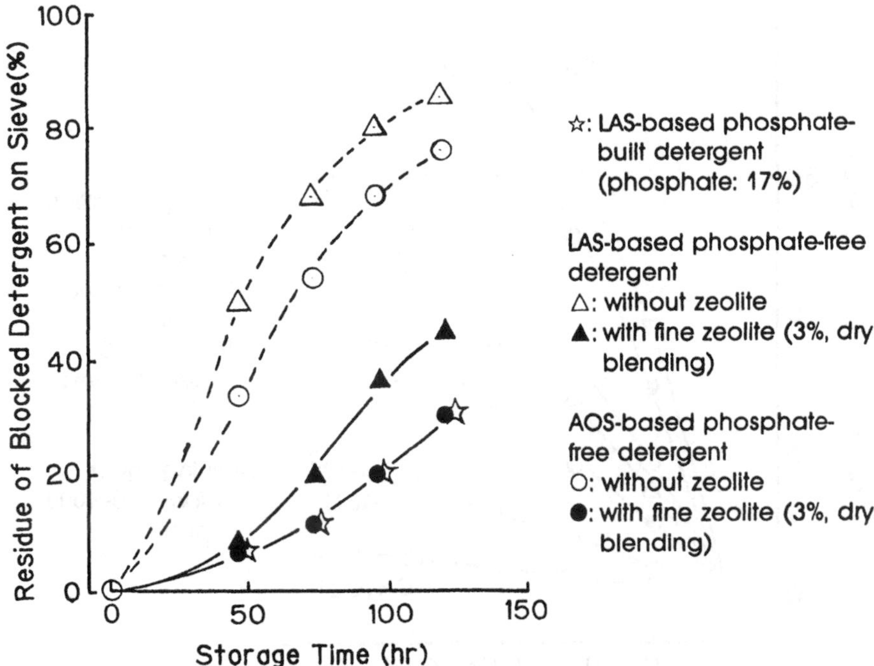

Figure 16 Caking of granular detergent by the storage. (From Ref. 56.)

isomer of AOS), and IOS were superior to that of LAS in both soft and hard water. Additionally, Odioso [66], investigating the dishwashing foam, of AOS, reported that of the preferred alkyl chain lengths, all appeared to be equal to or better than C_{11}–C_{14} LAS.

Figure 19 illustrates the comparative performances of various anionic

Table 7 Interfacial Tension of 0.045% Solutions of Surfactants versus Crisco at 60°C (dyn/cm)

Surfactant	300 ppm water hardness
LAS-average C_{11} side chain	0.57
Lauryl alcohol ether sulfate	0.25
Commercial paraffin sulfonate	0.56
C_{14} alkane-1-sulfonate	0.06
C_{14} hydroxyalkane-1-sulfonate	0.14

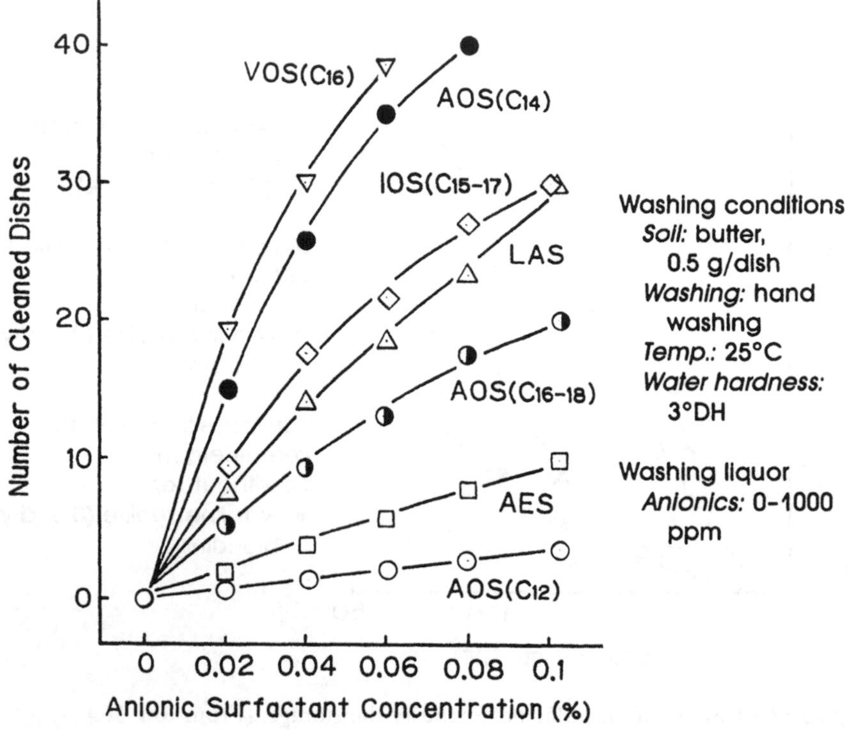

Figure 17 Effect of surfactant concentration on the number of cleaned dishes.

surfactants, and as shown, AOS and AES exhibited a synergism for soil removability and foamability [27]. Additionally, Okumura [15] investigated emulsification of liquid oils such as triolein as a factor greatly contributing to the detergent performance, and concluded, as shown in Figure 20, that the combination of AOS with amine oxide increases the emulsifying ability synergistically, leading to much better results.

Safety. In the case of dishwashing detergents, the potential risks conceivable via the oral route include chronic toxity induced by long-term intake of trace AOS residual on the article washed, as well as such specific toxicities as teratogenicity, or carcinogenicity. On the other hand, risks caused by the contact with hands are exemplified by skin roughness, primary skin irritation, and sensitization.

As for the possibility of chronic oral intake from washed tableware and vegetable, the fact is that the safety factor calculated from the maximum no-effect level and estimated human daily intake of AOS is greater than 100 times the concentration accepted as the criterion for safety of food ad-

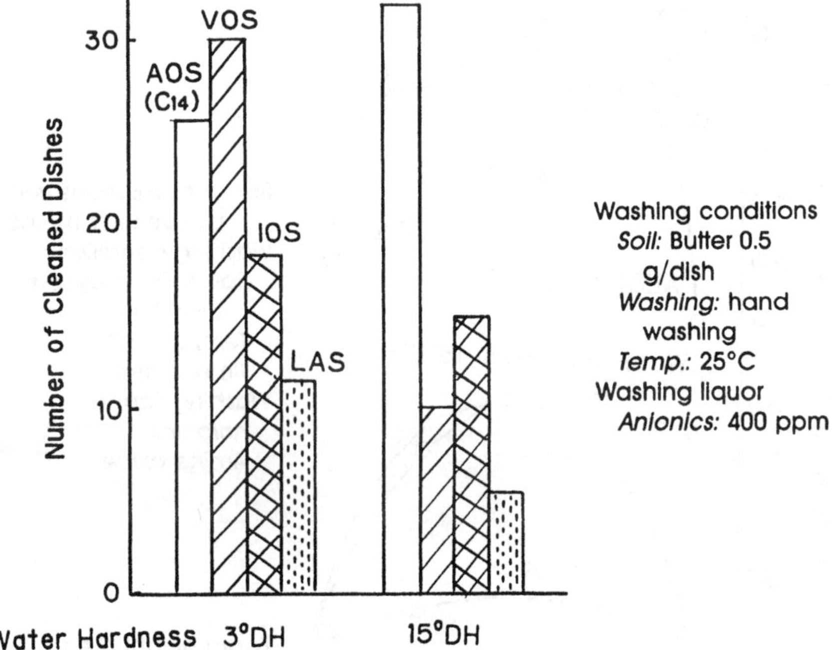

Figure 18 Effect of water hardness on the number of cleaned dishes of various anionic surfactants.

ditives by the World Health Organization. Specific toxicity evaluations (e.g., teratogenicity and carcinogenicity tests) resulted in the determination of no potential toxicity [55] for these materials.

As for the sensitization risks, the results obtained were negative in the maximization and intradermal challenge methods [55]. Additionally, a 9-month field test of AOS-based dishwashing detergent resulted clinically in no skin trouble [55].

Application of AOS in Dishwashing Detergents

Another important function required of dishwashing detergents is that they do not cause hand skin roughness. Clinically, the influence on the skin roughness is tested by primary skin irritation (patch) and skin roughness (dipping) methods. Such tests require numerous repetitions, however, and at present, the results obtained are qualitative.

Ohbu et al. [67] investigated the bovine serum albumin denaturing property, an indicator highly correlated with the clinical irritabilities noted above, and found that as shown in Figure 21, AOS and AES were

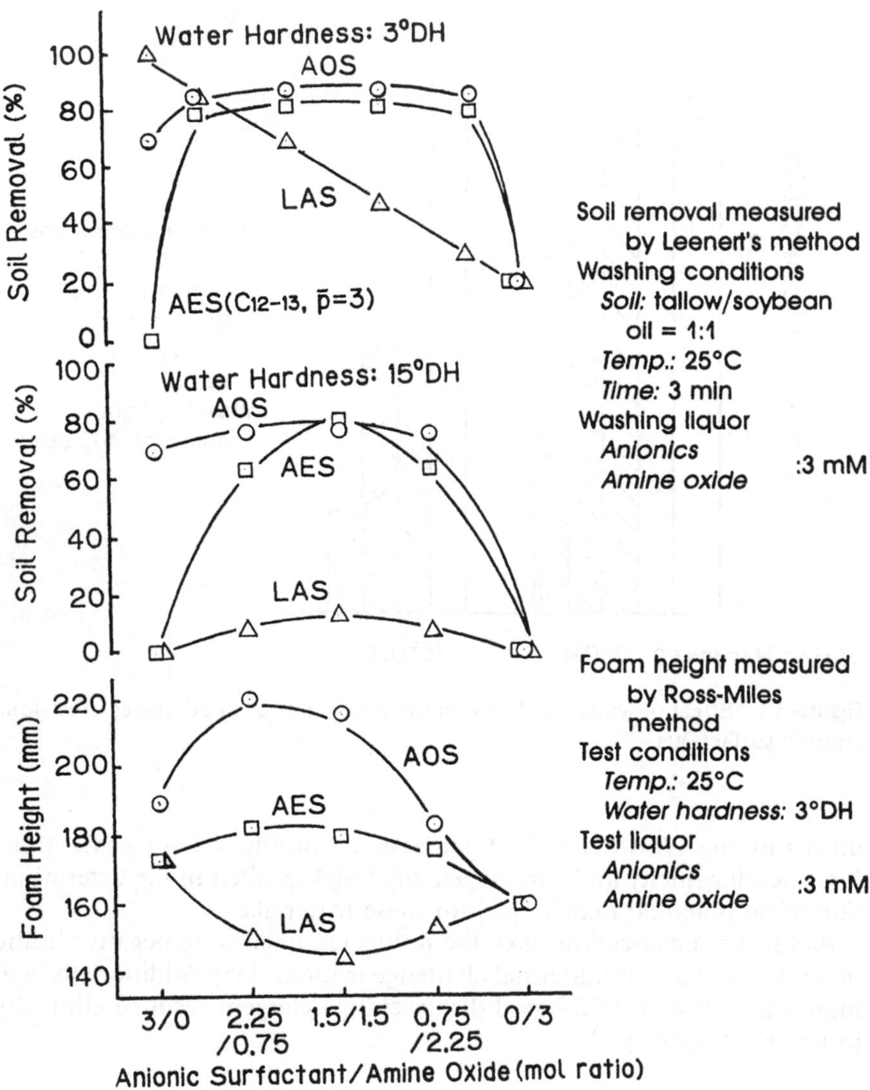

Figure 19 Soil removability and foam height of the binary system anionic surfactant and amine oxide.

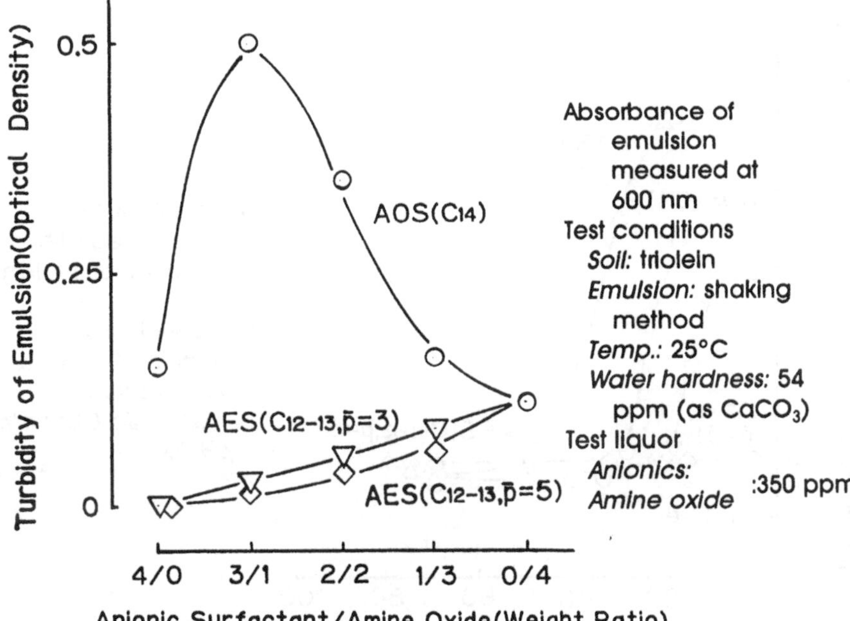

Figure 20 Effect of the binary system anionic surfactant and amine oxide on emulsification ability.

surfactants with less tendency to denature proteins. It was also determined that this property could be improved by the addition of amine oxide. This combination of AOS/amine oxide features superior soil removability, foamability, and mildness to the skin; some formulations are exemplified in patents for use in dishwashing detergents [68].

From the viewpoint of convenience, hand dishwashing detergents are mostly prepared in the form of liquid formulations. However, AOS formulations are relatively unstable compared to LAS or AES, in that they can become turbid at lower temperature or form a gelatinous precipitate during storage [69–71]. With this instability in mind, VOS or IOS, isomers of AOS having lower Krafft points, were used as dishwashing liquids and claimed in patents [72].

Dishwashing detergents other than those for hand washing include detergents for automatic dishwashers. AOS is suitable for use in hand washing detergents but is not appropriate for automatic dish washing because of its high foaming characteristics.

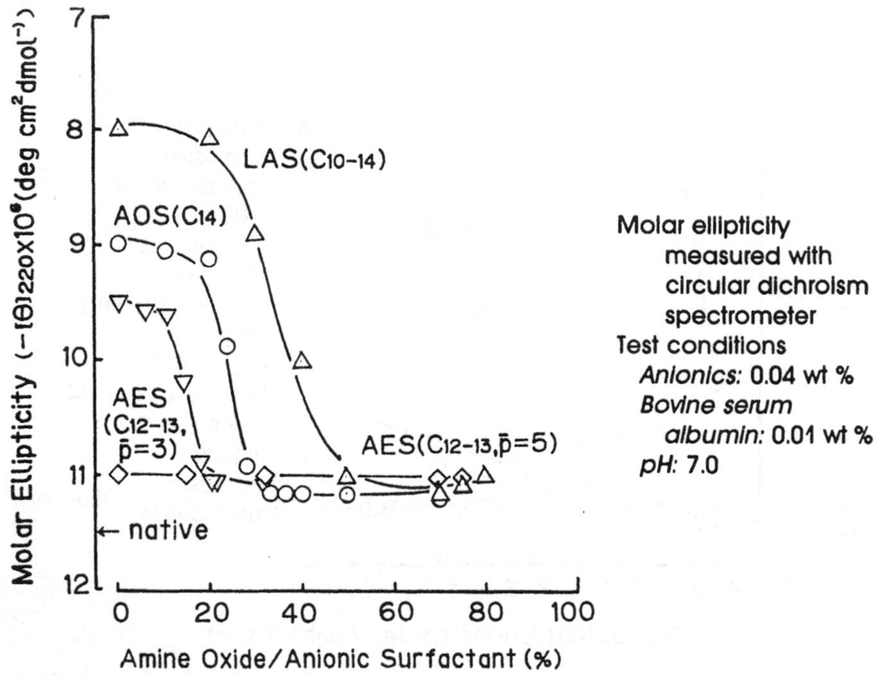

Figure 21 Denaturation property of the binary system anionic surfactant and amine oxide onto protein.

2.3 Other Applications

AOS is also utilized in smaller-volume applications, including rug shampoo and oven cleaners. As rug shampoos, there are patents wherein AOS has been claimed to provide good foam-stable liquid formulations, as well as cleaning-foam-generating aerosol preparations [73,74]. As cleaners for ovens, clear gel compositions and formulations are also claimed in patents to be effective [75, 76].

3 PERSONAL CARE PRODUCTS

3.1 Shampoo

Alcohol sulfate (AS) has been the key active in shampoos for many years. However, as high-quality AOS came to be produced successfully, cosmetic chemists have focused their attention on this new active. In addition to soil removability, foaming properties and mildness to skin, hair, and eyes are especially required in the case of shampoos. AOS exhibits superior performance in these areas. From the viewpoint of economics, superiority of AOS has increasingly been recognized, and recently it has come to be used widely as a shampoo base in the United States and Japan.

Basic Properties

Foamability. As has been seen, AOS not only exhibits high biodegradability and soil removability at higher water hardness, but C_{14} AOS also presents, as shown in Figure 22, a higher foamability in the presence of sebum than AS or AES [27]. Moreover, Rinso [77] reported that the foam generated gives a creamy, voluminous, and soaplike feeling.

As for the incorporation of lauric diethanolamide, a foam booster, Schoenberg [78] reported that both AOS and AS used at pH values of 4.0 and 7.0 are synergistic with this additive with respect to foam height enhancement (Figs. 23 and 24). This synergism was more pronounced with AOS than with AS. The author also investigated the combination effects observed with the simultaneous use of lauramine oxide and cocamidopropyl betaine, and obtained similar results.

Safety. Regarding the skin irritation potency of AOS, immersion and upper-arm patch tests revealed its mildness to the skin [79], and in a study

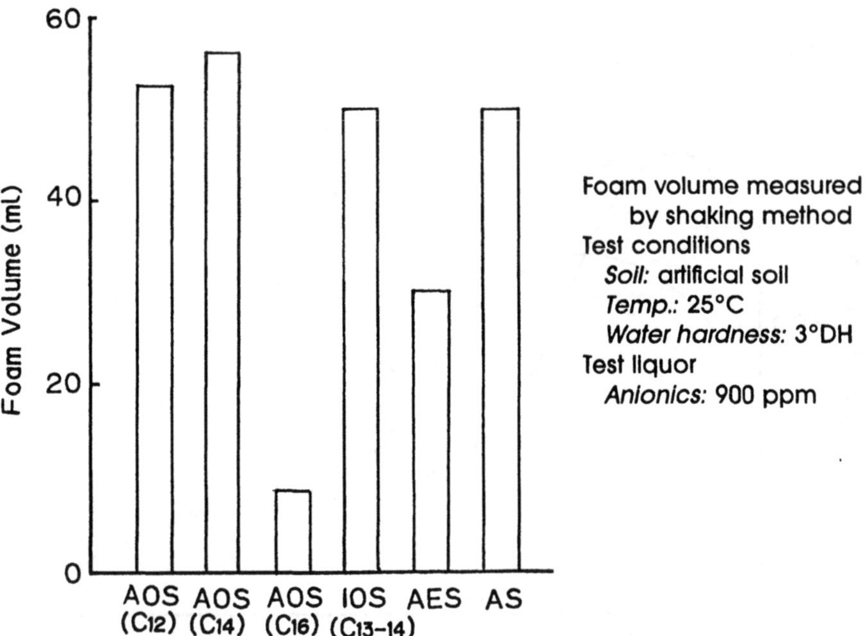

Figure 22 Foam property of various anionic surfactants.

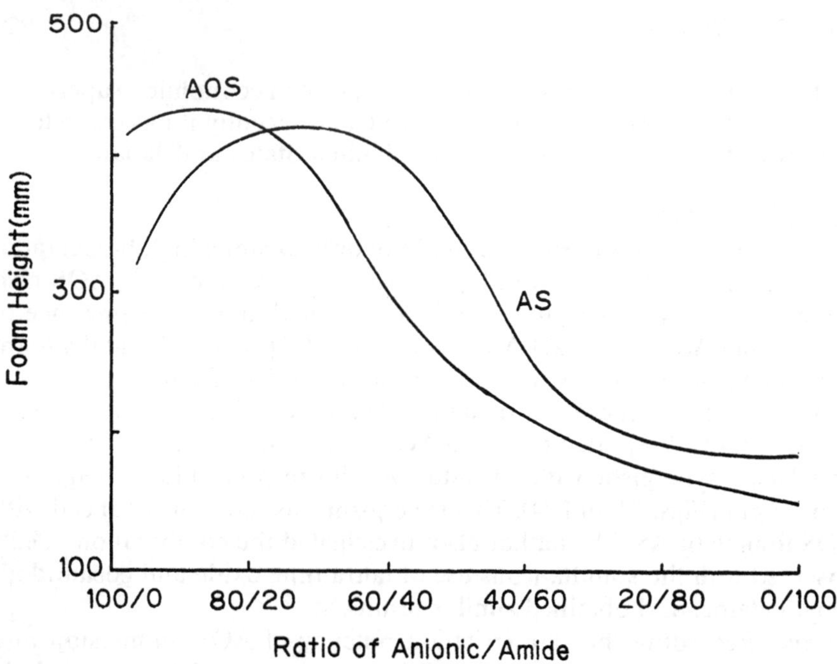

Figure 23 Foam height of the binary system AOS/lauramide and AS/lauramide at pH 7.0.

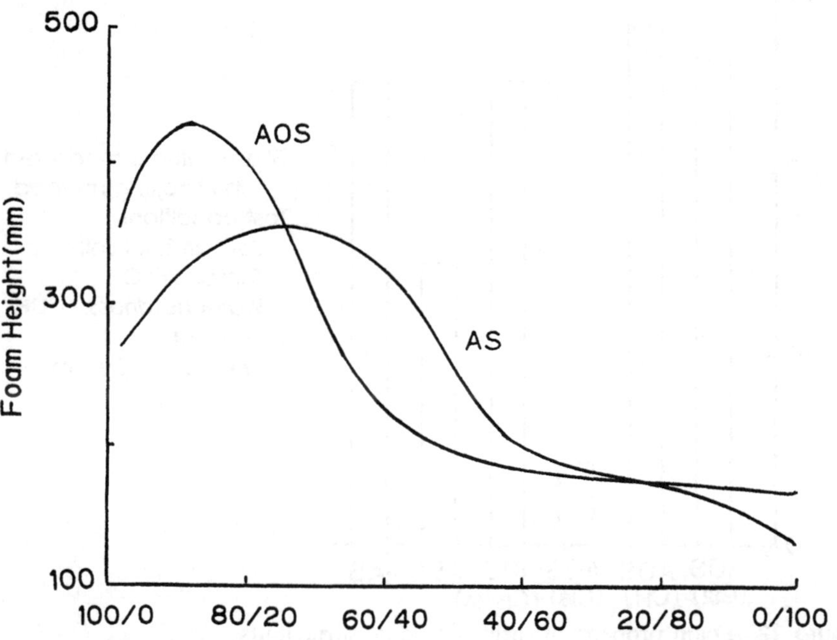

Figure 24 Foam height of the binary system AOS/lauramide and AS/lauramide at pH 4.0.

Table 8 Histopathological Observation in Guinea Pig Skin

Sample	Concentration (%)	Contact time (min)	Recovery time[a] (hr) 24	168
AOS (C_{14}–C_{19})	1	10	–	–
		120	–	–
	10	10	–	–
		120	–	–
	30	10	–	–
		120	+	–
LAS (C_{10}–C_{14})	1	10	–	–
		120	–	–
	10	10	+	–
		120	++	–
	30	10	++	–
		120	++	+
AS (C_{12}–C_{13})	1	10	–	–
		120	–	–
	10	10	+	–
		120	++	+
	30	10	++	+
		120	++	+

Source: Ref. 80.
[a]No reaction (–), slight reactions (+), and relatively heavy reactions (++) were observed, respectively.

using guinea pigs, AOS was proved to be milder than AS and LAS, as shown in Table 8. With respect to eye irritation potency, AOS exhibited an equal or lower level of potency compared to other surfactants, as Tables 9 and 10 clarify.

Application of AOS in Shampoo

Since AOS is stable over a wide pH range, as illustrated in Table 11, it is also suitable for use in slightly acidic shampoo bases. Hair conditioning function is a requisite for a shampoo, and in this respect, the combination of AOS with alkanolamide provides a composition that gives a good dried hair manageability [85].

Shampoos are mostly formulated as liquids from the viewpoint of convenience in handling and use. As stated before, however, AOS, has higher Krafft points than AES and its liquid state poses an instability problem at low temperature. To improve this instability of AOS in the form of liquid

Table 9 Evaluation of Rabbit Eye Irritation by Draze

Sample	Time after instillation (hr)	Concentration[a] (%)					
		0.01	0.05	0.1	0.5	1	5
AOS (C_{14}–C_{19})	24	0	0	0	0	2	14
	40	0	0	0	0	0	5
LAS (C_{10}–C_{14})	24	0	0	0	1	7	48
	40	0	0	0	0	2	25
AS (C_{12}–C_{13})	24	0	0	0	8	10	46
	40	0	0	0	0	2	23

Source: Refs. 81.
[a]Theoretical total maximum = 110.

preparations, its combination with alkane sulfonate (SAS) is recommended [15]. The low-temperature stability of AOS formulations can be improved by replacing one-third of the AOS with SAS. This results in a reduction in Krafft point without a sacrifice in the foamability and foam consistency of AOS (Fig. 25).

Table 10 Draze Irritation Scores (Dermal)[a]

Percent concentration	AOS	AES	LS	Ref.
1.0	0.2	0.1	0.2	83
2.0	0.0	—	—	84
5.0	0.3	—	—	84
10.0	0.5	—	—	84
20.0	2.9	—	—	84
28.0	—	5.0	—	83
30.0	—	—	5.2	83
36.8	4.5	—	—	83

Source: Refs. 82–84.
[a]Explanation of scores:
0.4 or less: nonirritant (83,84)
0.5: mild irritant (83,84)
2.9, 4.5: moderate irritant (83,84)
5.0, 5.2: primary irritant (83,84)

Table 11 Comparative Acid and Base Stability of Sulframin AOS, Sodium Lauryl Sulfate, and Sodium Lauryl Sulfoacetate (Percent Hydrolysis after 3 Months at 55°C)

	pH 2	pH 4	pH 10
Sulframin AOS	0	0	0
Sodium lauryl sulfate	100	>50 (sep'n)	0
Sodium lauryl sulfoacetate	100	>65	>20
Sodium lauryl ether sulfate	100	>50 (sep'n)	0

Source: Ref. 82.

3.2 Other Applications

Skin Products

Since AOS exhibits not merely mildness to the skin but also good foam properties, liquid hand soap formulations using AOS have enjoyed rapid acceptance in the U.S. market. A typical formulation is illustrated in Table 12.

Bath Products

AOS applications in bath products involve bubble bath preparations, bar soaps, and body shampoos. Since AOS possesses low hygroscopicity and

Table 12 Cold Blend Liquid Hand Soap[a]

	Parts by weight
Witconate AOS liquid	19.0
Witcamide 5133 (cocoamide DEA)	4.0
Oleic acid	0.4
Emcol NA-30[b] (cocoamidopropyl betaine)	2.2
Pearlescent agent	7.5
Preservative	0.2
Ammonium chloride (25% aq. soln.) to desired viscosity	4–7
Water, perfume, dye	q.s. to 100

Source: Ref. 86.

[a]Procedure:

1. Dissolve all raw materials in water with stirring. Adjust to pH 6.0–6.5 with acid (dilute phosphoric, citric, lactic, or hydrochloric acids).
2. Add salt for desired viscosity.

[b]Witco Chemical Corp., New York.

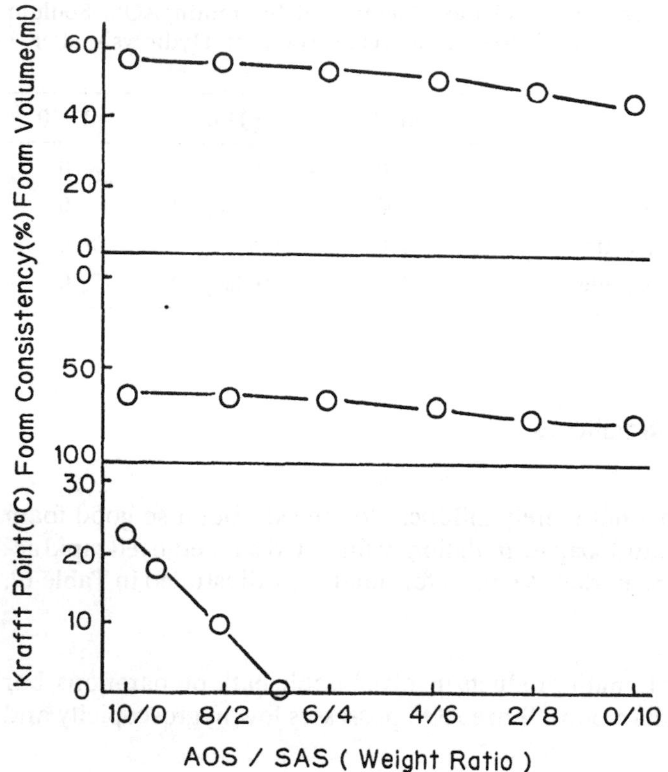

Foam volume measured by a shaking method
Foam consistency measured by a Stiepel flask method
Test conditions
Soil: artificial soil
Temp.: 25°C
Water hardness: 54 ppm (as $CaCO_3$)
Test liquor
Anionics: 9000 ppm

Figure 25 Effect of the binary system AOS and SAS on foam volume (upper), foam consistency (middle), and Kraft point (lower).

good flow properties, AOS in 90% active flake form has been applied to powdered bubble bath products [87]. As for AOS applications in bar soap products, several patents claim its use in this field with its high foamability, rinsability, and good shape retention [88]. Further, another patent includes its use in body shampoos, claiming its mildness to the skin and viscosity stability against temperature variation [89]. Another patent claims AOS application as a base in transparent gelled body shampoo composition, which is also stable at low temperature [90].

Toothpastes

Since AOS is a highly safe surfactant exhibiting high foaming characteristics and has practically no taste or odor, it is also suitable for use as a foaming agent in toothpastes [91].

REFERENCES

1. Tomiyama, S., *J. Jpn. Oil Chem. Soc., 19*(6): 359 (1970).
2. Yamane, I., *J. Syn. Org. Chem. Jpn., 38*(6): 593 (1980).
3. Yamane, I., *J. Am. Oil Chem. Soc., 55:* 81 (1978).
4. Toyoda, S., Ohgoshi, T., and Kitano, K., *Kagaku Kogaku, 46*(6): 298 (1982).
5. Toyoda, S., Ohgoshi, T., Kitano, K., and Miyawaki, Y., *Kagaku Sochi, 24*(6): 101 (1982).
6. Tomiyama, S., *Kagaku To Kogyo, 22*(3): 304 (1969).
7. Swischer, R.D., *J. Am. Oil Chem. Soc., 40:* 648 (1963).
8. Ohba, K., and Takei, R., *J. Jpn. Oil Chem. Soc., 30*(7): 450 (1981).
9. Masse, F. W. J. L., and Tilburg, R. V., *J. Am. Oil Chem. Soc., 60:* 1672 (1983).
10. Püshel, F., *Tenside, 4*(10): 320 (1967).
11. Yamane, I., Nagayama, M., Kashiwa, I., Kuwamura, H., Ando, S., and Mori, A., *Kogyo Kagaku Zasshi, 73*(4): 723 (1970).
12. Ogino, K., and Abe., M., *J. Am. Oil Chem. Soc., 52:* 465 (1975).
13. Yamane, I., *Kagaku To Kogyo, 30*(1): 23 (1977).
14. Tomiyama, S., Takao, M., Mori, A., and Sekiguchi, H., *J. Am. Oil Chem. Soc., 46:* 208 (1969).
15. Okumura, O., presented at the 76th AOCS Annual Meeting, Philadephia, May 1985.
16. Hunter, R. T., and Roga, R. C., *J. Am. Oil Chem. Soc., 46:* 199 (1969).
17. Spangler, W. G., Roga, R. C., and Cross, H. D., *J. Am. Oil Chem. Soc., 44:* 728 (1967).
18. Kemper, H. C., Martens, R. J., Nooi, J. R., and Stubbs, C. E., *Tenside, 12*(1): 47 (1975).

19. Crutchfield, M. M., *J. Am. Oil Chem. Soc., 55:* 58 (1978).
20. Tokiwa, F., and Imamura, T., *J. Am. Oil Chem. Soc., 46:* 571 (1969).
21. Kadoya, I., and Okumura, O., *Fragrance J., 6*(29): 61 (1978).
22. Yamane, I., Toyoda, S., Okumura, O., Nakazawa, T., and Ogawa, M., *J. Jpn. Oil Chem. Soc., 35*(3): 151 (1986).
23. Hoogerheide, I. C., *Fette Seifen Anstrichm., 70*(10): 743 (1968).
24. Liss, R. L., and Langguth, R. P., *J. Am. Oil Chem. Soc., 46:* 507 (1969).
25. Komaki, T., *Yushi, 23*(5): 113 (1970).
26. Sato, M., Sudo, M., and Minagawa, M., *Sen'i Syohikagaku Zasshi, 25*(10): 485 (1984).
27. Mori, A., and Okumura, O., CESIO World Surfactant Congress, Munich, vol. 2, p. 93, 1984.
28. Tomiyama, S., *J. Jpn. Oil Chem. Soc., 21*(6): 338 (1972).
29. Ohba, K., Sugiyama, T., Miura, K., Ishimatsu, T., Nishino, T., and Morisaki, Y., 7th International Congress on Surface Active Substances, Moscow, vol. 4, p. 190, 1976.
30. Acute toxicity of tetra decene sodium sulfonate, *Report*, Kitazato Univ., 1968.
31. Ohba, K., and Tamura, J., Acute toxicity of *n*-alpha-olefin sulfonates, *Agr. Biol. Chem., 31:* 1509–1510 (1967).
32. Ogura, Y., and Tamura, J. Pharmacological studies on surface active agents, III, Acute toxicity of household synthetic detergent, n-alpha-olefin sulfonate(AOS), *Ann. Rep. Inst. Food Microbiol., 20:* 83–87 (1967).
33. Webb, B.P., AOS—new biodegradable detergent, *Soap Chem. Spec.*, 61–62 (Nov. 1966).
34. Experimental data for one-month oral administration of tetra decene sodium sulfonate in the rat, *Report,* Kitazato Univ., 1968.
35. Experimental data for six- month oral administration of tetra decene sodium sulfonate in the rat, *Report*, Kitazato Univ., 1968.
36. Subacute toxicity for one-month oral administration of AOS-Mg, AOS-Na in the rat, *Report*, Yokosuka Inst. for Applied Pharmacology (Oct. 15, 1980).
37. Hunter, B., and Benson, H. G., Long-term toxicity of the sufactant alpha-olefin sulfonate (AOS) in the rate, *Toxicology, 5:* 359–370 (1976).
38. Chronic toxicity for six-month oral and percutaneous administration of AOS-Mg in the rat, *Report*, Yokosuka Inst. for Applied Pharmacology (Oct. 15, 1980).
39. Uchida, H., and Saski, A., The effect of the sysnthetic detergent into skin and organs in the mouse, *Yakubutsu Ryoho, 13*(2): 147–151 (1980).
40. Palmer, A. K., Readshaw, M. A., and Neuff, A. M., Assessment of the teratogenic potential of surfactants, II, AOS, *Toxicology, 3:* 107–113 (1975).

41. AOS-Mg: Segment II reproductive study in rats-oral administration, *Report 79/LIF036/328*, Life Science Research, June 20, 1979.

42. AOS-Mg: Effects upon the reproductive performance of rats treated continuously throughout two successive generations, *Report 80/LIF044/508*, Life Science Research, Dec. 4, 1980.

43. AOS-Mg: Effects of oral administration upon pregnancy in the rabbit, *Report 79/LIF040/381*, Life Science Research, Aug. 8, 1979.

44. Effect of AOS detergent on preganancy of the mouse (dermal application), *Report LFO/32/77265*, Huntington Research Centre, Apr. 21, 1977.

45. Sawano, J., Study on teratogenic effect of the percutaneous adsorption of "Detergents for kitchen use (AOS, AOS-S)," *Studies of the synthetic detergents,* Research Coordination Bureau, Japanese Science and Technology Agency, 1978, pp. 112–113.

46. Effect of AOS detergent on pregnancy of the rate (dermal application), *Report LFO/31/76860*, Huntington Research Centre, Dec. 14, 1976.

47. AOS-Mg: Segment II study of the effects of topical application upon reproduction in the rat, *Report 80/LIF038/056*, Life Science Research, Mar. 3, 1980.

48. AOS-Mg: Effect of topical application upon pregnancy in the rabbit, *Report 79/LIF42/258*, Life Science Research, May 14, 1979.

49. Inoue, S., On the relation between the binding of alpha- olefin sulfonate and its metabolite to serum albumin and these elimination, *Folia Pharmacol Jpn., 74*(2): 25 (1978).

50. Inoue, S., O'Grodnik, J. S., and Tomizawa, S., Metabolism of alpha-olefin sulfonate (AOS) in the rats, *Fund. Appl. Toxicol., 2:* 130–138 (1982).

51. Minegishi, K., Osawa, M., and Yamaha, T., Study of synthetic detergent. I, Skin adsorption of alpha-olefin sulfonate, Abstract of paper presented at the 96th Annual Meeting of the Pharmaceutical Society of Japan, vol. 3, 1976, p. 172.

52. Minegishi, K., Osawa, M., and Yamaha, T., Pecutaneous absorption of alpha-olefin sulfonate(AOS) in rats, *Chem. Pharm. Bull. (Tokyo), 25*(4): 821–825 (1977).

53. Tomizawa, S., and Inoue, S., The metabolic fate of alpha-olefin sulfonate(AOS) in rats. Abstract of paper presented at the 97th Annual Meeting of the Pharmaceutical Society of Japan, 1977/4, p. 251.

54. Yamaha, I., Minegishi, K., and Osawa, M., Metabolism of alpha-olefin sulfonate, *Studies of synthetic detergents*, Research Coordination Bureau, Japanese Science and Technology Agency, 1978, pg. 174–188.

55. Ohba, K., Mori, A., Nagai, T., Presented at the 76th AOCs Annual Meeting, Philadelphia, May 1985.

56. Okumura, O., Nakagawa, R., and Tsuruta, Y., *Gypsum Lime, 206:* 29 (1987).

57. Okumura, O., et al., Jpn. Kokai patent 29,302 (1974), to Lion Corp.: *Chem.*

Abstr., 81, 51553; Lion Corp., Jpn. Kokai patent 69,698 (1980): *Chem Abstr.*, 93, 170056; Lion Corp., Jpn. Kokai patent 37,095 (1983): *Chem. Abstr.*, 99, 214538; Lion Corp., Jpn. Kokai patent 37,098 (1983): *Chem. Abstr.*, 99, 214538; Rubinfeld, J., U.S. patent 3,980,588 (1976), to Colgate-Palmolive Co.: *Chem Abstr.*, 85, 162360; Sheldon, G. P., Ger. patent 2,242,157 (1973), to Colgate-Palmolive Co.: *Chem Abstr.*, 78, 161181; Yurko, J. A., Ger. patent 2,540,510 (1976), to Colgate-Palmolive Co.: *Chem Abstr.*, 81, 16634.

58. Fujii, T., et al., Jpn. patent 7,587 (1971), to Lion Corp.: *Chem. Abstr.*, 76, 47623; Okumura, O., et al., Jpn. Kokai patent 27,603 (1978), to Lion Corp.: *Chem Abstr.*, 89, 61381; Fukano, K., et al., Ger. patent 2,948,791 (1980), to Lion Corp.: *Chem. Abstr.,* 93, 1700065; Tomiyama, S., et al., U.S. patent 3,544,475 (1970), to Lion Corp.: *Chem. Abstr.*, 74, 100923.
59. Okumura, O., et al., Jpn. Kokai patent 105,914 (1977), to Lion Corp.: *Chem. Abstr.*, 88, 24522; Okumura, O., et al., Ger. patent 2,727,346 (1977), to Lion Corp.: *Chem. Abstr.*, 88, 63525; Nakamura, M., et al., Ger. patent 2,839,619 (1975), to Lion Corp.: *Chem. Abstr.*, 90, 170534.
60. Kashiwa, I., et al., Jpn. patent 33,488 (1975), to Lion Corp.: *Chem. Abstr.*, 84, 91980; Baurman, B., et al., Ger. patent 2,355,788 (1975), to Farbwerk Hoechst A.-G.: *Chem. Abstr.*, 83, 146544; Okumura, O., et al., Ger. patent 2,621,088 (1976), to Lion Corp.: *Chem. Abstr.*, 86, 18758; Ohgoshi, T., et al., Ger. patent 2,730,951 (1978), to Lion Corp.: *Chem. Abstr.*, 88, 75611; Nakamura, M., et al., Ger. patent 2,722,698 (1977), to Lion Corp.: *Chem. Abstr.*, 88, 39261; Farbwerk Hoechst A.-G., Fr. patent 2,250, 820 (1975): *Chem. Abstr.*, 84, 6869; Stepan, F. Q., U.S. patent 4,111,853 (1978), to Stepan Chemical Co.: *Chem. Abstr.*, 90, 89132.
61. Nakamura, M., and Toyoda, S., 4th International Drying Symposium, 1984, p. 403; Kawakami, A., et al., Jpn. Kokai patent 80,285 (1977), to Lion Corp.: *Chem. Abstr.*, 87, 169562; Lion Corp., Jpn. Kokai patent 192,499 (1982): *Chem. Abstr.*, 98, 217640; Lion Corp., Jpn. Kokai patent 192,500 (1982): *Chem. Abstr.*, 98, 217641.
62. Nobble, W. R., and Linfield, W.H., *J. Am. Oil Chem. Soc., 53:* 172 (1976).
63. Lion Corp., Jpn. Kokai patent 92,096 (1982): *Chem. Abstr.*, 97, 184377; Lion Corp., Jpn. Kokai patent 92,097 (1982): *Chem. Abstr.*, 97, 184378.
64. Unilever N. V., Neth. patent 6,717,002 (1969): *Chem. Abstr.*, 71, 114479.
65. Tuvell, M. E., Kuehnhanss, G. O., Heidebrecht, G. D., Hu, P. C., and Zielinski, A. D., *J. Am. Oil Chem. Soc., 55:* 70 (1978).
66. Odioso, R. C., *Detergent Age, 4*(2): 29 (1967).
67. Ohbu, K., Jona, N., Miyajima, N., and Kashiwa, I., *J. Jpn. Oil Chem. Soc., 29*(11): 866 (1980).
68. Lion Corp., Jpn. Kokai patent 4,694 (1984): *Chem. Abstr.*, 101, 25413; Lion Corp., Jpn. Kokai patent 12,997 (1984): *Chem. Abstr.*, 100, 194071; Miyajima, M., et al., Ger. patent 3,011,017 (1981), to Lion Corp.: *Chem. Abstr.*, 94, 210647.
69. Tokiwa, F., Jpn. patent 39,246 (1976), to Kao Corp,: *Chem. Abstr.*, 80, 61395.

70. Stephen K. C., et al., U.S. patent 3,970,596 (1975), to Colgate-Palmolive Co.: *Chem. Abstr*, 83, 81854.
71. Kuwamura, H., et al., Jpn. patent 48,766 (1976), to Lion Corp.: *Chem. Abstr.*, 75, 84605.
72. Ethyl Corp. Jpn. Kokai patent 78,706 (1974): *Chem Abstr.*, 84, 173795; Sweeney, A., U.S. patent 3,708,437 (1973), to Chevron Research Co.: *Chem. Abstr.*, 78, 60045; Bentley, F. E., Ger. patent 2,241,031 (1973), to Jefferson Chemical Co., Inc.: *Chem. Abastr.*, 78, 149209.
73. Hans M. H., et al., Ger. patent 2,408,895 (1975), to Chemische Werke Huels A.-G.: *Chem. Abstr.*, 83, 207847.
74. Koller, O., Ger. patent 2,146,373 (1973), to Farbwerk Hoechst A.-G.: *Chem Abstr.*, 78, 149210.
75. Watanabe, H., et al., U.S. patent 4,451,393 (1984), to Stepan Chemical Co.: *Chem. Abstr.*, 101, 56891.
76. Gilles V. M. L., et al., U.S. patent 4,477,395 (1984), to Miles Laboratories, Inc.: *Chem. Abstr.*, 101, 232252.
77. Rinso, R. R., *Soap Cosmet. Chem. Spec., 54*(6): 56 (1978).
78. Schoenberg, T. G., *Soap Cosmet. Chem. Spec., 56*(5): 54 (1980).
79. Knaggs, E. A., et al., Annual Meeting of the AOCS, May 1967.
80. Iimori, M., Iwata, E., and Sato, Y., *J. Jpn. Oil Chem. Soc., 20*(9): 584 (1971).
81. Iimori, M., Ogata, T., and Kudo, K., *J. Jpn. Oil Chem. Soc., 21*(6): 334 (1972).
82. Barker, G., Barabash, M., and Sosis, P., *Soap Cosmet. Chem. Spec., 54*(3): 38 (1978).
83. A study of the irritancy and sensitization potential of several detergent substances, *Paper*, Tulane Univ.
84. AOS and LAS: concentration vs irritation, Letter, Rinehart W. E. to File, July 28, 1972.
85. Baviak, S. C., U.S. patent 3,870,660 (1975), to Gulf Research and Development Co.: *Chem. Abstr.*, 83, 84711.
86. Friedman, S.K., Barker, G., and Mausner, M., *Soap Cosmet. Chem, Spec., 59*(9): 35 (1983).
87. Mausner, M., et al., U.S. patent 3,776,861 (1973), to Witco Chemical Corp.: *Chem. Abstr.*, 80, 97679.
88. Lion Corp., Jpn. Kokai patent 62,897 (1981): *Chem. Abstr.*, 95, 152506; Lion Corp., Jpn. Kokai patent 74,196 (1981): *Chem. Abstr.*, 95, 171472; Lion Corp., Jpn. Kokai patent 74,391 (1982): *Chem. Abstr.*, 97, 146562; Malhotra, V. N., et al., Ger. patent 2,335,171 (1974), to Unilever N.V.: *Chem. Abstr.*, 81, 4885.
89. Lion Corp., Jpn. Kokai patent 74,197 (1981): *Chem. Abstr.*, 95, 171490; Lion Corp., Jpn. Kokai patent 95,993 (1981): *Chem. Abstr.*, 95, 205805.
90. Lion Corp., Jpn. Kokai patent 36,599 (1981): *Chem. Abstr*,., 95, 117435.
91. Lion Corp., Jpn. patent 38,016 (1977): *Chem. Abstr.*, 88, 141507.

CHAPTER 8

Surfactants: Alkylates

GEORGE R. LAPPIN and JOHN D. WAGNER Ethyl Corporation, Baton Rouge, Louisiana

1 INTRODUCTION

Linear alkylbenzene (LAB) is a large-volume intermediate to linear alkylbenzene sulfonates (LAS), widely regarded as being the key active ingredient in modern detergents. Some sources estimate that (LAS) represents one-third of the active ingredients in detergents worldwide. It is used as a sulfonate in light-duty liquids, heavy-duty liquids, heavy-duty powders, general-purpose household cleaners, and various industrial and institutional cleaners. The largest area of LAS use is laundry detergents, which represents over 50% of U.S. consumption.

Less than 20% of LAB is produced from alpha olefins. Purchased alpha olefins may be economical for LAB production when a small volume of LAB is required. At large volumes, routes based on *n*-paraffins often are more attractive. There is a trend to lower paraffinic crude oils that may result in a shift in the LAB production economics to favor alpha olefins. Perhaps offsetting this effect, recent improvements in catalyst efficiency by UOP [1] have improved the economics of the paraffin route. The economics are also influenced by the differences in the cost of paraffins and the cost of ethylene.

Usage of alpha olefins to produce LAB varies from country to country depending on the volume of LAB used in the country and on economic

Table 1 Estimated Linear Alkylbenzene Annual Capacity by 1989

Region	Million pounds	Thousand metric tons	Percent
North America	790	359	17.0
South America	364	165	7.8
Western Europe	1782	674	31.9
Eastern Bloc	439	200	9.5
Africa/Middle East	409	186	8.8
Asia/Pacific	1164	529	25.0
Total	4648	2113	100.0

factors. In the United States the primary ongoing use of alpha olefins is to balance carbon numbers. When demand is high, usually during a strong economic period, demand for LAB, including exports, approaches or exceeds U.S. capacity; and then alpha olefins are used to obtain incremental capacity. In Europe, Central and South America, and Africa, alpha olefins and linear internal olefins are routinely used to produce linear alkylbenzene, often in units formerly producing branched alkylbenzene.

Table 1 shows the distribution of the 4.6 billion pounds (2.1 million metric tons) of LAB capacity by various world regions. Europe has the most capacity, and Asia/Pacific has the next-largest capacity.

In developing countries, alpha olefins are occasionally used to replace propylene tetramer to convert "hard" alkylate plants to "soft" alkylate plants at little or no capital cost. Thus alpha olefins offer an immediate solution to river and stream pollution problems. Longer term, the larger-volume producers of LAB tend to switch to paraffin-based LAB, but the smaller-volume producers may continue to use alpha olefins. Discussion of LAB in this book is limited primarily to its aspect as a market for alpha olefins.

2 HISTORY

Synthetic surfactants were developed early in the twentieth century, with the first being made during World War I in Germany. Being very harsh, they were used in industrial applications but not in household detergents. In 1933, Procter & Gamble developed the first synthetic household detergents and these were based on alcohol sulfates. Wartime shortages of tallow from cattle and of coconut oil from the Philippines slowed the development of these detergents. Alkylbenzene sulfonates based on kerosene petroleum fractions were used in the synthetic detergents that appeared after World War II. Propylene tetramer replaced these early kerosene fractions because of better economics. These early synthetics (of the late 1940s to early 1950s) performed better in hard water than the soaps used in most of the laundry products of the time.

One of the key actives in the early detergents was the propylene-based "hard" alkylbenzene sulfonates. By the mid-1960s, there was much concern about detergents causing foaming in sewage treatment plants and effluents. Photographs of rivers brimming from bank to bank with foam appeared in newspapers and magazines. The major soapers and alkylate suppliers identified the source of the problem as being caused by the multiple branching in the propylene-based alkylate which resisted biodegradation.

By 1965, the major soap and detergent producers had voluntarily switched to the more biodegradable alkylbenzene sulfonates based on linear paraffins in the United States and Europe, where high population density and high per capita detergent usage caused the greatest problems.

Since this book concerns alpha olefins, we will note that this also led to decisions to build various other linear organic raw material plants for detergents, including higher primary alcohol plants by Conoco (now Vista) in 1962 and by Ethyl in 1965 and the Gulf (now Chevron) alpha olefin plant in 1966. Biodegradable surfactants were considered to be a significant market for heavier olefins when the Ethyl alpha olefin plant was built in 1970 and were a key to Shell building olefin, alcohol, and LAB plants in Europe and an olefin/alcohol plant in the United States.

The detergent industry continues to change, driven by consumer habits, environmental concerns and regulations, research, and economics. Higher energy costs spurred development of detergents that perform at lower wash temperatures. Environmental regulations resulted in banning of phosphates in laundry products in some areas (e.g., the Great Lakes region of the United States). Low- or no-phosphate formulations represented new challenges because of cleaning deficiencies compared to the phosphate formulations. Use of polyester fibers in no-iron garments resulted in new cleaning challenges, as they do not release oily substances as easily as cotton.

Many of these changes resulted in less LAS being used in formulations, as mixtures of alcohol ethoxylates and alcohol ethoxy sulfates generally seemed better, especially at colder washing temperatures. This led to rationalization of supply in the United States as the original six suppliers of LAB were reduced to two by the end of 1985. Consumption of LAB in the United States peaked in 1978–1979, slumped in 1983 as Procter & Gamble reformulated some of its products away from LAB, and rebounded in 1985–1986 due in part to the popularity of Lever's Sunlight liquid detergent and other detergents with significant concentrations of LAS. Trend-line growth for LAB is forecast to be that of the U.S. gross national product, or about 1 to 2% annually [2].

Japan, South Africa, and other countries adopted soft alkylbenzenes soon after Europe and the United States. The process of conversion to soft alkylate has proceeded more slowly elsewhere, but several countries have converted recently or plan to convert by the end of this decade.

Alpha olefins are a good economic choice for LAB feedstock when the plant size is small, perhaps less than 50 to 100 million pounds per year. The olefins tend to cost more than the linear paraffins, which are extracted from kerosene. However, using alpha olefins means that the capital cost of conversion from propylene tetramer is lower than for paraffins and the conversion can be made quickly. Higher capital costs for the more expensive paraffin routes must be amortized, and these result in the alpha olefins being good choices for smaller plants. The paraffin route tends to be favored for totally new plants built for large and stable demands.

In the paraffin route based on dehydrogenation, linear odd and even internal olefins are produced. The resulting olefins must be linear, but they do not have to be 1-olefins, as rearrangement of the olefin bond by the alkylation catalyst occurs during the alkylation reaction. Thus identical products are produced no matter which type of linear olefin is fed to the alkylation reactor.

Linear alkylbenzenes are also produced by alkylation of benzene with secondary alkyl chlorides from partial chlorination of linear paraffins. The process is similar to the catalytic dehydrogenation method, in that after benzene alkylation, the unreacted paraffins are separated and returned to the chlorination reactor. About 37% of active U.S. LAB capacity in 1989 will use the chlorination-dehydrochlorination approach, and about 63% will use catalytic dehydrogenation, augmented at times with 1-olefin as indicated above.

The LAB price in the United States was in the vicinity of 10 to 11 cents per pound from 1965 through 1975, when crude oil was $3 per barrel. The U.S. price increased to the vicinity of 43 to 48 cents per pound in the early 1980s, when crude oil averaged about $28 per barrel. LAB prices tended to be lower in Europe and higher in Japan than in the United States. The economics of LAB production are directly dependent on energy values.

3 PRODUCERS

Although there have been many changes in recent years, Table 2 is believed to represent world producers by 1989. Little investment is required to change from hard alkylate to soft alkylate by using alpha olefins

Table 2 Estimated Linear Alkylbenzene Annual Capacity by 1989

Region/country	Company	Millions of pounds	Metric tons	Catalyst	Feedstock
North America					
United States	Monsanto	300	136	HF	Paraffin/olefin
United States	Monsanto	45	20	$AlCl_3$	Paraffin/olefin
United States	Vista	225	102	$AlCl_3$	Paraffin
United States	Vista	150	68	HF	Paraffin
Mexico	Pemex	70	32	$AlCl_3$	Olefin
Total North America		790	359		
South America					
Argentina	YPF	90	41	HF	Paraffin
Venezuela	Venoco	50	23	$AlCl_3$	Olefin
Brazil	Emca	30	14	$AlCl_3$	Olefin
Brazil	Deten	97	44	HF	Paraffin
Brazil	Deten	97	44	HF	Paraffin
Total South America		364	165		
Western Europe					
France	Shell	110	50	HF	Olefin
Germany	Huels	175	80	$AlCl_3$	Paraffin
Germany	Wibarco	77	35	$AlCl_3$	Paraffin
Germany	Deutsche Texaco	66	30	$AlCl_3$	Paraffin
Italy	Enichem Augusta	154	70	HF	Paraffin/olefins
Italy	Montedison	100	45	$AlCl_3$	Paraffin
Italy	SIR	220	100	$AlCl_3$	Paraffin
Spain	Petresa	330	150	HF	Paraffin
United Kingdom	Shell	250	114	HF	Olefin
Total Western Europe		1982	674		

(*continued*)

Table 2 *Continued*

Region/country	Company	Millions of pounds	Metric tons	Catalyst	Feedstock
Eastern Bloc					
Bulgaria	State	66	30	HF	Paraffin
Yugoslavia	Prva Iskra	66	30	HF	Paraffin
Yugoslavia	OHIS	22	10	$AlCl_3$	Olefin
Czechoslovakia	Slovnaft[a]	120	55	HF	Paraffin
USSR	State	165	75	HF	Paraffin
Total Eastern Bloc		439	200		
Africa/Middle East					
Egypt	Amreya Petrol	88	40	HF	Paraffin
Egypt	Suez Oil	44	20	HF	Paraffin
Iraq	Arab Det/Chem[a]	110	50	HF	Paraffin
Nigeria	NNPC[a]	66	30	HF	Paraffin
South Africa	Karbochem	35	16	HF	Olefin
South Africa	Shell	66	30	HF	Olefin
Total Africa/Middle East		409	186		

Asia/Pacific					
India	IPCL	110	50	HF	Paraffin
India	Tamil Nadu[a]	110	50	HF	Paraffin
India	Reliance[a]	110	50	HF	Paraffin
Australia	Shell	66	30	HF	Olefin
Taiwan	FUCC	99	45	HF	Olefin
Korea	Esoo	97	44	HF	Paraffin
Indonesia	P.T. Unggul I.C.[a]	132	60	HF	Paraffin
China	State	132	60	HF	Paraffin
Japan	Mitsubishi	77	35	HF	Paraffin
Japan	Nihon Atlantic	66	30	$AlCl_3$	Paraffin
Japan	Nippon Pet Det	77	35	HF	Paraffin
Japan	Nissan Nalkyen	88	40	$AlCl_3$	Paraffin
Total Asia/Pacific		1164	529		
Grand total		4648	2113		

[a]Announced.

and internal linear olefins. Some producers may not be completely converted to LAB.

4 PRODUCTION

The basic chemistry of LAB production from olefins is relatively straightforward. The olefin is attached at a secondary or tertiary carbon atom to benzene in the presence of either aluminum chloride or hydrogen fluoride used as the Friedel-Crafts alkylation catalyst. The amount of dialkylate formation is reduced by using excess benzene and by controlling the temperature and catalyst concentration. When the reaction is complete, the catalyst is either deactivated through hydrolysis or separated for recycle. The products are separated by fractional distillation. Benzene is removed first, unreacted olefins are removed second, and the mono linear alkylbenzene product is distilled overhead in a third distillation column. The paraffins, olefins, and benzene are recycled. The dialkylbenzene bottoms from the last column are sent to storage for sale, usually to lubricant detergent producers.

R'CHR"

$RCH{=}CH_2$ + [benzene ring] → [benzene ring with R'CHR"]

1-olefin benzene LAB

Some of the heavier alkylate by-products are formed by:

1. Dialkylation, which occurs by alkylation of LAB with a second olefin molecule:

R'CHR" R'CHR"

$RCH{=}CH_2$ + [LAB ring] → [ring] CHR'R"

1-olefin LAB Dialkylbenzene

2. Oligomerization of the olefins followed by benzene alkylation:

$$2\ RCH{=}CH_2 \longrightarrow \text{Dimer, Trimer, etc.}$$

1-olefin

Dimer + benzene → dimer alkylbenzene [R = secondary or tertiary alkyl-1]

Some heavier alkylate is formed, with three and more units of the 1-olefin attached to the benzene ring.

Components of the alkylate bottoms all appear to perform similarly in lube oil detergents and other applications for the bottoms products, whether arising from 1-olefin or 1-olefin oligomer alkylation.

5 CARBON NUMBER EFFECTS

As discussed in the following paragraphs, linear alkylbenzene sulfonates with alkyl chain lengths in the range of 10 to 15 carbon atoms are used in household detergents. Higher carbon numbers in the range C_{16}-C_{24} are used in lubricant detergents in the form of alkylbenzene sulfonates. The alkylbenzene sulfonate derivatives of the C_{16} and higher carbon numbers have low water solubility and are well suited for functions needed in lubricants and in cutting fluids. Lubricant producers need oil solubility and avoid water solubility, so they do not use C_{14} and lighter alkylbenzene sulfonates. Household detergent producers need water solubility, so they avoid C_{16} and higher alkylbenzene sulfonates. In enhanced oil recovery, the C_{16} alkylbenzene sulfonate is often ideal, as it has the proper interfacial properties.

Light-duty liquids for use in light household chores such as dishwashing are produced using LAB with chain-length carbon numbers between C_{11} and C_{12}. Heavy-duty laundry applications require LAB with C_{12}-C_{14} chain lengths. Various carbon number distributions have been studied in different liquid and powder detergents. Since the uses are so diverse, optimization on one formula for all uses is not possible. Mixtures are preferred in detergents to be able to handle any of several end uses in which a particular formulation may be employed. Ethyl has developed a 10% C_{10}, 75% C_{12}, 15% C_{14} mixture that appears to satisfy the requirements of both light- and heavy-duty applications. Several companies have worked to develop optimal mixtures for their own applications. Some of these efforts are reported below.

5.1 Ethyl Corporation

Ethyl has prepared information to assist companies in their conversion from propylene tetramer to even-carbon-number alpha olefins. Ethyl [3] lists these advantages for using an alpha olefin blend containing 10% C_{10}, 75% C_{12}, and 15% C_{14}.

No plant modifications are required.

No light ends are formed during alkylation, as there is no olefin degradation compared to the fragmentation of the branched alkyl groups in tetramer-based alkylation.

Yields are higher with the alpha olefins.

Alpha olefins or internal olefins give the same phenyl isomer distribution.

The alkylbenzene product column takes C_{14} alkylbenzene overhead. A typical analysis for alkylate based on Ethyl's C_{10}–C_{14} olefin is shown in Table 3. This carbon distribution was tested in several sizes of equipment, including commercial units in which large quantities were produced. The commercial yield was very good, as 1 lb of alpha olefins yielded 1.33 lb of linear alkylbenzene. Detergency results showed the sulfonate of the alpha olefin mixture (LAS) to be equivalent or superior to the sulfonate of the propylene tetramer-based alkylbenzene (also called hard alkylate). The olefin-based LAS gave similar performance in dishwashing tests using Crisco as a soil (Table 4). Ross-Miles foam heights were slightly higher for the alpha olefin-based LAS (Table 5). Measurements of reflectance in laundry tergotometer tests showed equivalent performance using dust sebum on cotton (Table 6). The olefin-based LAS appeared to clean polyester/cotton better than did the tetramer-based product (Table 7). Detergency results may vary depending on the test conditions, the type of soil, and other ingredients in the formulation, so it cannot be said that a given compound will always outperform another. The detersive efficiency of one formulation is shown in Table 8.

Results will vary with application and with carbon number distribution. Selection of the 10% C_{10}, 75% C_{12}, and 15% C_{14} blend was made primarily

Table 3 Analysis of LAB Produced from Alpha Olefins

Alpha olefin feed composition (wt %)	
C_{10}	10
C_{12}	75
C_{14}	15
Alkylate	
Produced in commercial plant	
Molecular weight approx.: 245	
Percent moisture: 0.02	
Bromine number: 0.003	
Phenyl isomer distribution (wt %)	
2	20
3	18
4	17
5	21
6, 7	24
Dialkylbenzene: none detected	
Diphenyl alkylate: none detected	
Indanes: none detected	
Tetralins: none detected	
LAS sodium salt slurry	
Sulfonated in pilot plant SO_3 unit and neutralized with sodium hydroxide	
Active (wt %)	52.2
Free oil (wt %)(AI)	1.8
Na_2SO_4 (wt %)	2.8
Total solids (wt %)	56.1
pH (1% solution): 8.5	
Klett color (5% AI): 50	

on the basis of overall optimum performance when using the even-carbon alpha olefins available from the Ethyl process. Mixtures generally give better overall performance than do single-component systems. If only one alpha olefin were to be chosen, it would be C_{12}. The Ethyl results and recommendations are consistent with the work of others as discussed herein.

5.2 Vista

Matheson and Matson, researchers at Conoco (now Vista) [4] reported on the effect of carbon chain and phenyl isomer distribution on use properties of linear alkylbenzene sulfonate, a comparison of "high" and "low" 2-phenyl LAS homologs. This work includes the effect of odd and even car-

Table 4 Alkyl Carbon Number Optimization for LAS in Light-Duty Detergents

Formulation	wt %
LAS	30
STPP	5
Silicate	8
Na_2SO_4	40
Na_2CO_3	5
Water	12
Total	100

Conditions

Miniplate test: *J. Am. Oil Chem. Soc., 43:* 576 (1966)
Water hardness: 200 ppm as $CaCO_3$
Temperature: 24°C
Soil: Crisco

	Plates washed
Ethyl C_{10}–C_{12}–C_{14} LAS (10:75:15)	24
Propylene tetramer alkylate sulfonate	24

Table 5 Performance in Light-Duty Detergents
Foam properties: Ross-Miles mm foam height at 40°C; 150 ppm and 300 ppm water hardness (Ca^{2+}/Mg^{2+} = 3/2)

	LAS based on Ethyl's olefins		Sulfonate from hard alkylate	
Concentration (g/L)	Initial	After 5 min	Initial	After 5 min
150 ppm hardness				
0.5	80	45	70	30
1.0	130	100	120	90
1.5	160	130	155	115
2.0	165	130	150	120
2.5	170	150	165	140
300 ppm hardness				
0.5	60	45	50	30
1.0	100	70	90	60
1.5	160	130	135	110
2.0	165	165	150	150
2.5	170	170	150	150

Table 6 Alkyl Carbon Number Optimization for LAS in Heavy-Duty Detergents

Formulation	wt %
LAS	20
STPP	25
Silicate	5.3
Na_2SO_4	44
Carboxymethylcellulose	1
Water	4.7
Total	100

Conditions:
Terg-O-Tometer test
Water hardness: 200 ppm as $CaCO_3$
Temperature: 24°C
Cloth: dust sebum on cotton

	Reflectance Units	
	Unformulated	Formulated
Ethyl C_{10}-C_{12}-C_{14} LAS (10/75/15)	71.4	78.3
Propylene tetramer alkylate sulfonate	71.2	75.5

bon numbers that are available from paraffin-based processes. As is usual, the performance of each carbon number varies with the hardness level of the water. Results show light-duty performance peaking between C_{11} and C_{12} for several applications under hard water conditions. With no water hardness, the peak shifts to C_{13}. For heavy-duty applications, the C_{12}, C_{13}, and C_{14} chain sizes are effective.

The 2-phenyl content varies with the type of alkylation catalyst, as HF produces 19% 2-phenyl and $AlCl_3$ produces 29% 2-phenyl. The authors reported that the 2-phenyl content had little effect on LAS performance in both light-duty and heavy-duty detergent applications, and the carbon-number chain size is far more important.

5.3 Colgate-Palmolive

Rubenfield, Emery, Cross, Anstett, and Munger, researchers at Colgate-Palmolive [5–8], reported in several papers in 1964–1966 on their studies to identify the effect of the mass spectral characteristics on the performance of sulfonates in heavy-duty formulations and light-duty formulations. This work was done to support their switching from the propylene oligomer-based detergents to the linear alkyl-based detergents.

Table 7 Performance in Heavy-Duty Detergents

Detersive efficiency (Terg-O-Tometer)
110°F, dust sebum polyester/cotton with permanent-press finish 4030
150 ppm and 300 ppm water hardness (Ca^{2+}/Mg^{2+} = 3/2)

Formulation	wt %
LAS	20
STPP	28
Silicate	9
Na_2SO_4	36
Na_2CO_3	1
Water	6
Total	100

Concentration (g/L)	LAS based on Ethyl's olefins	Sulfonate from hard alkylate
150 ppm hardness		
0.5	64.5 ± 0.4	61.9 ± 0.3
1.0	72.4 ± 0.2	64.2 ± 0.4
1.5	77.9 ± 0.3	69.1 ± 0.6
2.0	82.7 ± 0.2	71.1 ± 0.2
2.5	83.3 ± 0.2	75.0 ± 1.0
300 ppm hardness		
0.5	64.1 ± 0.4	62.2 ± 0.6
1.0	67.5 ± 0.5	64.1 ± 0.5
1.5	72.1 ± 0.4	66.4 ± 0.4
2.0	76.4 ± 0.3	70.9 ± 0.6
2.5	79.9 ± 0.5	73.8 ± 0.4

They immediately found that they could not interchange the two materials. They found that an alkylate with molecular weight range 231 to 241 gave the same performance in light-duty detergents. They recommended a 2-phenyl concentration in the range 15 to 25%. They recommended a minimum of 95% of C_{10}, C_{11}, and C_{12}. In heavy-duty applications, they recommended an alkylate with a molecular weight in the area of 258 to 266 and with a minimum of 70% in the C_{12} and C_{14} range, with a maximum of 5% C_{11}, 20% C_{12}, and the remaining 5% C_{15}.

5.4 Chevron

Sweeney and Olson [9], researchers at Chevron, reported the following for heavy-duty detergents:

A peak in performance was observed in the side-chain range C_{11}–C_{17}.

The peak shifted to lower molecular weights as water hardness increased, especially at low detergent concentration.

LAS (from linear alkylbenzene) performance is generally equal to PPABS (their abbreviation for alkylbenzene sulfonate based on propylene oligomers), except in dishwashing foam in soft water.

The LAS peaks are shifted to one-carbon-number-lower molecular weight than the PPABS; therefore, best overall light-duty performance is obtained for LAS, whose average molecular weight corresponds to a 12.5-carbon chain average, whereas PPABS is best with an average 13.5-carbon side chain.

For heavy-duty use, peak performance came from between C_{12} and C_{14} alkylbenzene sulfonate.

High 2-phenyl content (made from $AlCl_3$) has better solubility than random isomer content.

6 BIODEGRADABILITY

Sedlak and Booman recently reported on LAS biodegradability [10]. This study of the Enid, Oklahoma, wastewater treatment plant was sponsored by the U.S. Soap and Detergent Association (SDA). The study demonstrated the high removal of LAS and alcohol ethoxylates (AE) in an activated sludge plant. Only about 1% of the LAS, and AE left the plant in the effluent. About 3% of the LAS left with the sludge. Based on other studies, the small amount leaving the plant in the effluent would be expected to further biodegrade, as long as aerobic conditions prevail. This article contains good references, including other SDA studies and reports by Arthur D. Little, Inc.

REFERENCES

1. Vora, B. V., et al., *Hydrocarbon Process.*, 86–90 (Nov. 1984).
2. Carson, H. C., *HAPPI*, 34–42, 92 (June 1986).
3. Tuvell, M. E., Ethyl Corporation private communication (1986).
4. Matheson, K. L., and Matson, T. P., *J. Am. Oil Chem. Soc., 60:* 9 (1983).
5. Rubenfield, J., Emery, E. M., and Cross H. D., *J. Am. Oil Chem. Soc., 41:* 822 (1964).
6. Anstett, R. M., Munger, P. A., and Rubenfield, J., *J. Am. Oil Chem. Soc., 43:* 25 (1966).

7. Rubenfield, J., Emery, E. M., and Cross, H. D., *Ind. Eng. Chem. Prod. Res. Devel. Q., 4*(1): 33 (1965).
8. Rubenfield, J., and Cross, H. D., *Soap Cosmet. Chem. Spec., 4:* 1 (1967).
9. Sweeney, W. A., and Olson, A. C., *J. Am. Oil Chem. Soc., 41:* 815 (1964).
10. Sedlak, R. I., and Booman, K. A., *Soap Cosmet. Chem. Spec.* (Apr. 1986).

CHAPTER 9

Surfactants: Amines

JAMES E. BORLAND and JOE D. SAUER Ethyl Corporation, Baton Rouge, Louisiana

1 INTRODUCTION

Fatty amines of the alkyldimethylamine and dialkylmethylamine types may be prepared from several different feedstocks. The total world production of these compounds (from all routes) is estimated to be about 58 million pounds per year. Of this total, over 39 million pounds per year of amine are derived directly from alpha olefins. A portion of the remaining production can be considered to be obtained indirectly from alpha olefins as well, since substantial amounts of tertiary amine are prepared from synthetic fatty alcohols, which are based on alpha olefin raw materials. In this chapter, we deal with long-chain alkyl tertiary amines, their properties, their derivatives and reactions, and their applications.

2 GENERAL STRUCTURE, COMPOSITION, AND PHYSICAL PROPERTIES

Amines are classified as primary, secondary, or tertiary, depending on the number of organic groups attached to the nitrogen atom. They are derivatives of ammonia in which one, two, or three of the hydrogen atoms have been replaced by alkyl groups, resulting in the primary, secondary, and tertiary amine, respectively (see Fig. 1).

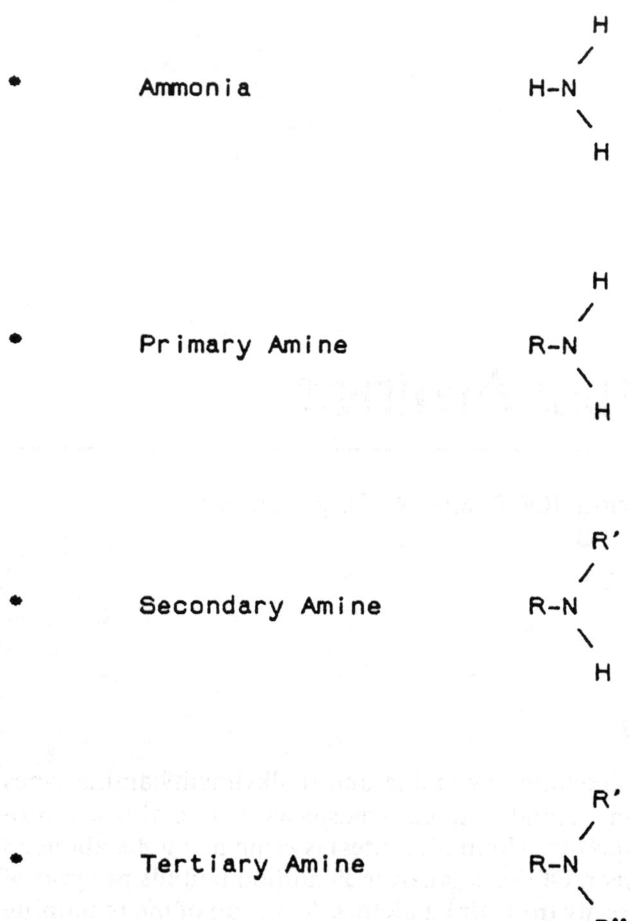

Figure 1 General structures of amines.

2.1 Long-Chain Alkyldimethylamines

Alkyldimethylamines of industry interest are generally tertiary amines containing two methyl groups, along with a third alkyl group, C_8–C_{18}.

Alkyldimethylamine

$$CH_3-(CH_2)_n-CH_2-N\begin{matrix} CH_3 \\ \\ CH_3 \end{matrix}$$

a tertiary amine where n = 6 through 16.

2.2 Di-Long-Chain Alkylmethylamines

Dialkylmethylamines are generally tertiary amines with one methyl group, along with two even-carbon-numbered alkyl groups, C_8 and/or C_{10}.

Dialkylmethylamine

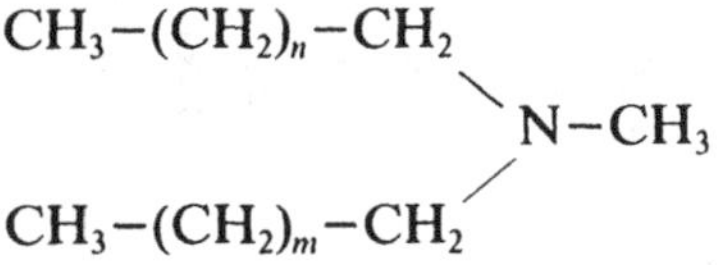

$$\begin{array}{l} CH_3-(CH_2)_n-CH_2 \\ \qquad\qquad\qquad\qquad \searrow N-CH_3 \\ \qquad\qquad\qquad\qquad \nearrow \\ CH_3-(CH_2)_m-CH_2 \end{array}$$

a tertiary amine where n, m = 6 or 8.

The tertiary amines described above are clear liquids at room temperature and have the characteristic odor of fatty amines.

Reactions of the monoalkyldimethylamines and the dialkylmethylamines can lead to a broad range of derivatives having diverse applications. For example, tertiary amines containing one and two long-chain alkyl groups are recognized by the regulatory agencies (EPA, FDA) for suitability in the manufacture of quaternary and oxide derivatives registered for microbiocidal and surfactant applications.

High-quality tertiary amines are characterized by low color (5 to 10 APHA typically) and high tertiary amine content. Typical analytical data for olefin-based tertiary amines are listed in Tables 1 and 2. Extensive physical property data for alkyldimethyl- and dialkylmethylamines are given in Figures 4 through 12.

3 COMMERCIAL ROUTES TO LONG-CHAIN ALKYLMETHYLAMINES

In addition to the synthesis method based on alpha olefins, several alternative routes to the tertiary alkylmethylamines are utilized in industrial practice today. These routes, based on well-known chemistry, are described briefly below. The four major competing schemes are outlined in Figures 2.

Each of the routes obviously has advantages and disadvantages. For example, product from the olefin route (Fig. 2a) inherently contains about 3% linear secondary alkyl groups due to Markovnikov addition of the HBr to the olefin raw material. The isomeric amines are found to be somewhat less reactive than those containing linear primary alkyl groups in some derivatizations. The alcohol route (via alkyl chloride, Fig. 2c) results in high-quality primary alkylamine products with only trace amounts of the isomeric amines. However, it has the disadvantages of corrosivity of the intermediates and the higher cost of the fatty alcohol raw material when

Table 1 Typical Analytical Values for Olefin-Derived Alkyldimethylamines

Typical values	Alkyl chain length in alkyldimethylamine					
	C_8	C_{10}	C_{12}	C_{14}	C_{16}	C_{18}
Carbon number distribution (wt %)						
C_8 (octyl-)	99.3					
C_{10} (decyl-)	0.7	98.8				
C_{12} (dodecyl-)		1.2	97	2		
C_{14} (tetradecyl-)			3	97	1	
C_{16} (hexadecyl-)				1	97	4
C_{18} (octadecyl-)					2	92
C_{20} (eicosyl-)						4
Tertiary amine (wt %)	>97	>97	>97	>97	>97	>97
Primary and secondary amine (wt %)	<0.1	<0.1	<0.1	<0.1	<0.1	<0.1
Water (wt %)	<0.1	<0.1	<0.1	<0.1	<0.1	<0.1
Color (APHA)	<10	<10	<10	<10	<10	<10
Amine value (mg KOH/g)	352	300	258	229	206	187
Specific gravity, 25/25 (°C)	0.765	0.778	0.788	0.794	0.800	0.807
Freezing point (°C)	−57	−35	−22	−6	8	21
Boiling point (°C)	195	234	271	302	331	347
Formula weight	157	185	213	241	269	297
Theoretical amine value	357.3	303.2	263.4	232.8	208.6	188.9
Density, 25°C (g/mL)	0.763	0.775	0.786	0.792	0.798	0.805
Flash point (°F)						
Tag closed-cup method	148	196	—	—	—	—
Pensky-Martens method	—	—	237	270	287	325

Table 2 Typical Analytical Values for Olefin-Derived Dialkylmethylamines

Typical values	Alkyl groups present in amine	
	Didecylmethyl	Octyldecylmethyl
Carbon number distribution (wt %)		
C_8;C_8 (dioctyl-)	<0.1	25
C_8;C_{10} (octyl-; decyl-)	<0.1	50
C_{10};C_{10} (didecyl-)	99.9	25
C_{10};C_{12} (decyl-; dodecyl-)	0.1	
C_{10};C_{14} (decyl-; tetradecyl-)	<0.1	
Tertiary amine (wt %)	97.5	97
Primary and secondary amine (wt %)	0.05	0.1
Water (wt %)	0.09	0.1
Color (APHA)	15	15
Amine value (mg KOH/g)	178.3	196.2
Specific gravity, 25/25 (°C)	0.810	0.801
Freezing point (°C)	−6.3	−38.6
Boiling point (°C)	~370	~329
Formula weight	311	282
Theoretical amine value	180.4	199.3
Density, 25°C (g/mL)	0.808	0.799
Flash point (°F); Pensky-Martens method	>200	330

compared to alpha olefin. Direct amination of an alcohol (Fig. 2c), on the other hand, requires catalyst systems and the more expensive fatty alcohol, but is capable of producing high-quality product. The acid/ester route (Fig. 2b) historically has given very poor quality tertiary amines, due in part to the by-products inherent in the process and in part to traces of amide and nitrile intermediates typically remaining in the final product. This route may enjoy, at times, a low-cost raw material position compared to the other paths shown. However, fatty acid cost, from the volatile natural product markets, typically fluctuates (occasionally dramatically) from year to year, sometimes making it the highest-cost route.

4 LITERATURE OF TERTIARY AMINES

The general chemistry of tertiary amines has been studied extensively and numerous syntheses and reactions have been recorded in the literature

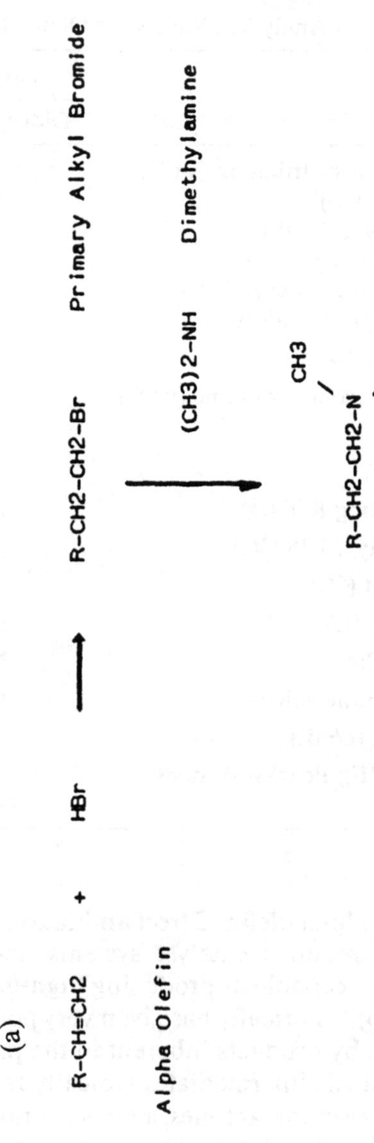
(a)
R-CH=CH2 + HBr
Alpha Olefin
R-CH2-CH2-Br
Primary Alkyl Bromide
(CH3)2-NH
Dimethylamine
CH3
R-CH2-CH2-N
CH3
Alkyldimethylamine (ADA)

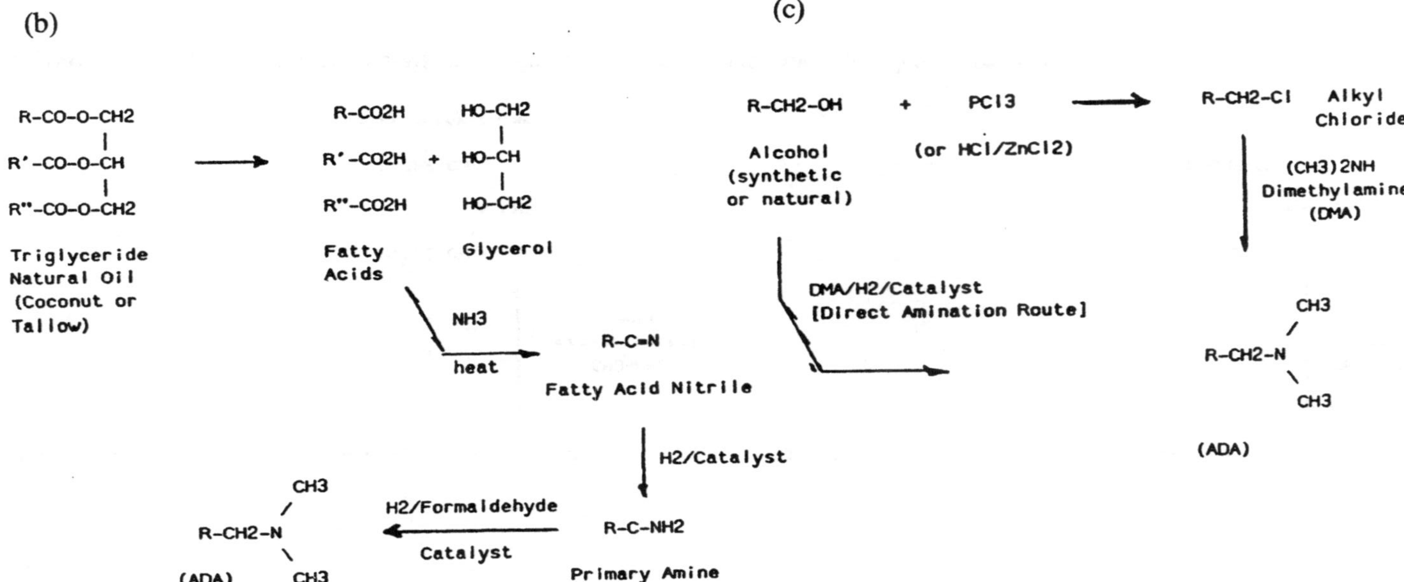

Figure 2 Commercial routes to alkyldimethylamines (ADA): (a) olefin feedstock; (b) fatty acid feedstock; (c) alcohol feedstock via alkyl chloride/direct animation.

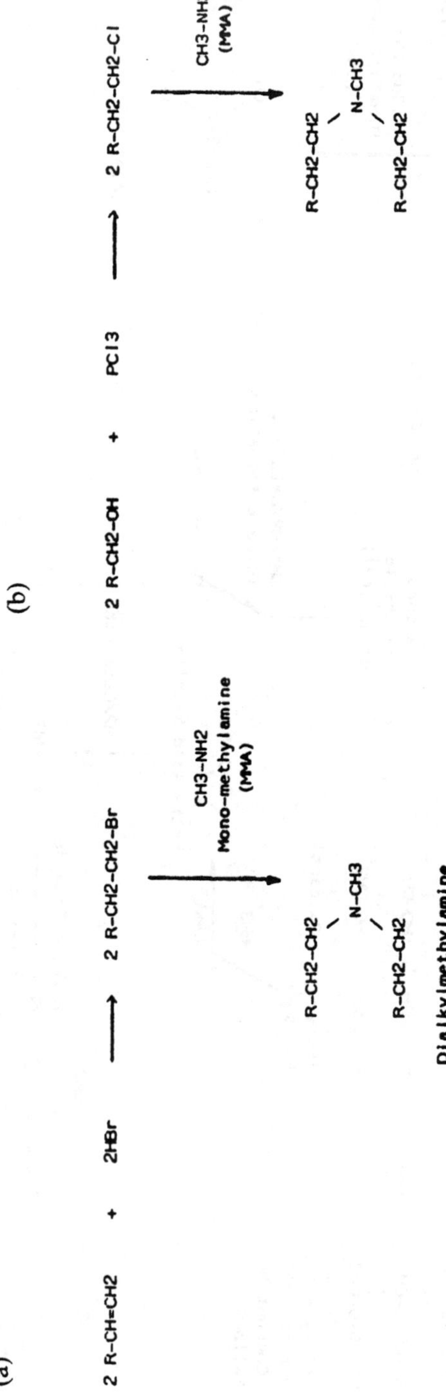

Figure 3 Commercial routes to dialkylmethylamines: (a) olefin feedstock; (b) alcohol feedstock.

and in patents. This body of literature is diffuse, with no one general reference or pertinent set of review articles covering the broad scope of amines and their derivatives. Information for this chapter was obtained in a literature search covering the period 1937–1982, in which 3100 citations were found. Because it is not practical to cite all these references in this survey, a summary of application references is given later in Table 4.

5 DERIVATIVES AND REACTIONS

The commercially important reactions of alkyldimethyl- and dialkylmethylamines are described briefly below.

5.1 Quaternary Ammonium Halides

Quaternary ammonium halides are among the most important commercial derivatives of tertiary amines. These materials are tetrasubstituted ammonium salts with four organic groups attached directly to the nitrogen by C–N covalent bonds. The alkyl groups can contain a wide variety of substituents, including heteroatoms such as O, N, S, Si, and so on, but the most common "quats" contain only simple alkyl, aralkyl, or aryl groups.

Nomenclature

Quaternary ammonium compounds are typically named as substituted nitrogen systems. According to nomenclature rules, the substituent groups may be listed alphabetically and without prefixes such as di-, tri-, etc. (as in *Chemical Abstracts*) or by increasing size and complexity. Based on accepted nomenclature practices, both of the names below would be acceptable for the structure shown:

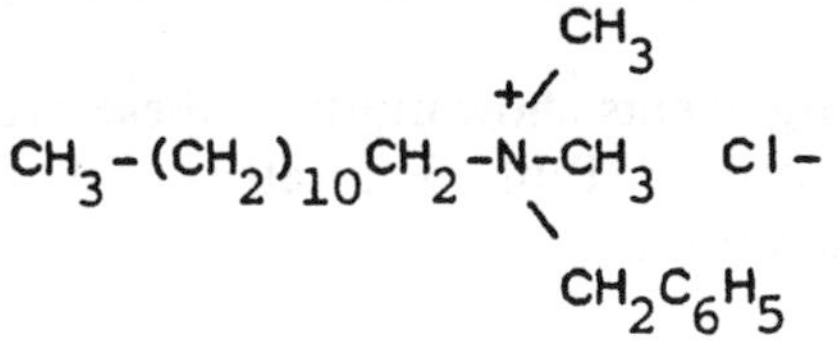

$$CH_3-(CH_2)_{10}CH_2-\overset{+}{N}(CH_3)_2(CH_2C_6H_5)\quad Cl^-$$

Benzyl(dodecyl)dimethylammonium chloride; or

Dimethyl(benzyl)dodecylammonium chloride

Preparation

A tertiary amine reacts with an "alkylating agent" to form the quaternary ammonium product as shown by the equations.

```
     CH3                                              CH3
    /                                               +/
R-N              +        R'X      ------>      R-N----CH3    X-
    \                                               \
     CH3                                             R'
```

Alkyldimethylamine

```
     R'                                               R'
    /                                               +/
R-N              +        R''X     ------>      R-N----CH3    X-
    \                                               \
     CH3                                             R''
```

Dialkylmethylamine

Many different products are available, since the identity of R, R′, and R″ can be changed readily to suit the needs of the application (i.e., water solubility, melting point, microbiological activity, etc.).

In commercial practice, these reactions are generally carried out in stainless steel, special alloy, or glass-lined reactors, depending on the corrosivity of the particular reagents employed. Although many of the quaternizations involve relatively high-boiling precursors, it is possible to employ alkylating agents with high vapor pressures, such as methyl chloride. Thus in some cases the use of pressure equipment is required for the derivatization step. Although these reactions are normally carried out in solution, typically in ethyl or isopropyl alcohol and water to afford 50–80% "active" final products, it is possible to isolate 100% active, solid, crystalline quats in some instances.

Although many different alkylating agents show utility in these preparations, the most common are listed in Table 3, together with the quaternary compounds obtained therefrom.

Properties

Changes in the structure of a quaternary ammonium compound can result in profound alterations in the physical properties of the material. Low-molecular-weight quats are highly water soluble and exhibit low solubilities in common nonpolar organic solvents such as ether or benzene. As the molecular weight of the quat increases, by increasing the

Table 3 Quaternary Compounds via Common Alkylating Agents and Tertiary Amine R_3N

Alkylating agent	Quaternary compound
Benzyl chloride ($Ph-CH_2-Cl$)	$R_3^+N-CH_2-C_6H_5$ Cl^-
Ethylbenzyl chloride ($CH_3CH_2-Ph-CH_2-Cl$)	$R_3^+N-CH_2-C_6H_4-CH_2CH_3$ Cl^-
Dichlorobenzyl chloride ($Cl_2-Ph-CH_2-Cl$)	$R_3^+N-CH_2-C_6H_3-Cl_2$ Cl^-
Methyl chloride (CH_3Cl)	$R_3^+N-CH_3$ Cl^-

length and/or number of nonmethyl alkyl chains attached to the nitrogen atom, water solubility decreases and organic-soluble character becomes more dominant. This trend may continue until the "heavy" quats exhibit "dispersability" and not true "solubility" in aqueous systems.

The quaternary ammonium compounds differ from most common surfactants in that the hydrophilic portion of the molecule is positively charged (cationic) [most commonly used surfactants are either negatively charged (anionic) or neutral (nonionic)].

Some of the most striking properties of the quaternaries are in their biocidal activities. A great many of the materials exhibit marked biocidal activity against a full range of pathogens, including gram-positive bacteria, gram-negative bacteria, viruses, and, fungi. This activity apparently derives from surface activity in the quats, allowing for interaction with the organism's cell wall, which interferes with normal metabolic activity. Fortunately, the bacteriostatic action is observed at very low concentrations which appear to be innocuous to higher forms of life.

Other advantageous properties of these compounds include:

Good thermal stability
Low corrosiveness
Low toxicity
Long shelf lives
General antistatic properties

Additionally, dialkyl quats exhibit the following advantageous properties:

Higher hard water tolerance than alkyldimethylbenzylammonium chlorides, referred to as benzalkonium chlorides, at the same use level
Greater biocidal activity in the presence of proteinaceous materials than benzalkonium chloride quats

Some disadvantages that should be noted include:

Monoalkyl quat biocidal properties are inhibited or reduced by the presence of organic matter.
Quats are incompatible with anionic surfactants (they frequently form insoluble complexes).
Quats are ineffective against some strains of bacteria (notably tubercle bacillus).

A survey of end uses for quaternary ammonium compounds is given in Table 4.

Efficacy Data

As mentioned earlier, quaternary-based disinfectants are effective against a wide range of pathogens. A brief survey of the literature for industrial products provides the following list of common organisms which are controlled by quaternary systems.

Bacteria, Gram Negative:

Enterobacter aerogenes
Escherichia coli
Klebsiella pneumoniae
Nisseria gonorrhea
Proteus mirabilis
Proteus morganii
Proteus vulgaris
Providence spp.
Pseudomonas aeruginosa
Pseudomonas fragi
Pseudomonas pseudomallei
Salmonella choleraesuis
Salmonella enteritides
Salmonella gallinarium
Salmonella paratyphi A
Salmonella shotmulleri
Salmonella typhimurium
Salmonella typhosa
Serratia marcesens
Shigella flexneri type II
Shigella sonnei
Vibro cholerae

Bacteria, Gram Positive:

Citrobacter freundii
Citrobacter diversus
Corynebacterium diphtherias
Diplococcus pneumonia
Enterococcus aerogenes
Listeria monaestaens
Myxobacteria spp.
Staphylococcus albus
Staphylococcus aureus
Staphylococcus citrens
Staphylococcus epidermis
Streptococcus pyrogenes
Streptococcus viridans

Fungi:

Aspergillus niger
Candida albicans
Microsporum audouni
Trychophyton interdigitale
Trychophyton mentagrophytes

Viruses:

Adenovirus type IV
Feline pneumonitis
Herpes simplex type I
Influenza A (Japan)
Influenza A2 (Aichi)
Influenza A2 (Hong Kong)
Parinfluenza (Sendai)
Polio virus
Reovirus
Respiratory syncytial virus
Vaccinia

5.2 Amine Oxides

Amine oxides are the N-oxides of tertiary amines. Although both aromatic and aliphatic amine oxides are known, based on whether the nitrogen atom is part of an aromatic ring system or not, only the aliphatic type will be discussed in this section. Typically, the amine oxides of commercial utility contain one long-chain alkyl group on the nitrogen.

Preparation

The most widely utilized route to amine oxides involves oxidation of tertiary fatty amines with hydrogen peroxide. This reaction is usually carried out in a solvent system including one or more of the components: water, light alcohols, acetone, or acetic acid. The commercial preparation commonly is done in a stainless steel or glass-lined reactor. Since the reaction is often exothermic, provisions for cooling the reaction mixture to maintain a temperature in the range 60 to 80°C must be provided. A problem associated with this process is the difficulty in handling 35 to 70% aqueous hydrogen peroxide. Thus adequate engineering safeguards must be applied.

When isopropyl alcohol or IPA/water solvent systems are utilized, final amine oxide concentrations of 70% are possible. In water or other solvent systems, active levels above ~35% may lead to gel formation and attendant handling problems.

The general reaction is

$$\text{R-N(CH}_3)_2 + H_2O_2 \longrightarrow \text{R-N(CH}_3)_2\text{→O} + H_2O$$

Alkyldimethylamine → Alkyldimethylamine oxide

The bond between the nitrogen and the oxygen atom might be called a coordinate covalent bond, with the electron density shifted toward the more electronegative oxygen. At neutral pH, no formal charge is present and these species would behave as nonionic surfactants.

Properties

Aliphatic amine oxides behave as typical nonionic surfactants in neutral and/or alkaline aqueous systems. At low pH (<3), the predominant species present is the cationic form, $R_3N^+OH\ X^-$, which, as such, may be incompatible with anionic surfactants because of formation of insoluble precipitates. Owing to their surfactant properties, these materials are reported to be wetting agents, cleaning agents, dispersants, and foam boosters.

Not only do these compounds have good surfactant properties, they also impart beneficial properties to many formulated products designed for skin contact. In this regard, aliphatic amine oxides are typically described as having good emollient properties, anti-irritant characteristics, viscosity builder properties, low toxicity, and compatibility with anionic, cationic, and nonionic species. It is not surprising that these materials find applications in light-duty dishwash detergents, bar and liquid soaps, shampoos, and bubble baths. A more complex examination of amine oxide end uses is available in the section of this chapter dealing with applications for amines and amine derivatives (Table 4).

5.3 Betaines

The term "betaine" arose through reference to the trimethylglycyl group.

```
       CH3  O
     +/     ||
CH3-N-CH2-C-O-
      \
       CH3
```

Contemporary references have extended this definition to the extent that a betaine can be any amine, with two methyl groups and an alkyl chain attached directly to the nitrogen, that has been quaternized with sodium chloroacetate.

```
           CH3  O
         +/     ||
R-CH2-N-CH2-C-O-
          \
           CH3
```

Preparation

Betaines can be prepared readily from alkyldimethylamines. In a typical procedure, sodium chloroacetate (20 to 25% molar excess) is added to a warm suspension of fatty amine in a solvent such as water. The reaction is normally completed in a few hours to yield around 30% active solutions of the desired surfactant.

$$R{-}N(CH_3)_2 + Cl{-}CH_2{-}C(=O){-}O^-\ Na^+ \longrightarrow R{-}N^+(CH_3)_2{-}CH_2{-}C(=O){-}O^- + NaCl$$

Properties

Betaines are surfactants possessing a broad range of useful properties that can be tailored to some degree by alteration of the alkyl group carbon chain length. Typically, increasing the chain length decreases the foaming properties and increases the detergency of the surfactant. Although rarely used alone, these compounds are valuable in formulated products for the following characteristics:

They behave as foam boosters and stabilizers.
They have high solubility throughout the pH range.
They reduce the irritation potential of anionic surfactants.
They contribute substantivity and conditioning to hair.

These compounds have the ability to act as cationic surfactants in acid solution and as amphoteric surfactants in alkaline solution. For this reason, betaines are compatible with most common surfactant systems employed today.

5.4 Amine Salts

Since amines are organic bases, they will react rapidly with acids (organic and inorganic) to form salts.

$$R_3N + HX \rightarrow R_3N^+H \quad X^-$$

The individual properties of these salts will depend on the nature of both of the reactive parent compounds. Relative solubilities and melting points can vary widely with structure.

6 TOXICOLOGY

Alkyldimethylamines and dialkylmethylamines ranging in alkyl chain carbon length from C_8 through C_{18} have relatively low levels of toxicity.

Based on the results of animal studies, these fatty amines are considered only slightly toxic both orally and dermally. Irritancy tests on rabbits show that these tertiary amines are corrosive to the eye and initially produce moderate skin irritation followed by corrosivity at 48 hrs. Transient contact with human skin can cause a rash or dermatitis, possibly delayed. Prolonged skin contact can cause a more severe effect such as a delayed burn.

The toxicology of the various derivatives known for these fatty amines depends entirely on the nature of the specific derivative. Many of these compounds find usage in personal care applications such as cosmetics and hair care items, while others may exhibit some corrosive and/or toxic nature equal to, or in some cases, surpassing the parent amines. Information on particular derivatives should be consulted for further details.

7 HANDLING AND STORAGE PRECAUTIONS

Alkyldimethyl and dialkylmethyl fatty amines are organic bases and should be treated with caution. Good industrial hygiene and safe working practices must be observed. Care should be taken to avoid ingestion; contact with eyes, skin, and clothing should be avoided by wearing protective clothing, including gloves and goggles.

These amines can be stored in carbon steel vessels. Precautions should be taken to avoid contact with moisture, carbon dioxide, and copper compounds in general. Various properties relating to storage and handling of these tertiary amines are detailed in Tables 1 and 2 as well as in Figures 4 through 12.

8 APPLICATIONS OF TERTIARY AMINES

Commercial uses of tertiary amines have become widely diversified. These compounds find utility on an as-is basis, as components of formulated products and as feedstocks for various specialty chemical derivatives. Although it is not possible to list all current applications, Table 4 provides a rough "key-word" survey of general applications for tertiary amines.

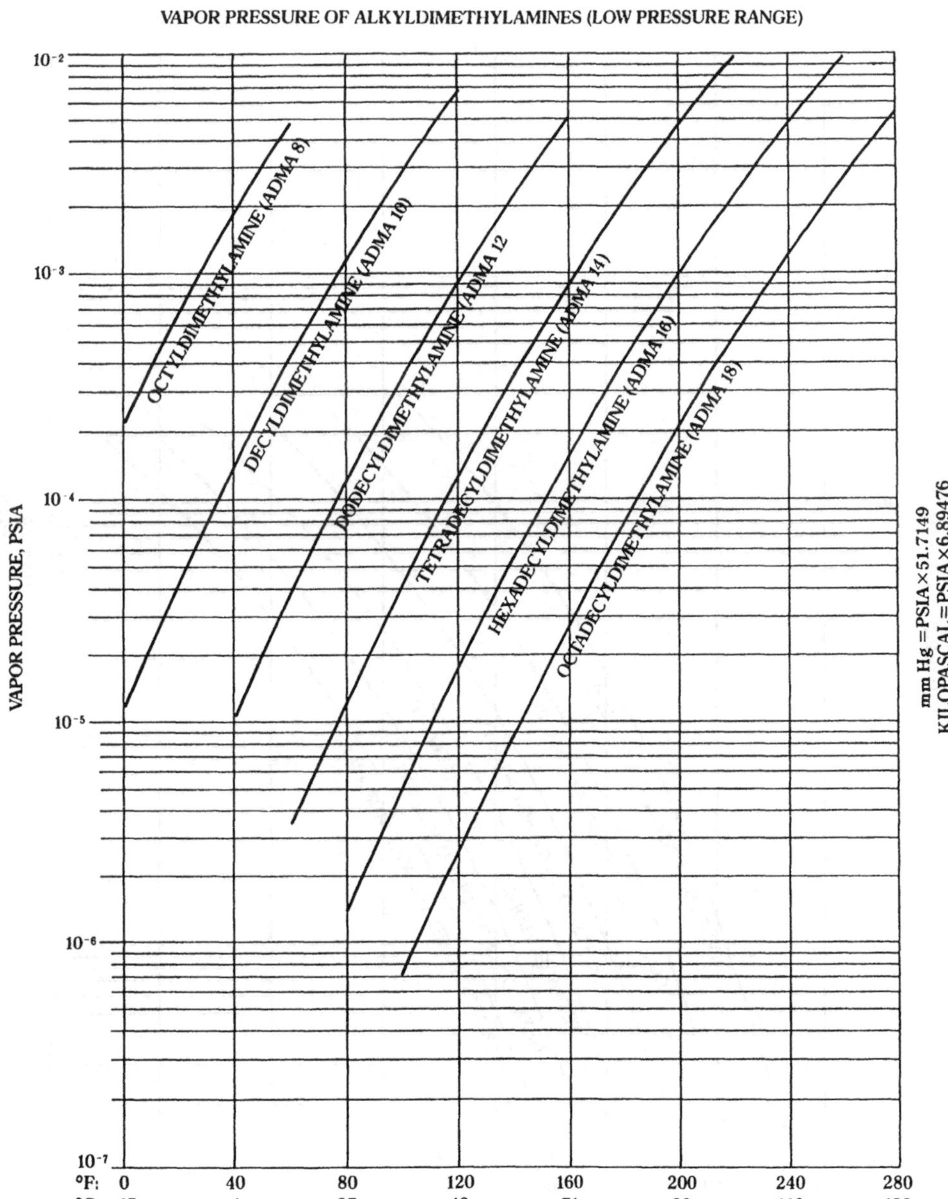

Figure 4 Vapor pressure of alkyldimethylamines vs. temperature: low range.

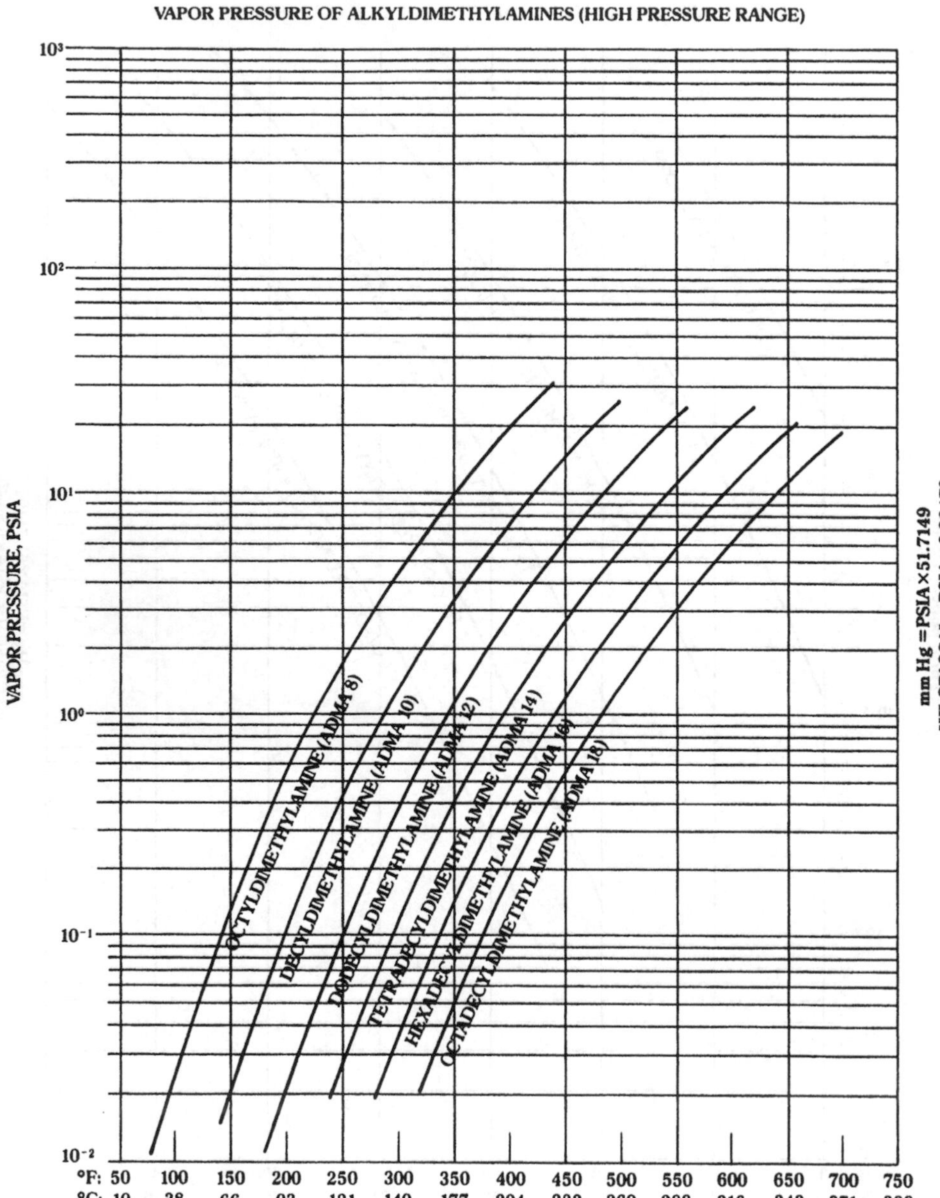

Figure 5 Vapor pressure of alkyldimethylamines vs. temperature: high range.

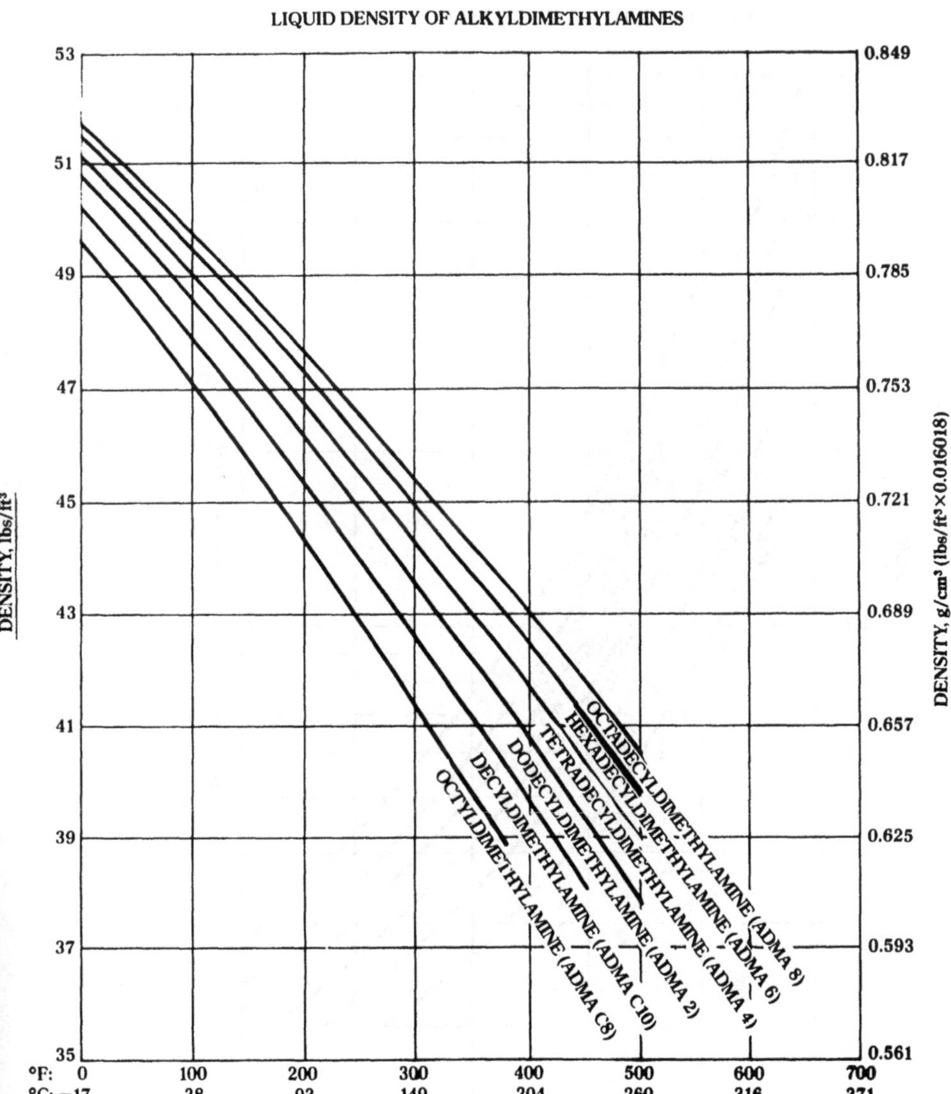

Figure 6 Liquid density of alkyldimethylamines.

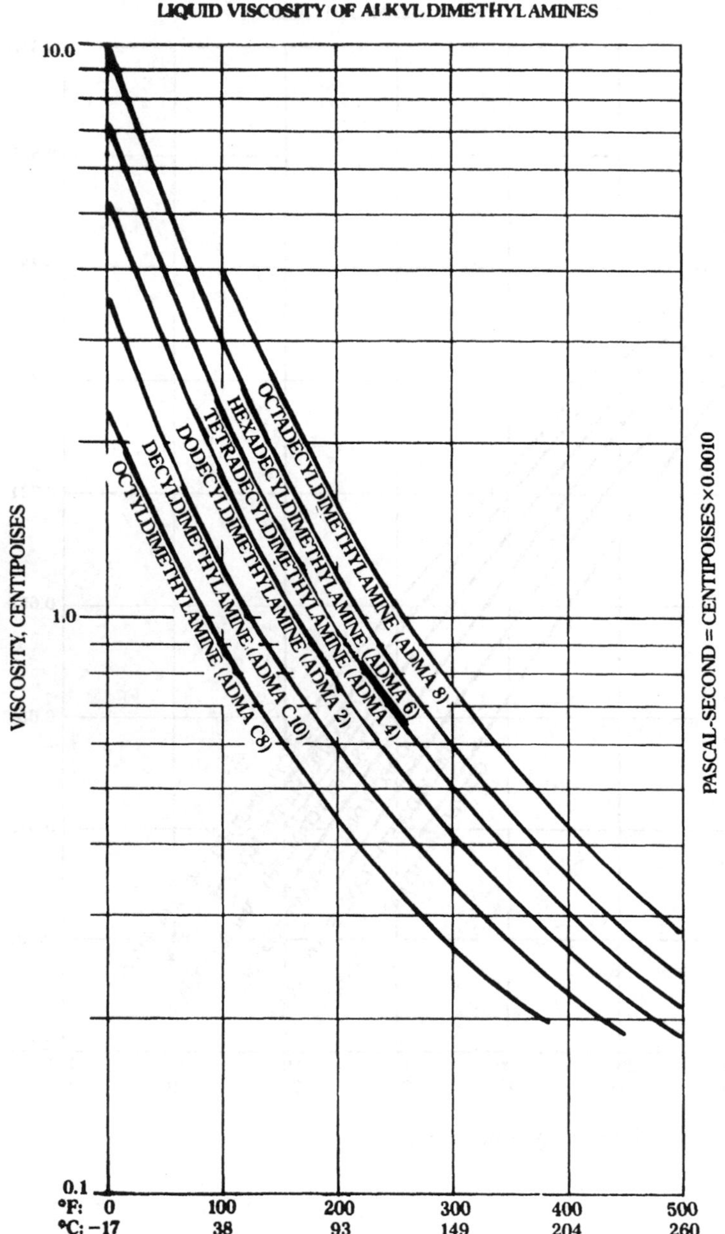

Figure 7 Liquid viscosity of alkyldimethylamines.

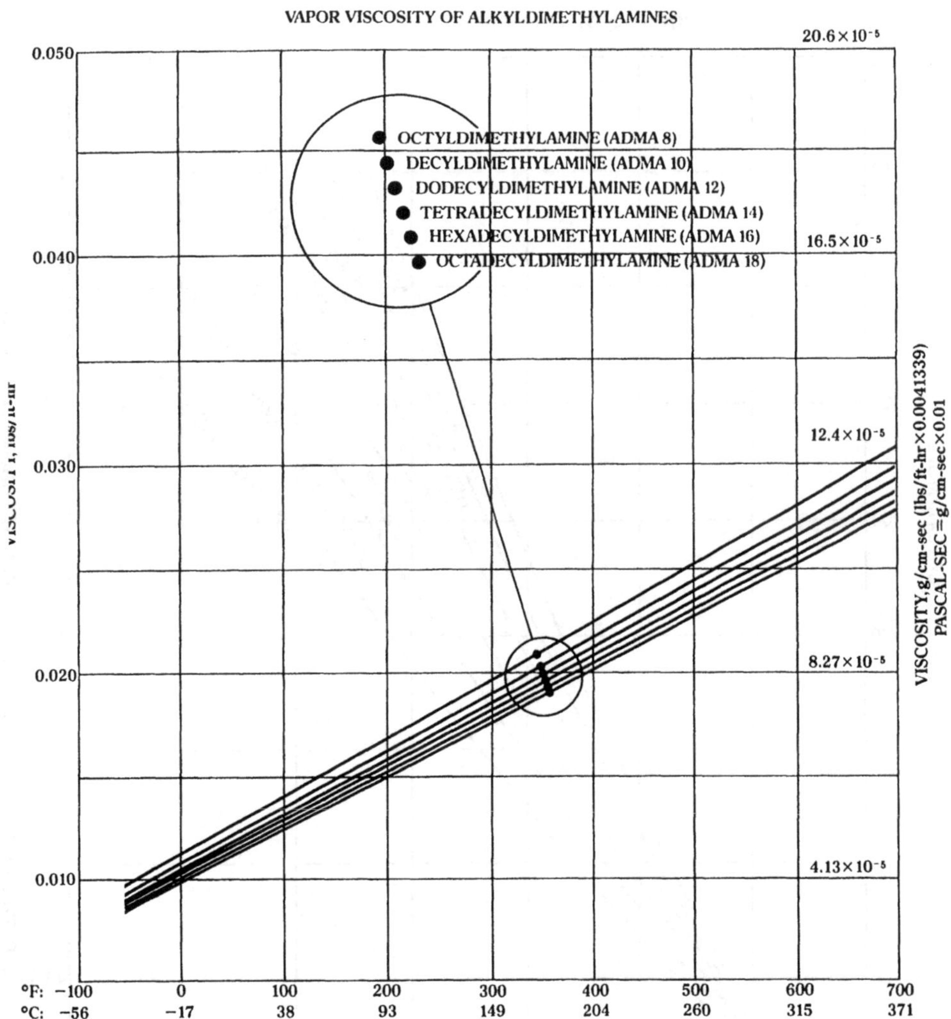

Figure 8 Vapor viscosity of alkyldimethylamines.

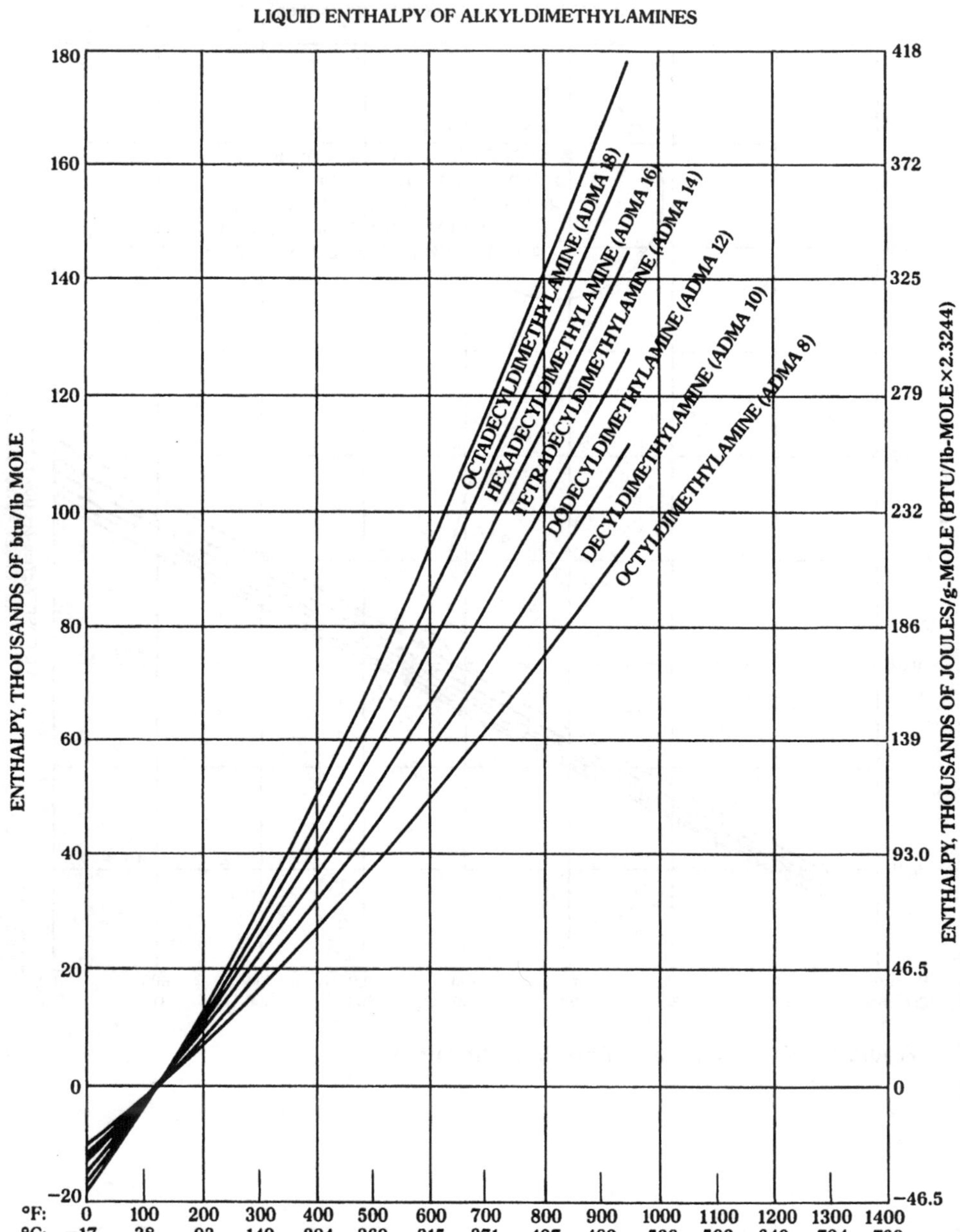

Figure 9 Liquid enthalpy of alkyldimethylamines.

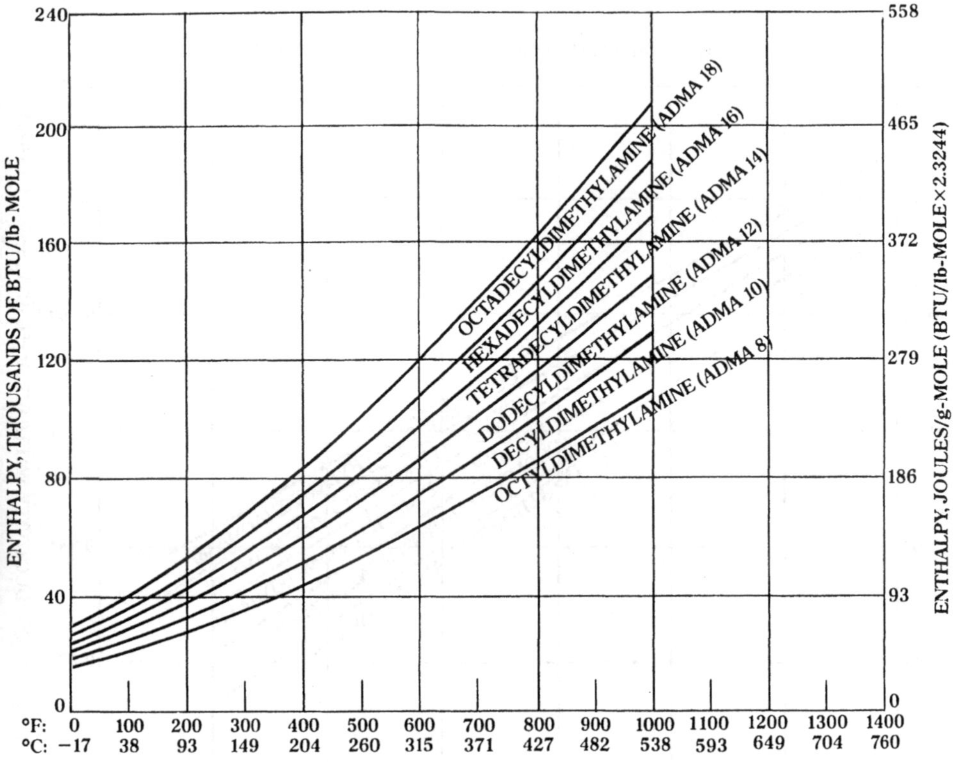

Figure 10 Ideal gas enthalpy of alkyldimethylamines.

Table 4 End Uses of Tertiary Amines and Derivatives

Alphabetical listing by end-use	Derivative classification	Comments, additional information
Acid extractions	Amine, as is	Formation of salts with various acids, organic and/or inorganic, for removal and/or purification
Acid pickling baths	Quat	As corrosion inhibitors in baths used to remove mill scale and oxides from iron and steel fabrication pieces
Adhesives	Amine, as is Quat	(*See* Asphalt)
Aerosol sprays	Quat	Used as active ingredient in aerosol disinfectants, cleaners, and deodorizers; also used in formulation of antiseptic sprays

(*continued on page 280*)

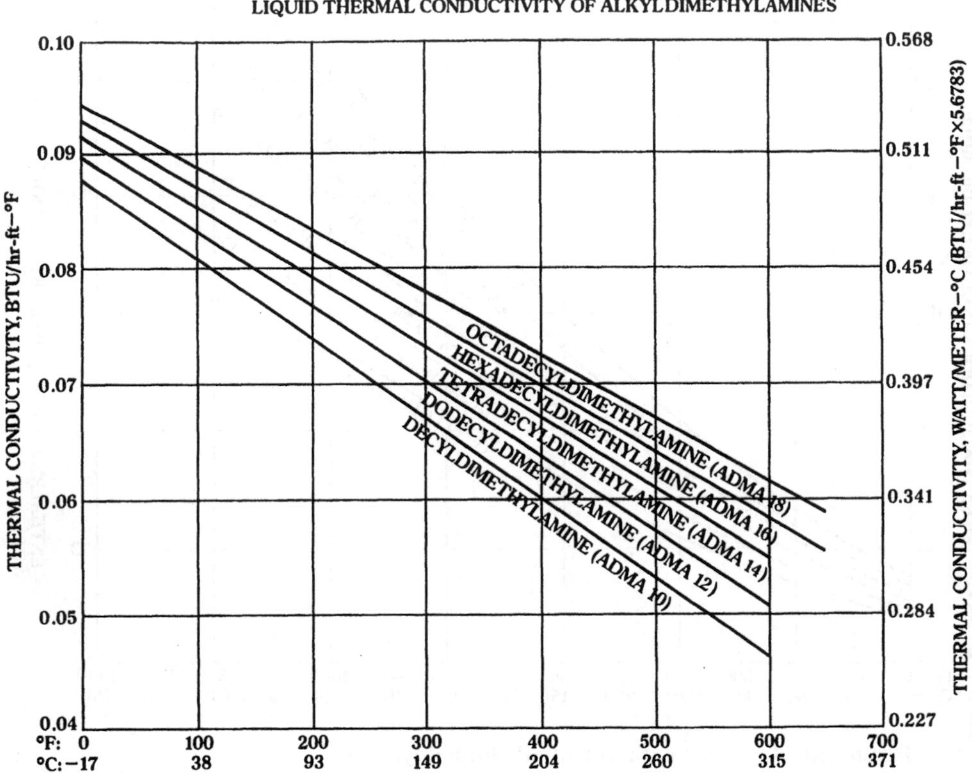

Figure 11 Liquid thermal conductivity of alkyldimethylamines.

Table 4 *Continued*

Alphabetical listing by end-use	Derivative classification	Comments, additional information
Agricultural applications	Quat	Utilized as surfactants and emulsifiers for pesticides; has value as adjuvant and synergist to toxicant, thereby reducing cost of per-acre treatment; shows insecticidal and larvicidal effects; utilized as growth regulator; active as fungicide
Air deodorizers	Quat	Active ingredient used to kill microorganisms causing odor buildup (*see* Bactericides)

(continued on page 282)

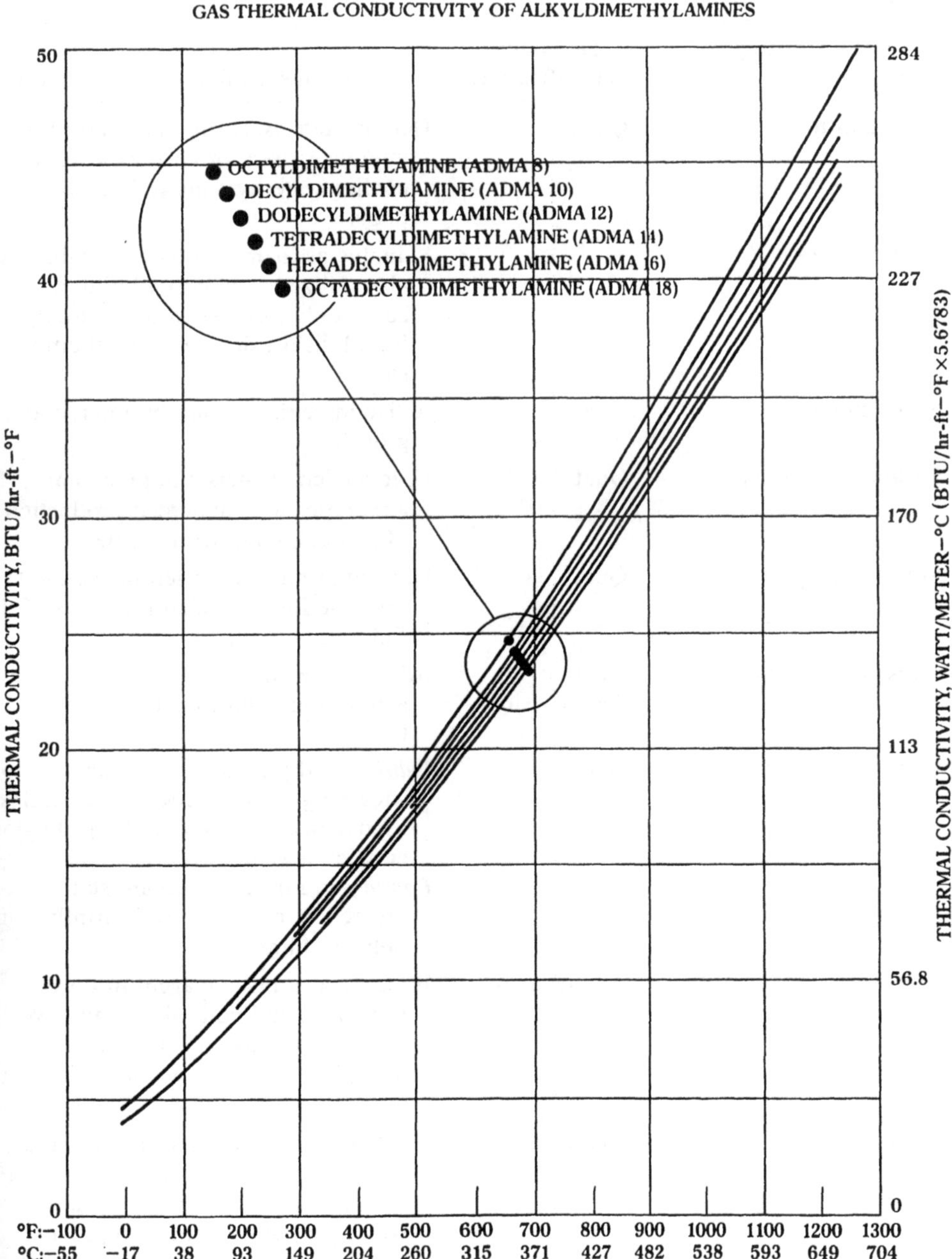

Figure 12 Gas thermal conductivity of alkyldimethylamines.

Table 4 *Continued*

Alphabetical listing by end-use	Derivative classification	Comments, additional information
Algae control	Quat	Demonstrate especially high hard water tolerance; used for algae control in cooling water systems and in swimming pools
Amine oxides	—	Valuable as foam booster and stabilizer; also utilized as viscosity builder; commonly formulated in light-duty liquid dishwash detergents, shampoos, etc.
Antibacterial scrubs	Quat	Active ingredient to inhibit microbial growth
Antibiotic recovery	Quat	Used as demulsifiers and precipitant auxiliaries in commercial production of penicillin and tetracycline
Anticaking agents	Quat	Utilized as additive in fertilizers, etc. to promote free-flow characteristics of the finished powders
Antiseptic creams	Quat Amine oxide	Additive for inhibition of growth and/or action of microorganisms on living tissue
Antistatic agents	Quats	*Fabrics:* in dry-cleaning formulations for wide range of synthetics and fiberglass used in manufacture/handling of nylon acrylics, acetates, wool, etc. *Printing:* in inks to afford antistatic protection, especially with printing on polymer films
	Amine, as is	Effective as antistatic agent in polypropylene and both high- and low-density polyethylene at use levels of 0.5 pph resin; also used as ingredient in furniture paste wax
Asphalt	Amine, as is Quat	Used as anchoring agent for bituminous solutions, cut-back bitumen, and hot bitumen; these additives improve coating on wet substrate to provide better adhesion; also used as emulsifier in the asphalt mixes

Table 4 *Continued*

Alphabetical listing by end-use	Derivative classification	Comments, additional information
Bactericides	Quat	Active ingredient, microbiocidal activity against a broad range of pathogens, including bacteria (gram negative and gram positive), fungi, and viruses
Bar soaps	Amine oxide	Used as foam booster, foam stabilizer, emollient, wetting agent, lime soap dispersant, viscosity builder, antistat, anti-irritant
Betaines	—	Prepared from tertiary amine and sodium chloroacetate; used as detergent sanitizer, shampoo, hair conditioner, bubble bath, textile surfactant; exhibit wide range of properties, including wetting agent, antistat, dispersant, emulsifier, anti-irritant
Bubble baths	Amine oxide Betaine	Active ingredients impart good foaming, emolliency, anti-irritant properties
Buffering agents	—	
Car-drying agent	Quat	Added to final rinse in automatic car wash to accelerate air drying of the washed surface
Catalysts (urethane foam)	Amine, as is	Tertiary amines are effective accelerators and curing agents in urethane foams
Cationic flocculation	Quat	Additive for water treatment; also used in mineral froth flotation operations
Cements	Amine, as is	Used as additive during the growing of the clinker or after preparation of the cement to increase storage life of the finished material
Corrosion inhibitors	Quat Amine, as is	Tert amines and quats are effectively used to inhibit corrosion in a large variety of systems; inhibition is observed at very low concentrations; a few examples include marine tanks, bilges, oil pipelines, oil storage tanks, furnace-oil additives, component of water-soluble cutting oils, metal cleaning compounds, metal pickling baths, etc.

(continued)

Table 4 *Continued*

Alphabetical listing by end-use	Derivative classification	Comments, additional information
Cosmetics	Quat Amine oxide Betaines	Useful additives for imparting emolliency, substantivity, anti-irritant properties, increasing foaming characteristics; also shows the benefits expected of a good microbiocide; routinely formulated in conditioners, cream rinses, protective hand creams, color rinses, etc.
Crude-oil pipelines	Quat	Corrosion buildup in pipelines can limit production, cause equipment failure, and contaminate crudes; quats find utility as corrosion inhibitor and production enhancers
Cutting oils	Quat	Additive use to impart corrosion resistance, surfactant properties, etc.; anti-irritant and noncorrosive to workers in contact
Dairy industry	Quat	Used as microbiocide to control bacteria growth in equipment; can also be applied directly to a cow's udder
Dehydrating agents	Amine, as is	Anthracene oil is a lubricant in gas-holding tanks; this oil absorbs water over a period of time but can be effectively dehydrated by adding a tertiary amine
Deodorants	Quat	Used as microbiocide to control growth of odor-causing organisms; used both as surface cleaners and in aerosol form
Detergents	Quat Amine oxide Betaine	Used as cationic and nonionic surfactants; also used as foam boosters, microbiocides, etc. in formulated packages
Dewatering agents	Quat	Formulated with mineral oil and naphtha, 1–5% active-level quat can act to disperse water on surfaces (especially useful on metals)
Disinfectants	Quat	Active ingredient; has microbiocidal activity against a broad range of pathogens, including bacteria (gram negative and gram positive), fungi, and viruses

Table 4 *Continued*

Alphabetical listing by end-use	Derivative classification	Comments, additional information
Dispersing agents	Quat	Cationic surfactants find use in the paint industry as additives to increase pigment hydrophobicity; this improves the oil-wetting character of the pigment and increases dispersability
Dye-leveling agents	Quat	Agent used to increase migration ability of polar dyes; a main objective in dying is that of producing finished materials which are uniform in coloration
Electroplating bath additives	Quat	Quat additives increase brightness of copper plating; similar additives have been used to control foam levels and decrease fume emissions in chrome plating operations
Electrostatic films	Quat	Used to regulate conductivity levels of electrostatic paints and films
Emollients	Amine oxide	Active as emollient in bar soaps, shaving soaps, after-shave lotions, etc.
Emulsifiers	Amine oxide	Emulsifiers promote and/or stabilize emulsions; fatty amine derivatives are useful in this regard in a wide range of formulations, including paints, inks, waterless hand cleaners, lotions, detergents, etc.
Epoxy-resin curing agents	Amine, as is	Tertiary amines are effective curing agents for epoxy resins; they also have commercial value as additives for electrophoretic coating baths containing epoxy resin to decrease the amount of bath liquid adhering to coated objects after removal
Fabric softeners	Quat	These materials find wide application as textile softening agents; can also be used in home laundry as dryer-applied softening agents
Fertilizers	Amine, as is Amine salts	Used in fertilizer manufacture as anti-caking agents

(*continued*)

Table 4 *Continued*

Alphabetical listing by end-use	Derivative classification	Comments, additional information
Firefighting foams	Amine oxide	Used to enhance water retention of high-water-content foams in firefighting formulations; also good foam booster, stabilizer
Flocculating agents	Quat	Used as flocculating agent/manufacturing aid in production of silica and hydrated silica
Flushing agents	Quat Amine oxide	"Flushing" is a method used to enhance and preserve the initial high fineness (small particle size) of pigments. These additives find wide application in the paint industry
Foam stabilizers	Amine oxide	Provides foam booster/stabilizer properties to formulations such as bubble baths, bath oils, dishwash concentrates, hair-coloring systems, softeners, cleaners, etc.
Food packaging	Quat	EPA registered quats are utilized in slimicides to be added to process water used in the production of paper and paperboard that contact food
Fuel additives	Amine, as is	Tertiary amines impede sludge formation, disperse sediments, and reduce corrosion in various fuel systems
Fuel oil dehazing	Quat	Addition of small amounts (~25 ppm) of quat can effectively demulsify water and other suspended matter, resulting in treated fuel stocks that burns cleaner
Fungicides	Quat	These materials are effective in formulations designed to control fungus and wood rot in the wood industry
Gasoline additives	Amine, as is	Used directly or as derivative (nature unspecified) as gasoline additives to impart several different characteristics to the blend; uses mentioned include carburetor detergent, anti-icers, corrosion inhibitors, fuel distribution aids

Table 4 *Continued*

Alphabetical listing by end-use	Derivative classification	Comments, additional information
Gel formation additives	Amine oxide Betaine	These surfactants impart several properties to formulated systems, especially enhanced foaming ability and built viscosity
	Quat	Gelling agent for butadiene-styrene systems
General emulsion manufacture	Quat	Used as general cationic surfactant-emulsifier in wide variety of industrial applications: paints, shampoos, hand cleaners, cosmetics, lotions, creams, etc.
Germicides	Quat	Active ingredient with microbiocidal activity against a broad range of pathogens, including bacteria, fungi, and viruses
Hair conditioners	Quat	Provides good "wet combing" and manageability
	Amine oxide	Acts as foam booster/stabilizer; contributes to substantivity; builds viscosity
Hand cleaners	Amine oxide	Used as emulsifier in waterless hand cleaners; contributes emolliency
Hardeners—epoxy resins	Amine, as is	Tertiary amines are used as accelerator, hardening agent for epoxy resins
Herbicides	Quat	Used as emulsifiers, solubilizing agents, adjuvants, synergists for compounding herbicidal formulations and sprays
Hide preservation agents	Quat	Used in leather processing to prevent mildew and mold
Ink antistatic agent	Quat	Additive in printing inks to give antistatic properties, useful when printing plastics and films for better ink distribution
Inks	—	—
Insecticides	Amine, as is	Fatty amines have good potential for use against insecticide-resistant mosquitoes; tertiary amines are used in emulsions as treatment for endoparasitic and ectoparasitic nematodes in soil

(*continued*)

Table 4 *Continued*

Alphabetical listing by end-use	Derivative classification	Comments, additional information
Latex emulsions	Quat	Used as emulsifier
Latex foam rubber production	Quat	This additive reduces the stability of a latex emulsion and controls the rate of gellation; by hastening the gelation process, the finished product exhibits less shrinkage
Leather dying aids	Quat	Increased affinity for dyestuffs can be imparted to leather goods by "drumming" with a quat solution
Leather processing aids	Quat	Benzalkonium quats are used in the leather industry to decrease mildew of the hides
Leather softeners	Quat	Quat dispersions have been utilized to soften and "wet" out sheepskins after prolonged storage resulted in dried, stiffened skins
Lignin recovery	Quat	Water treatment additive utilized to remove lignin from papermill wastewater (lignins are highly colored components of wood, released when the wood is converted into pulp)
Liquid dishwash formulations	Amine oxide	Used to promote foam stability, increase viscosity, enhance emolliency, and increase mildness to human skin
Liquid fertilizer clarification	—	—
Liquid starch stabilizers	Quat	Lowers initial viscosity of the starch anc stabilizes the product against significan viscosity changes; provides superior surface lubrication to eliminate stickin to hot irons and increase speed and eas of ironing
Lubricants	Quat Amine, as is Amine oxide	These materials are used in the manufacture of extreme, high-pressure lube tertiary amines have also been used during manufacture (drawing step) of aluminum tubing
Metal cleaning/polishing emulsions	Quat	Used as a wetting agent in metal-cleanii formulations

Table 4 *Continued*

Alphabetical listing by end-use	Derivative classification	Comments, additional information
Metal extraction	Amine, as is Quat Betaine	Used in froth flotation purification and ore beneficiation processes; possible uses include foamers, froth modifiers, collecting agents, etc.
Metal purification	—	—
Metal treatment	Amine, as is Quat	Fatty amines and quats are used in the metal treatment area as corrosion inhibitors, dewatering agents, lubricant additives, etc.
Microbiocides	Quat	Active ingredient showing a broad spectrum of activity against a variety of pathogens, including bacteria, viruses, and fungi
Mineral acid inhibitors	Amine, as is	Can be used as organic base to neutralize mineral acids (form fatty salts); often used in corrosion inhibitor packages, etc.
Mineral separation/ enhancement	Quat	Increases settling rates and improves clarity of supernatant water
Mining (flotation agents)	Amine, as is Quat Betaine	Used in froth flotation purification and ore beneficiation processes as foamers, froth modifiers, collecting agents, etc.
Mold release agents	Amine, as is	Used as mold release agent for synthetic and natural rubber
Mothproofing	Amine, as is	Tertiary amines are active ingredients (used with a dry-cleaning solvent) to impart mothproofing to fabrics
Neutralization of acids	Amine, as is	Used as organic base to neutralize acids and form fatty amine salts; often used in corrosion-inhibitor packages, etc.
Oil drilling	Amine, as is Amine oxide Quat	Find use in drilling muds and fluids; used as water treatment chemicals in secondary oil recovery; used to combat sulfate-reducing bacteria producing corrosive hydrogen sulfide
Ore flotation	Quat	Used in flocculation/flotation as a general method of liquid/solids separation; flotation is recommended with slimy ores or where cationic activity is desired

(*continued*)

Table 4 *Continued*

Alphabetical listing by end-use	Derivative classification	Comments, additional information
Paint components	Amine oxide Quat Amine, as is	Used as leveling agents and thixotropic agents in paints; quats regulate conductivity in electrostatic paints; amines make it possible to paint on wet surfaces by displacing the water
Paper (antistat)	Quat	Used to increase paper conductivity and decrease static charge buildup
Paper (manufacturing)	Quat	Used as general softener in manufacture of tissue paper
Paper (softener)	Quat	Additive allows low-grade wood pulp to be softened to approach the texture of expensive stocks
Petroleum applications	Quat	Corrosion inhibitors, emulsion breakers, general surfactant properties
Pharmaceuticals	Quat	Used as process aid to prevent slime and soluble protein carryover into finished antibiotics after the fermentation step is complete
Phase transfer catalysts	Quat	Used in side range of multiphase systems to catalyze reactions
Photographic chemicals	Amine, as is	Tertiary amines are used as stabilizers and sensitizers in photographic films and plates
Pickling inhibitors	Quat	Anticorrosion agent; keeps fresh metal surfaces clean and bright while acids attack mill scale
Pigment dispersion agents	Amine, as is	Amine additives decrease overall grinding times in pigment production; as a consequence, production capacity can be increased
Pigment wetting agents	Quat	—
Plastic applications	Amine, as is Quat	As processing aids, antistat
Polymer applications	Amine, as is Quat	Quat provides antistat properties; fatty amines are used as a variety of processing aids such as polypropylene polyethylene—anti-stat; solvent polymerization of formaldehyde—initiator; polylactones—accelerator;

Table 4 *Continued*

Alphabetical listing by end-use	Derivative classification	Comments, additional information
Polymer applications (*continued*)		polyamide—corrosion inhibitor; polyester—stabilizer; polyurethane—foamer, catalyst, accelerator; polycaprolactam—dye modifier
Polyurethane foam manufacture	Amine, as is	Used as catalyst and accelerator in production of polyurethane; also used as a foamer
Premoistened towelettes	Quat	Used as surfactant, wetting agent, microbiocide in disinfectant premoistened towelettes
Quaternary ammonium salts	—	Currently, the most important application of quaternary ammonium salts revolves about their microbiocidal activity; other important industrial users include: textile—antistats, softeners, dye levelers; mining—froth flotation agents/modifiers; plastic—antistats, flocculating agents; metal—anticorrosion agents; etc.
Rubber curing agent	Amine, as is	Used as curing agent for some rubbers; for example, addition of 4% *t*-amine to butyl rubber formulation can result in the product hardening in thin layers or in bulk applications
Rubber molding aids	Amine, as is	Acts as good processing aid, imparting several properties, such as internal mold release; antisulfur bloom, etc. (*see* Rubber processing/manufacture)
Rubber processing/ manufacture	Amine, as is	Utilized for several different reasons: as internal mold release agent, as antiblocking agent, as antisulfur bloom agent, as processing aid, as gloss agent, as ozone scavenger, etc.
Saltwater lubricants	Quat	Used as corrosion inhibitor in heavy, calcium-base greases used in marine applications
Sanitizers	Quat	Used extensively in formulated sanitizers in hospitals, institutions, industry, and homes (*see* Bacteriacides)

(*continued*)

Table 4 *Continued*

Alphabetical listing by end-use	Derivative classification	Comments, additional information
Secondary oil recovery	Quat	Used as surfactant, etc. in secondary oil recovery, especially, water flooding
Sewage treatment	Quat	Used in sewage treatment as flocculant
Shampoo	Amine oxide Betaine	Used as foamer, foam booster, foam stabilizer; also to impart good manageability to hair (antistat)
Silica compounding	Quat	Used as flocculating agents in manufacture of silica and hydrated silica
Slimicides	Quat	Utilized in water treatment, paper manufacture, etc. to control slime and fungal growth
Stabilizers	Amine, as is	*t*-Amines act as stabilizers for 1,1,1-trichloroethane to inhibit metal-induced decomposition; also used as stabilizers in polyester manufacture, etc.
Sugar decolorizing agents	Quat	Fatty amine quats find application in the sugar industry as clarifying and decolorizing agents for raw cane sugars
Surface coatings	Quat	Addition of quats to several different types of paint produces material having better wetting capacity and adhesion for both wet and dry surfaces
Swimming suits	Quat	Illinois has legal ordinance requiring sterilization of suits worn by swimmers; disinfection can be accomplished with quat solutions with no adverse effects on fabric
Textiles	Quat	Used to prevent static electrical charge accumulation; also used as dye levelers
Thickeners	Amine oxide Betaine	Used to build viscosity of surfactant formulations
Toilet-bowl cleaners	Amine oxide Quat	These additives increase viscosity to keep cleaner in contact with vertical surfaces during bleaching process; also as active bactericide
Wastewater treatment	Quat	Useful in water treatment as algae control agent, slime control agent, and as corrosion inhibitor

Table 4 *Continued*

Alphabetical listing by end-use	Derivative classification	Comments, additional information
Waterless hand cleaner	Amine oxide	Utilized to stabilize emulsion; also impart emolliency
Waterproofing	Amine salt	Polyester fabric treated with a polysiloxane consisting of a *t*-amine salt can provide water repellance without causing discoloration
Wood-pulp processing	Amine oxide	Purified wood pulp can be treated with cationic-form amine oxide to hasten breakdown during shredding; this improves handling characteristics in the xanthate or similar processes
Wood carbonizing	Amine salt	Fatty amine salts are effective surfactants in the wool carbonizing process, where aluminum chloride is the carbonizing agent

CHAPTER 10
Oil Field Applications

PATRICK C. HU Ethyl Corporation, Baton Rouge, Louisiana

1 INTRODUCTION

Alpha olefin derivatives are either being used or considered for use in every phase of oil production operations, starting with well drilling and ending at transporting crude oil to the refinery. Based on their functionalities, use of alpha olefin derivatives in oil field applications can be classified into eight categories: emulsifiers, demulsifiers, foamers, defoamers, dispersants, corrosion inhibitors, scale inhibitors, and biocides. Nearly all the applications of alpha olefin derivatives in oil field operations can be classified as surfactant in nature.

This chapter is intended to familiarize the reader with alpha olefin derivatives, which are applicable in every phase of oil field operations, where a significant amount of alpha olefin derivatives are or may be used. In the discussion of each phase, the mechanical processes and the functionalities demanded of the chemical additives are introduced. Naturally, in the content of this chapter it will only be possible to develop a very simplistic outline for each complex operation.

2 ENHANCED OIL RECOVERY

When a hole is drilled through the overlying impervious strata and the pressure on the reservoir is released in the immediate vicinity of the well bore, the high pressure in the reservoir causes both oil and connate water to migrate to the bore hole, where the oil is recovered. The well is considered to be in its primary oil production stage. Once the primary oil production of a field has slowed below the point of economic operation, it is necessary to move to secondary oil recovery. The most commonly used secondary oil recovery process in the United States several decades ago was water flooding. To perform water flooding either new wells are drilled or some of the production wells are converted to injection wells where water is pumped in to push oil to the production wells. Since the porosity of various sections of a formation may vary and water has a higher mobility than that of oil in place, inevitably the water front overruns some of the oil and breaks through into the production wells, leaving a substantial amount of oil trapped in the formation. It is estimated [1] that on an average, two-thirds of the original oil in place was retained in the formation rock after water flooding.

Removal of the residual oil in place after water flooding requires the use of tertiary treatment such as with heat or chemical additives. Stimulation of oil production by secondary or tertiary methods can be referred to as enhanced oil recovery (EOR) methods. Due to economic considerations, more and more EOR methods are undertaken when an oil field is still in its primary production stage. It is estimated [2] that about 60 billion barrels of additional oil from known reservoirs in the United States are recoverable using existing EOR technologies.

All EOR methods are sensitive to environmental elements or factors that commonly exist in oil reservoirs. For any EOR process to be effective, all the environmental elements must be carefully considered [3], so that the heat or the chemical added can reach the oil-bearing region, dislodge the oil, and push it to the recovery wells.

2.1 Micellar Flood

Technology using a concentrated aqueous surfactant solution to dislodge oil from a formation is referred to as micellar flood. The surfactant solution displaces the oil by wetting the formation rock and promotes oil/water emulsification by reduction of oil/water interfacial tension. The surfactant slug is formulated based on the available information relating to the characteristics of formation, connate water, and oil in place. To improve efficiency, a surfactant slug often contains electrolytes, commonly inorganic alkali metal salts and cosurfactants such as low-molecular-weight alcohols, ethoxylated alcohols, and ethoxylated alcohol sulfonates. To improve sweep efficiency, the surfactant solution is commonly pushed through the formation by a polymer slug at a rate of a few feet per day. During propagation through a formation, a surfactant slug may encounter connate water containing high levels of alkali and alkaline earth metal inorganic salts which can precipitate anionic surfactants. Due to their nature, surfactants also tend to be adsorbed on the formation rock. Chromatographic separation of surfactant components can be a serious problem. To achieve good oil recovery efficiency the composition of the surfactant slug should retain its integrity throughout the process. In addition to forming ultralow interfacial tensions at the oil/water interface, an effective surfactant should also have good brine tolerance, thermal stability, and low adsorbtivity by the formulation rock.

Numerous surfactants [4–7] have been evaluated for micellar floods. Among all the surfactants studied, alkylaryl sulfonates have received most attention by investigators in the field, due to their excellent thermal stability, availability in large quantity, and low cost. It has been demonstrated [8–10] that depending on the characteristics of the crude oil, alkylaryl sulfonates of various molecular weights and branching content may be used to formulate a surfactant slug producing the required ultralow interfacial tension with the oil under simulated formation conditions. Compared with other types of surfactants available, alkylaryl sulfonates are generally very sensitive to the brine content encountered by a surfactant slug. Consequently, it is often necessary to add a low-molecular-weight alcohol such as isobutyl alcohol or ethoxylated alcohol to the system. One effective micellar system that functions well in a high brine environment can be formulated by combining alkylaryl sulfonates with ethoxylated alcohol sulfonates. Other types of surfactants that have received attention are alpha olefin sulfonates, petroleum sulfonates, lignosulfonates, and ethoxylated alcohols and phenols. Recent work at the University of Texas in Austin showed [11,12] that effective micellar slugs can be formulated using alpha olefin sulfonates.

2.2 Alkaline Flood and Surfactant-Assisted Alkaline Flood

Alkaline flood [13–16] and surfactant-assisted alkaline flood [17–21] are similar to micellar flood in principle. However, instead of depending on injected surfactant as in micellar flood to move oil in a formation, alkaline flood utilizes in situ-generated surfactant resulting from neutralization of petroleum acids present in the crude oil. Many crude oil candidates for enhanced oil recovery by alkaline flood produce their lowest interfacial tension at very low concentration of alkali when the petroleum soap is the only surfactant present in the system. To maintain the optimal salinity, a low-concentration alkaline slug must be used. Alkali consumption by formation rock makes propagation through porous media of such dilute alkaline solutions prohibitively slow. Also, a dilute alkaline solution is often ineffective in controlling the divalent ions (such as Ca^{2+} and Mg^{2+}) present in the formation. Without effective control of divalent ions, surfactant loss by precipitation represents an additional difficulty.

To expand the application scope and recovery efficiency of alkaline flood, additional surfactant is often injected along with alkali. This method is known as surfactant-assisted alkaline flood or cosurfactant-enhanced alkaline flood. The surfactant selected for use with in situ-generated petroleum soap should form a system with sufficiently high tolerance to electrolytes to permit use of the desired concentration of alkali. Generally, surfactant cost for surfactant-assisted alkaline flood is significantly lower than that for a typical micellar flood, and the oil recovery efficiency of a surfactant-assisted alkaline flood is expected to be higher than that of an alkaline flood.

Potential surfactant candidates for surfactant-assisted alkaline flood are internal olefin sulfonates, alkylaryl sulfonates, alcohol sulfates, ethoxylated alcohol sulfates, ethoxylated alkylphenol sulfates, alcohol ethoxylates, ethoxylated phenols, petroleum sulfonates, and ethoxylated alcohol sulfonates. Due to their limited hydrolytic stability, sulfate types of surfactants should not be selected for applications where elevated temperature may be present.

2.3 Miscible and Immiscible Floods

Decreasing the viscosity of crude oil and increasing bulk and relative permeability to increase oil flow at formation pressure are the principles of using fluids for miscible or immiscible floods [22–29]. Miscibility with reservoir oil is the criterion for referring to the process as miscible or immiscible. Fluids commonly used are carbon dioxide, nitrogen, air, liquefied natural gas, or enriched gas.

From the pore-scale point of view, a dense gas or fluid is an ideal displacement medium for many crude oils, because it can generate a miscible transition zone with the oil such that the interface is eliminated, to result in nearly complete oil displacement. Unfortunately, even when conditions for miscibility are met, high sweep efficiency is not often achieved, due principally to the nonuniformity of the flow pattern. Because the dense gas has a viscosity much lower than the oil and water exposed to the same pressure gradient, the displacement front is very susceptible to the growth of viscous fingers in which injected gas or fluid moves faster than the oil and water. When dispersion alone is incapable of smearing the fingers on a typical reservoir scale, early breakthrough occurs and sweep efficiency decreases. The fingering problem can be alleviated using a water alternated with gas (WAG) process. However, the large quantity of water injected with such a procedure could isolate reservoir oil from contact with the fluid. Also, the solubility of some of the fluid in water could increase the overall fluid consumption to make the process less economically attractive.

The idea of using foams for mobility control is not new. It is known that flow impedance by foam in a porous medium is several hundred times higher than can be achieved by either consituent alone. Recently, it was demonstrated [29] that flow impedance by using foam is generated by the distortion of the numerous interfaces present in foam during the propagation of the foam through porous media. Consequently, it is possible to use a surfactant foam to control the flow pattern of a fluid to improve sweep efficiency. In addition to the other desired surfactant characteristics commonly required for other EOR operations, the surfactant selected for this application must produce stable foams. Alcohol sulfates, ethoxylated alcohol sulfates, ethoxylated alkylphenol sulfates, alkylaryl sulfonates, alpha olefin sulfonates, ethoxylated alcohol sulfonates, and ethoxylated alkylphenol sulfonates are all potential candidates for this application. However, the limited hydrolytic stability of sulfates in acidic media may exclude the use of sulfate-type surfactants in applications where carbon dioxide or sour hydrocarbons are present.

2.4 Steam Injection [30–36]

Injection of steam into an oil-bearing formation has been by far the most widely used enhanced recovery technique and is expected to remain so in heavy oil reservoirs for years to come. This process has the advantages over others of low investment, a rapid return on investment, and a large ultimate recovery. It is estimated that over 70% of the oil recovered by EOR techniques is by steam injection. There are two types of steam injection currently used: cyclic steam injection (huff and puff) and steam drive.

In cyclic injection [32], steam is first injected into an oil well for 2 to 3 weeks. The well is then capped for a few days, known as the soaking period, when steam is condensed and heats the oil-bearing formation close to the well bore to lower viscosity and improve flow toward the well bore for recovery when the well is uncapped. When oil recovery rates decrease to an unacceptable level, the process is repeated.

Subsequent cyclic steam injections increase the water-to-oil ratio. Once the water-to-oil ratio reaches a certain limit, economic considerations dictate the use of steam drive because a greater steam penetration can be achieved. Both cyclic steam injection and steam drive are most effective for moderately viscous to viscous crudes since very viscous crude oil and tar may not be appreciably mobilized with steam [37].

In addition to reservoir heterogenieties, severe differences in density and mobility between steam and crude oil cause undesirable steam penetration patterns. As a result, good sweep efficiency is rarely attained. The low steam density, relative to that of crude oil, tends to result in steam rising to the upper region of the reservoir, a phenomenon known as gravity override. The extremely high mobility ratio between steam and crude oil also tends to result in steam channeling. Consequently, in steam injection some of the oil-bearing regions may never be reached by steam.

Foam can be generated in situ when steam is injected into a formation along with a surfactant and an optional gas. The lighter and more mobile foam migrates and fills the paths of least resistance to block the steam and results in directing the steam sweep to the desired zones. Additional benefits can be realized after the steam foam condenses. The presence of surfactant in condensed steam tends to simulate a surfactant flood which not only accelerates oil flow but also alters the permeability profile to assist steam in reaching more inaccessible regions of the formation. In addition, the surfactant solution can mobilize heavy crude blocking pores near the well bore to result in greater permeability with resulting higher oil production rates.

In addition to having good foam characteristics, surfactants must also be stable at steam temperatures up to a few months. Hence sulfonates, such as alkylaryl sulfonates, alpha olefin sulfonates, or dimer alpha olefin sulfonates, are more suitable for many field applications. Due to their excellent brine tolerance, ethoxylated alcohol sulfonates and ethoxylated alkylphenol sulfonates are more effective in field applications where high electrolyte content is desired or naturally present as connate water.

3 WELL STIMULATION

The objective of well stimulation is to increase permeability to improve the rate at which oil will flow out of the reservoir. Treating the formation with acid [37–40] and inducting fractures by hydraulic pressure are some methods used. The reduction in permeability may have been caused by precipitation of minerals, improper well finishing, or heavy oil residuals blocking the pores near the well bore.

For an acidization well stimulation to be effective, the acidic fluid must etch flow channels by solubilization or remove solid debris which impedes flow to create or restore flow patterns that extend as far as possible from the well bore. Depending on job requirements, various types of acid at various concentrations are used. Ninety percent of all acid used in the oil field is hydrochloric acid. An acid concentration of 3 to 15% is typical. Due to concentration gradient, acid has a tendency to produce sludge deposits, which may reduce the permeability improvement created by the acid. Consequently, a surfactant is commonly added to the acid as a dispersant to prevent sludge deposition. The most common antisludge agent is alkylbenzene sulfonic acid.

Surfactants such as dodecylbenzene sulfonic acid are also used as wetting agents to enhance the penetrating power of the acid sludge. It is often desirable to achieve acidizing in a formation at a point far from the injection point at the well bore. In such cases it would be desirable to inhibit the acid reaction rate such that the acidizing solutions may reach the distant formation before being spent. To accomplish this, an acidified emulsion system may be used. Acidified emulsion is generally comprised of acid, oil, and a surfactant such as alpha olefin sulfonic acid, dodecylbenzene sulfonic acids, and amine salts such as dodecylamine, tetradecylamine, and dodecyldimethylamine hydrochlorides. One approach involves the use of an aqueous acid solution as an emulsion in oil. The acid becomes reactive as the emulsion breaks. The acid in oil emulsion gives the longest retardation period of acid activity and minimum corrosion damage to the equipment used. Surfactants such as ethoxylated alkylphenol, ethoxylated alcohol, and alkylamine are used as emulsifiers.

Nitrogen foam/acid stimulation [38–40] is a relatively new technique. The technique requires the injection of an acidic foam into a formation. As compared to using an aqueous acid solution, acidic foam well stimulation has the advantages of better sweep efficiency, less water loss, less residual acid, and a longer reaction period. A considerable amount of foam acidization is carried out in shallow and water-sensitive formations. Surfactants such as ethoxylated alcohol sulfonates, alkylarylsulfonic acids, and fluorosurfactants are used as foaming agents for acidic foam well stimulation.

In the area of hydraulic fracturing, an acid cleanup is often conducted prior to fracturing. After the acid work a fracture is initiated by applying hydraulic pressure and the formation is propped open with sand or other suitable material. During the fracturing process it is known that good fracture propagation can be achieved by the addition of quaternary ammonium surfactant such as dialkyldimethylammonium chloride, alkyltrimethylammonium chloride, or alkyldimethylbenzylammonium chloride into the hydraulic fluid. The hypothesis behind the use of quaternary ammonium surfactant is that the surfactant improves the wetting characteristics of the hydraulic fluid and the surfactant adsorbs rapidly on the formation. Once a fracture is initiated, the surfactant solution penetrates rapidly into the microscopic voids created and a monolayer of the surfactant is adsorbed on the newly created surface to form a barrier. The wetting and adsorption phenomena tend to cause propagation of the fracture. After the hydraulic fracturing, pressure must be maintained to keep the fractures open, and sand is pumped into prop open the fractures. Surfactant foam or a viscous fluid is used to transfer sand. When a surfactant foam is used, surfactants such as ethoxylated alcohol sulfates, alkylaryl sulfonates, and alpha olefin sulfonates are used as the foaming agent.

4 DRILLING FLUIDS

In any drilling activity the debris or chips generated at the bottom of a borehole need to be removed in order to maintain drilling efficiency. The solution to the removal problem is the application of drilling fluid [41] in the form of mud or foam. Drilling fluids are held in storage tanks on the rig. Fluids are pumped down through the hollow drill string using high-pressure positive-displacement pumps. At the bottom of the string, the fluid passes out through ports in the bit and returns up to the surface in the annulus formed between the borehole wall and the drill string, carrying with it the cuttings created by the bit.

In addition to debris removal, drilling fluid should have other characteristics to facilitate drilling operations. For example, the drilling fluid desirably lubricates and cools the drill string and bit during drilling operations, keeps cuttings in suspension, provides well bore stability, prevents fluid from entering reservoir formations, prevents formation fluids from entering the well bore, provides corrosion protection, and controls formation pressures.

In the following discussion, drilling fluid is classified in three categories: foam, oil-based drilling mud, and water-based drilling mud. Foam is produced as a surfactant solution and is injected into an airstream as it enters a well [42–47]. The foam provides a low-density fluid with high lifting capacity for cuttings. It improves suspension, separates drill cuttings to prevent aggregation, and helps remove water and oil contaminants by utilizing the numerous plateau borders as suction centers. The rheology of a foam is described as Bingham plastic. It has the property of keeping cuttings suspended during a pumping failure caused by a power loss or mechanical problems. Severe restarting difficulties can occur when the cuttings suspended in drilling fluid are allowed to settle and pack at the well bottom. The type and concentration of surfactant selected as a foamer are chosen to provide a stable foam under exposure to the specific fluid contaminants encountered. Surfactants such as ethoxylated alcohol sulfate or alpha olefin sulfonates are commonly used as the foamers.

Among the three types of drilling fluids, water-based drilling muds are the most widely used. A water-based drilling mud [48] is defined as a drilling fluid in which water is the continuous phase, maintaining a suspension of solids and dissolved materials. Only small quantities of olefin derivatives are used in water-based drilling muds. Occasionally, ethoxylated alkylphenols or ethoxylated alcohols are used as dispersants for solid materials and alkylaryl sulfonates are employed as defoamers. Some formations contain substantial hydratable clay minerals. Clays swell upon hydration and cause reduction in permeability. Clay swelling can also increase the drag on the totating drill pipe, resulting in reduction in drilling efficiency.

The higher drilling temperatures found in deeper wells causes rapid degradation of drilling mud, and frequent reconditioning of the drilling mud is required to retain the necessary functionalities. Under these conditions, the use of an oil-based drilling mud is often desirable. In oil-based drilling muds [49], oil is the continuous phase. Similar to water-based drilling muds, the oil-based muds may contain both suspended solids and liquids as well as dissolved materials. To prepare a homogeneous and functional oil mud, surface-active agents are often required. For example, bentonite clay is not wet by oil. For bentonite clay to function as a viscosity modifier in oil-based drilling muds, it is treated with a quaternary ammonium surfactant such as dialkyldimethylammonium chloride or alkyltrimethylammonium chloride. The quaternary ammonium species incorporates itself into the structure of bentonite clay by replacing some of the metal ions commonly found in it to make the clay wettable by oil.

Asphalt is commonly incorporated into oil-based muds for viscosity modification at elevated temperature. Nonionic surfactants such as ethoxylated alkylphenols or ethoxylated alcohols are used to wet the asphalt. Although an oil-based drilling mud needs no water to function, it must function in the presence of water. During a drilling operation the drilling mud will encounter formation water. To maintain the functional efficiency of the drilling mud, the water must be emulsified into oil to form a water-in-oil emulsion. Nonionic surfactants such as ethoxylated alkylphenols and ethoxylated alcohols are incorporated in oil-based drilling muds to facilitate emulsification.

5 EMULSIFIER AND DEMULSIFIER

An increasing number of oil fields are producing both crude oil and water. The water that is produced with the crude oil comes either from the formation or from injected water or a combination of the two sources. The injection water comes from water flood, alkaline flood, micellar flood, or condensed steam from steam flood. In the absence of a surface-active agent, water and oil are incompatible and tend to form discrete phases. In the presence of surfactants, stable emulsions often form. Surfactants adsorb at oil/water interfaces and form a boundary coating. The adsorption of surfactant decreases oil/water interfacial energy to enhance the efficacy of emulsion formation. The formation of a protective boundary layer resists the coalescence of emulsion droplets. Furthermore, solids species which are often present in crude oil may also attach to the oil/water interface and form a solid barrier, preventing coalescence breaking of the emulsion.

The presence of water in oil increases crude oil transportation cost. Also, the presence of inorganic ions such as Ca^{2+} and Mg^{2+} in water can poison refinery catalysts as well as corrode equipment. Water-in-oil emulsions tend to have higher apparent viscosities than that of the crude oil. Increasing viscosity causes a reduction in pumping efficiency during transportation in the pipe. The viscosity of an oil-in-water emulsion is often less than the viscosity of the oil. Therefore, sometimes during the transportation of a viscous crude oil in a pipeline, surfactant and water are added deliberately to create a less viscous oil in water emulsion for improved pumping efficiency.

One procedure often used for separating the water from the production crude oil is the addition of a demulsifier. Supposedly, a demulsifier enters the water/oil interfacial region and alters the characteristics of interfaces to accelerate coalescence and emulsion breaking. However, the crude oil/water emulsion system is highly complexed and closely related to the petroleum production. Efficacy of demulsifiers is highly dependent on the specific system involved. There is no single unique surface-active agent that works as a demulsifier in all oil/water emulsion systems. Since the most common types of emulsifiers found in a crude water emulsion are anionic, it is not surprising that cationic polyelectrolytes are found to be effective demulsifiers in many cases. Typical examples of products being evaluated as demulsifiers are ethoxylated alcohols, ethoxylated alkylphenols, dialkyldimethylammonium chlorides, propylene oxide/ethylene oxide adducts, cationic compounds, or phenolic resins.

6 CORROSION AND SCALE INHIBITORS

Corrosion inhibitors are used in nearly every phase of oil production to protect drilling equipment, pipelines, and facilities. Corrosion is caused by exposure of metal surfaces to acids, bases, corrosive gases such as carbon dioxide and hydrogen sulfide, or bacterial wastes in the presence of water. The evolution of modern, high-efficiency boilers led to an increase in the incidence of corrosion failures and scaling associated with generating steam for steam feed. Water fed into a boiler to generate steam inevitably contains some dissolved carbon dioxide, carbonate, and/or bicarbonate which decomposes to carbon dioxide. When the steam condenses, the carbon dioxide redissolves in water to form carbonic acid, which promotes corrosion.

Chemical treating agents for condensate usually consist of neutralizing amines, film-forming amines, and hydrazines. Numerous amines can be used for pH control in condensate systems. An amine is selected based on consideration of stability and vapor pressure or distribution ratio of the amine in the vapor phase to amine in the liquid phase during steam generation. A continuous release of a suitable amount of neutralizing amine can be achieved for each boiler operation. Aliphatic amines and their derivatives are commonly used as neutralizing amines.

Filming amines [50] function by forming a protective barrier on metal surfaces to block the attack by corrosive elements. Almost 90% of the film-forming corrosion inhibitors are aliphatic amines, alkylimidazolines, or quaternary ammonium compounds. Dependent on the specific application, film-forming amines of various types and concentrations are used to protect both downhole and above-surface equipment. In downhole applications, film-forming amines can be incorporated into water, steam or drilling mud to protect metal surfaces against excessive corrosion. Quaternary ammonium compounds are more water soluble and therefore more readily removed from metal surfaces. However, in solution they function as biocidal agents to control the generation of corrosive wastes from bacteria.

Chemicals are used to prevent or minimize the deposition of scale from water as a result of changing environmental conditions. Scale of calcium carbonate, calcium sulfate, and barium sulfate has been encountered both downhole and at the surface. The formation of scale reduces oil production rates from a well, transporting efficiency of a pipeline, and operational efficiency of a boiler.

The function of most scale inhibitors is basically that of adsorption on the microcrystalline nuclei to inhibit crystal growth to full-fledged scale crystals [51,52]. These inhibitors are usually phosphorus compounds in several forms: organic phosphate esters, organic phosphates, and organic aminophosphates. Depending on the specific system, scale inhibitors of various types, concentration, and form are selected. Organic aminophosphates are intended to provide not only scale control but also corrosion control. The phosphate ester portion of the molecule is the same as in the organic phosphate ester. In addition, the amino group in the molecule enables it to serve the corrosion control function. The presence of some bacteria may cause serious material problems which plague production operations.

7 SUMMARY

All the chemicals identified for various oil field applications can be derived by utilizing an alpha olefin as one of the starting raw materials. However, not all the chemicals mentioned require an alpha olefin as the starting material. Economic considerations sometimes dictate the use of different raw materials. Almost all the alpha olefin derivatives mentioned in this chapter are surface active in nature and may be classified as surfactants, with the exception of alkylamines. When primary, secondary, or tertiary amines are used as a film-forming amine for corrosion inhibition of metal surfaces, the amines are not surfactants. On the other hand, when the amines are used to prepare acidic emulsions, the amines are protonated and function as surfactants or, more precisely, emulsifiers.

Most surfactants used in oil field applications are also used in many household laundry and personal care products, with the exception of ethoxylated alcohol sulfonates and dimer alpha olefin sulfonates.

Ethoxylated alcohol sulfonates can be represented by using the general formula $RO(CH_2CH_2O)_xR'(SO_3M)_y$, where R is an alkyl or alkaryl group, R′ is alkylene, and M is an alkali metal ion. Ethoxylated alcohol sulfonates have high brine tolerance, comparable to that of ethoxylated alcohol sulfates. However, being sulfonates rather than sulfates, they are stable against thermal hydrolytic degradation.

Dimer alpha olefin sulfonates are prepared by reacting alpha olefins with sulfur trioxide, followed by dimerization and neutralization. Dimer alpha olefin sulfonates are tolerant to high brine concentrations without suffering precipitation. They have higher thermal stability than that of alpha olefin sulfonates.

Both ethoxylated alcohol sulfonates and dimer alpha olefin sulfonates are currently available only in developmental quantities from various sources. The utility of both surfactants has been demonstrated in laboratory and/or field test. However, due to their exceptionally high tolerance to brine and to their good thermal stability, it is expected that both will find applications in the areas mentioned.

In oil field operations, one must deal constantly with complex and ill-defined systems. Consequently, the perception of chemical needs may differ from operator to operator. Olefin derivatives, summarized below according to applications, are either being used or being considered for use.

Micellar Floods

Alpha olefin sulfonates
Alkylaryl sulfonates
Ethoxylated alcohols
Ethoxylated alcohol sulfonates

Alkaline- and Surfactant-Assisted Alkaline Floods

Alpha olefin sulfonates
Internal olefin sulfonates
Alkylaryl sulfonates
Alcohol sulfates
Ethoxylated alcohol sulfates
Ethoxylated alcohols

Miscible and Immiscible Floods

Alpha olefin sulfonates
Alkylaryl sulfonates
Ethoxylated alcohol sulfates
Ethoxylated alcohol sulfonates

Steam Injections

Alpha olefin sulfonates
Alkylaryl sulfonates
Dimer alpha olefin sulfonates
Ethoxylated alcohol sulfonates

Well Stimulations

Primary, secondary, and tertiary alkylamines
Alpha olefin sulfonates
Alkylaryl sulfonates
Ethoxylated alcohol sulfates

Drilling Fluids

Alkyl phosphates
Alkylaryl sulfonates
Alkylamines
Ethoxylated alcohols
Alpha olefin sulfonates
Ethoxylated alcohol sulfates

Emulsifiers/Demulsifiers

Quaternary ammonium halides
Imidazolines
Alkylaryl sulfonates
Alpha olefin sulfonates
Ethoxylated alcohols
Ethoxylated alcohol sulfates

Corrosion and Scale Inhibitors

Primary, secondary, and tertiary alkylamines
Alkyl phosphates
Imidazolines

REFERENCES

1. H. K. Van Poollen and Associates, Inc., *Fundamentals of Enhanced Oil Recovery*, PennWell Books, Tulsa, Okla. 1980.
2. Wilson L. A., Jr., in *Improved Oil Recovery by Surfactant and Polymer Flooding*, (D. O. Shah and R. S. Schechter, eds.), Academic Press, New York, 1977, pp. 1–54.
3. Geffen, T. M., *World Oil*, 53–57 (Mar. 1975).
4. Graciaa, A., Lachaise, J., Sayous, J. G., Grenier, P., Yiv, S., Schechter, R. S., and Wade, W. H., *J. Colloid Interface Sci., 93*(2): 474–486 (1983).
5. Wilson, P. M., Murphy, L. C., and Foster, W. R., *Paper SPE 5812*, presented at the SPE Symposium on Improved Oil Recovery, March 1976.
6. Bousaid, I. S., U.S. patent 4,230,182 (Oct. 1980).
7. Shupe, R. D., Maddox, J., Jr., U.S. patent 4,269,271 (May 1981).
8. Puerto, M. C., and Gale, W. W., paper presented at the SPE Symposium on Improved Oil Recovery, Mar. 1976.
9. Doe, P. H., El-Emary, M., Wade, W. H., and Schechter, R. S., *J. Am. Oil Chem. Soc., 55*(5): 513–520 (1978).
10. Doe, P. H., El-Emary, M., Wade, W. H., and Schechter, R. S., *J. Am. Oil Chem. Soc., 55*(5): 505–512 (1978).
11. Lalane-Cassou, C., Schechter, R. S., and Wade, W. H., *Am. Chem. Soc. Div. Pet. Chem. Prepr., 29*(4): 1187–1192 (1984).
12. Barakat, Y., Fortney, L. N., Lalane-Cassou, C., Schechter, R. S., Wade, W. H., Weerasooriya, U., and Yiv, S., *Soc. Pet. Eng. J., 23*(6): 913–918 (1983).
13. Wasan, D. T., Shah, S. M., Aderangi, N., Chan, M. S., and NcNarmar, J. J., *Soc. Pet. Eng. J., 18*(6): 409–417 (1978).
14. Ehrlich, R., and Wygal, R. J., Jr., *Soc. Pet. Eng. J., 17*(4): 263–270 (1977).
15. Ramakrishanan, T. S., and Wasan, D. T., *Soc. Pet. Eng. J., 23:* 602 (1983).
16. Trujillo, E. M., *Soc. Pet. Eng. J., 23:* 645 (1983).
17. Chang, H. L., U.S. patent 3,977,470 (1976).
18. Southwick, J. G., *Paper SPE 12771*, presented at the California Regional Meeting of Society of Petroleum Engineers, Long Beach, Calif., 1984.
19. Sydansk, R. D., *Soc. Pet. Eng. J., 22*(4): 453–462 (1982)
20. Quarterly Report for July–Sept. 1985, *Report NIPER-113*, National Institute for Petroleum and Energy Research, Department of Energy, Oct. 1985.

21. Nelson, R. C., Lawson, J. B., Thigpen, D. R. and Stegemeier, G. L., *Paper SPE/DOE 12672*, presented at the SPE/DOE 4th Symposium on Enhanced Oil Recovery, Apr. 1, 1984.
22. Spivalk, A., and Chima, C. M., *Paper SPE 12667*, presented at the California Regional Meeting of Society of Petroleum Engineers, Wilmington Field, Calif., 1984.
23. Bernard, G. G., Holm, L. W., and Jacobs, W. L., *Soc. Pet. Eng. J, 5:* 295–300 (1965).
24. Minssieux, L., *J. Pet. Technol., 26:* 100–108 (1974).
25. Raza, S. H., *Soc. Pet. Eng. J., 10:* 328–336 (1970).
26. Albrecht, R. A., and Marsden, S. S., *Soc. Pet. Eng. J., 10:* 51–55 (1970).
27. Holm, L. W., *J. Pet. Technol., 22:* 1477–1506 (1970).
28. Orr, F. M., Heller, J. P., and Taber, J. J., *J. Am. Oil Chem. Soc., 59*(10 (1982).
29. Hu, P. C., Tuvell, M. E., and Bonner, G. A., *Paper SPE/DOE 12660*, presented at the 4th SPE/DOE Symposium on Enhanced Oil Recovery, Tulsa, Okla. 1984.
30. Dilgren, R. E., Deemer, A. R., and Owens, K. R., paper presented at SPE Annual Meeting, Tulsa, Okla., Apr. 4–7, 1982.
31. Doscher, T. M., *J. Pet. Technol., 34:* 1533 (1982).
32. Blair, C. M., Jr., Schribner, R. E., and Stout, C. A., *Paper SPE/DOE 10700*, presented at the SPE/DOE 3rd Joint Symposium on Enhanced Oil Recovery, Tulsa, Okla., 1982.
33. French, T. R., Broz, J. S., Lorenz, P. B., and Bertus, K. M., *Paper SPE 15052*, presented at the 56th California Regional Meeting of the Society of Petroleum Engineers Oakland, Calif., 1986.
34. Duerksen, J. H., *Paper SPE 12785*, presented at the Califorina Regional Meeting of the Society of Petroleum Engineers, Long Beach, Calif., 1984.
35. Dilgren, R. E., Deemer, A. R., and Owens, K. B., *Paper SPE 10774*, presented at the California Regional Meeting of the Society of Petroleum Engineers, San Francisco, 1982.
36. Dilgren, R. E., and Owens, K. B., *J. Am. Oil Chem. Soc., 59*(10): 818A–822A (1982).
37. Neuman, C. H., *J. Pet. Technol., 37:* 163–169 (1985).
38. William, G. F., and Loy, D. R., *J. Pet. Technol. 37:* 89–97 (1985).
39. Crowe, C. W., U.S. patent 3,917,536 (1975).
40. Young, D. C., and Maly, G. P., U.S. patent 4,101,425 (1978).
41. Schweizer, R. G., U.S. patent 3,726,796 (1973).
42. Mitchell, B. J., *Oil Gas J., 69*(36): 96–100 (1971).
43. Anderson, G. W., *World Oil, 173* (4): 39–42 (1971).

44. Garavini, O., Radenti, G., and Sala, A., *Oil Gas J., 69*(33): 82–84, 89–90 (1971).

45. Hutchison, S. O., Anderson, G. W., and McKinnell, J. C., U.S. patent 3,486,560 (1970).

46. Arthur, R. P., Chocola, L. R., Shore, A., and Shore, S., U.S. patent 3,394,768 (1968).

47. Cooper, L. W., Hook, R. A., and Payne, B. R., *World Oil, 184*(5): 95–96, 100, 102, 106 (1977).

48. Nelson, R. C., *J. Am. Oil Chem. Soc., 59*(10): 823A–826A (1982).

49. Hayes, J. B., Haws, G. W., and William, G. B., U.S. patent 4,012,329 (1977).

50. Beecher, J., *Power*, 74–77 (July 1981).

51. Adamson, A. W., *Physical Chemistry of Surfaces*, Wiley, New York, 1976.

52. Mysels, K. J., *Introduction to Colloid Chemistry*, Interscience, New York, 1967.

CHAPTER 11

Fatty Acids

JERRY D. UNRUH and JAMES R. STRONG Celanese Chemical Company, Inc., Corpus Christi, Texas
CHRISTINE L. KOSKI Celanese Chemical Company, Inc., Dallas, Texas

1 INTRODUCTION

Fatty acids are considered to be carboxylic acids with six or more carbon atoms (C_6–C_{24}). Of these, heptanoic (C_7) and pelargonic (C_9, nonanoic) acids became commercially available from Celanese Chemical Co. in 1980 from alpha olefins via hydroformylation (oxo) of 1-hexene and 1-octene, respectively. In turn, these olefins are readily available from the Ziegler ethylene-growth process, wax cracking, and the Shell SHOP process. These synthetic fatty acids are ideal as replacements for, or as additions to, naturally derived fatty acids in the 5- to 10-carbon range. Prior to the oxo process, the only commercial routes to fatty acids were from natural fats and oils.

Naturally derived heptanoic acid is obtained from caster oil. The castor oil is pyrolyzed to give ricinoleic acid, which is then processed to yield undecanoic acid and heptanal. The heptanal can be converted to heptanoic acid, and the undecanoic acid is used in the production of nylon-11. Naturally derived pelargonic acid is coproduced along with azelaic acid through the ozonolysis of oleic acid. In essence, both naturally derived heptanoic and pelargonic acids are by-products of processes that yield other primary products, and their availability is dependent on the need for the primary products.

Whereas the price of naturally derived heptanoic and pelargonic acids may be affected by natural causes, such as weather patterns, pricing of the synthetic fatty acids is tied to hydrocarbon pricing. Therefore, the economic advantage of one source over the other will change from time to time. The main differences between the synthetic and naturally derived heptanoic and pelargonic acids are factors such as consistency of availability, purity, and types of impurities inherent in the modes of manufacture.

There are many applications that benefit from the consistent quality and availability of synthetic heptanoic and pelargonic acids. These include synthetic lubricants, alkyd resins for coatings, plasticizers, high-water-based cutting fluids, metal salts for grease thickeners and paint driers, and many others.

2 MARKET ANALYSIS

2.1 Supply and Production

As stated above, Celanese is the only producer of C_7 and C_9 fatty acids derived from alpha olefins. Competitive by-product acids are available in the United States. Emery produces pelargonic acid (C_9) as a by-product of the ozonolysis of oleic acid to dibasic azelaic acid. ATOCHEM imports heptanoic acid (C_7), from France, which is a by-product of the nylon-11 process, using castor bean oil as its raw material. The capacities, or imports to the United States in 1985, are listed in Table 1.

2.2 Markets

The end uses for heptanoic and pelargonic acids are quite diverse. Figure 1 is a pie chart showing relative end uses for both acids. Detailed descriptions of the major end uses are discussed later in this chapter.

Table 1 By-Product Acid Capacities or Imports, 1985

Producers	Acid	Millions of pounds capacity/imports
Celanese Chemical Co., Inc.	C_7 or C_9	40
Emery Industries	C_9	10–19
ATOCHEM (France)	C_7	4–5

Source: Celanese Chemical Company estimates.

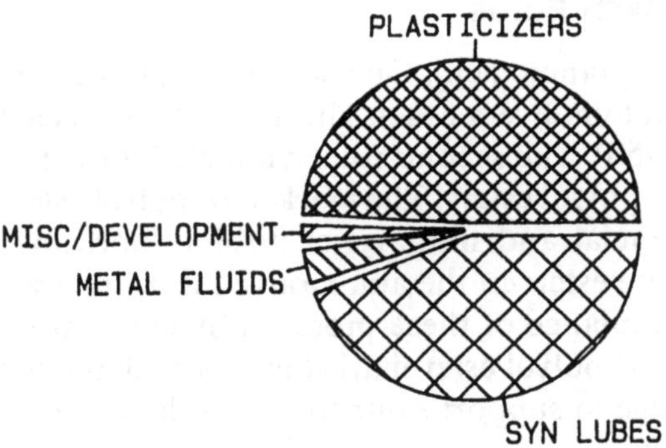

HEPTANOIC ACID

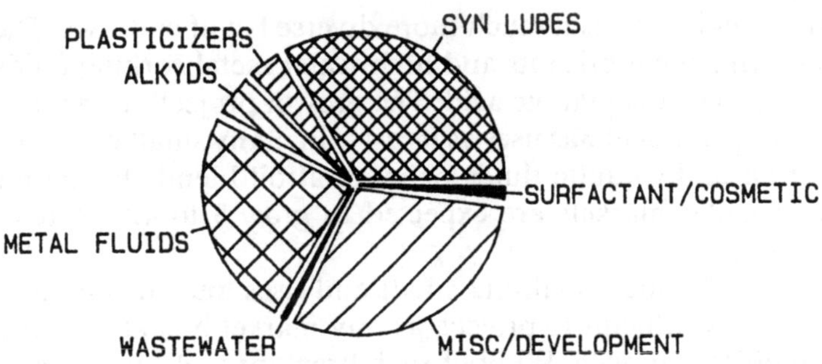

PELARGONIC ACID

Figure 1 Relative end use of heptanoic and pelargonic acid. (From Celanese Chemical Company estimates.)

2.3 Projected Market Growth Rates

There are two major uses for heptanoic acid. One is in vinyl plasticizers that are used primarily in the automotive market. This market is expected to grow 3 to 4% per year with GNP. The second is in synthetic lubricants, where heptanoic acid is used in polyol esters. The market for polyol esters is primarily for use in commercial and military jet turbine lubricants. There is a small market for these esters in the automotive lubricant area, but there has been limited acceptance of these products by automakers and the public. The growth of the polyol ester market is expected to track GNP, unless automakers change to support synthetics or unless there is an elevation of military activity.

The use of heptanoic acid in high-water metalworking fluids has grown in excess of 20% over the last several years. The amount of acid used in these products is small; therefore, a dramatic change from the traditional oil-based fluids would be required before there would be significant market impact.

Pelargonic acid (C_9) has a much more diverse list of end uses. The two major uses, synthetic lubricants and high-water metalworking fluids, are much the same as in heptanoic acid. The growth projections are similar; however, pelargonic acid has uses as its ester in many small markets, such as in water-based hydraulic fluids or mineral oil blends for traditional use. These smaller markets are expected to grow 3 to 4% greater than GNP.

Growth rates for the remaining small end uses and the development end uses are very difficult to project, as any market breakthrough could result in significant growth. Without such breakthroughs these small use areas are expected to track or fall below GNP.

3 USES AND APPLICATIONS OF SYNTHETIC FATTY ACIDS

3.1 Synthetic Lubricants

In the past decade, synthetic lubricants (syn-lubes) have been recognized as the solution to many critical service problems. These synthetic lubricants are generally composed of one or more of the following: (1) diesters of diacids such as adipic acid, (2) esters of polyols such as pentaerythritol, or (3) polyalpha olefins. One of the advantages of synthetic lubricants is the ability to tailor their properties to specific uses by proper selection of ingredients. Because of the consistent purity of synthetic fatty acids produced from alpha olefins, they are especially suitable where dependable quality and availability are required. The carboxylic acids used to manufacture the polyol esters are usually C_5–C_{10} acids, with most based on C_7, C_8, or C_9 acids. Polyol esters are discussed in more detail below.

Polyol esters are produced by the reaction between a polyol, such as trimethylolpropane or pentaerythritol, and a carboxylic acid, usually in the 5- to 10-carbon length. Both heptanoic and pelargonic acids have found use in these polyol esters. Trimethylolpropane and pentaerythritol esters of these acids are soluble in poyalphaolefins, diesters, and naural mineral oils, increasing their flexibility in the synthetic lubricant field [1].

Some of the advantages of polyol esters [2] are wide operating temperature range (−75 to 600°F), low volatility, good oxidative stability, high solvency and dispersibility, and greater seal-swell characteristics than diesters. Some typical uses of polyol esters as synthetic lubricants include jet aircraft engine and gas turbine lubricants, two-stroke engine lubricants, gear lubricants, and automotive crankcase lubricants.

The increasingly severe performance demands on lubricants by jet aircraft engines are typical of needs that polyol esters can satisfy. Polyol esters have, for the most part, replaced diesters in this application. Polyol esters offer, as well as low viscosity, greater thermal stability than diesters, owing to the stability of the neopentyl structure. Diesters are subject to thermal decomposition to an olefin and a carboxylic acid as shown in Figure 2.

In the polyol esters (see Fig. 3), the beta carbon from the alcohol used to prepare the ester does not have an attached hydrogen that can be abstracted as does the alcohol used in the diester. Polyol esters are hindered in such a way that thermal decomposition is unlikely and, therefore, the thermal stability of the molecule is increased [3,4].

3.2 Plasticizers

A plasticizer is a nonvolatile organic compound incorporated into a polymer or plastic to increase flexibility, increase internal lubricity, soften, and in other ways modify the polymer's properties. The purpose may be to improve either the workability of a plastic in a process or the properties of the product produced.

Fatty acid diesters of glycols offer properties similar to those of esters of dibasic acids [5]. Due to the relatively high volatility of some of these es-

$$R'\text{-}CH_2\text{-}CH_2\text{-}O\text{-}\overset{\overset{\displaystyle O}{\|}}{C}\text{-}R \xrightarrow[\text{heat}]{} R'\text{-}CH{=}CH_2 + HO\text{-}\overset{\overset{\displaystyle O}{\|}}{C}\text{-}R$$

Figure 2 Thermal decomposition of a diester.

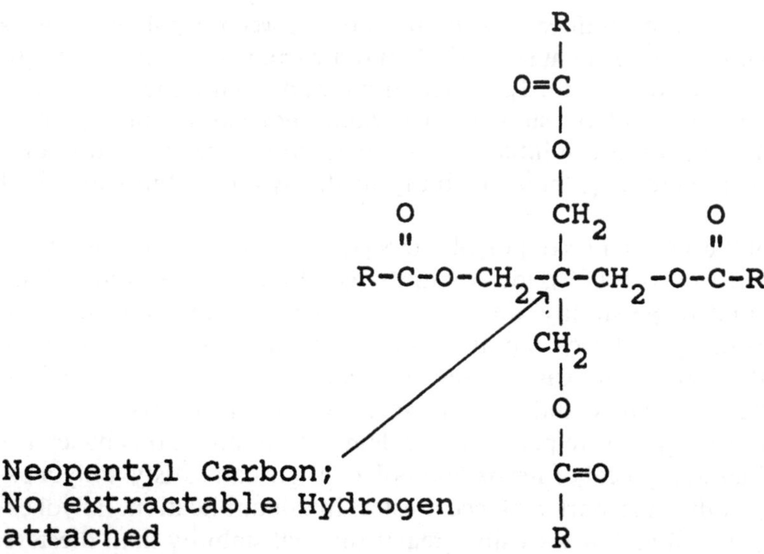

Figure 3 Polyolester molecule.

ters, especially the monoesters, they are often used as a secondary plasticizer. Esterification of either heptanoic or pelargonic acid with higher alcohols and diols yields esters that are useful in the production of synthetic rubber and vinyl plasticizers. The heptanoic and pelargonic acid esters of triethylene glycol are useful as specialty vinyl plasticizers, with excellent low-temperature properties and stability to heat and ultraviolet light. [1,6]. A glycol diheptanoate is in use as a plasticizer in poly(vinyl butyral), which has widespread use in manufacture of automobile safety glass. A glycol dipelargonate has been used in nitrile-type rubbers. Polyol esters of fatty acids in this range are frequently used as plasticizers in electrically insulating compounds.

3.3 Metal Working Fluids

Heptanoic acid's odd-carbon content differs from the even-carbon-numbered "natural" fatty acids, such as capric and caprylic acids, providing it with some unique performance characteristics (e.g., greater stability, better low temperature fluidity, and lower melting point). One developing use for heptanoic acid is in the production of lubricity additives and extenders for synthetic high-water-content metalworking fluids (HWCF). The triethanolamine (TEA) salt of heptanoic acid is particularly suited to this application, due to its (1) high degree of water stability (2) synergistic effect when used with polyglycols as cutting fluids (3) low foaming characteristics (4) and low cost [7].

One major advantage of the heptanoic acid/TEA salt over other fatty acid salts is its tolerance to hard water. Many fatty acid salts begin precipitating in water with as little as 50 ppm calcium. Heptanoic acid/TEA salt is stable in water containing over 700 ppm calcium. This property allows the use of tap water for diluting these HWCFs [7,8].

In the past, one shortcoming of synthetic HWCF compared to oil-based systems has been lubricity. Data published by Celanese Chemical Company demonstrate that the use of an amine salt of heptanoic acid helps overcome this lubricity problem [7,8]. In addition to enhancing lubricity, the amine salt of heptanoic acid functions as a low-cost extender or diluent for other cutting fluids.

3.4 Alkyd Resins

Alkyd resins are essentially oil-modified polyester resins, the oil being fatty acid triglycerides or fatty acids. They have wide application in the coatings field, either alone or blended with cross-linking agents such as melamines or urea-formaldehyde resins. An alkyd resin formulation includes a glycol or polyol or combination thereof, a diacid or its anhydride, and an oil or fatty acid. If an oil is used, the production process has two steps. The first is an alcoholysis reaction between the oil and the polyol or glycol to convert the oil's triglycerides to monoglycerides, and the second is the esterification reaction utilizing the diacid or anhydride. If a fatty acid is used, the reaction can be performed in one step, with all reactants (polyol/glycol, diacid/anhydride, fatty acid) added at the beginning. A wealth of literature is available on general alkyd resin technology [5,9–11], so we will concentrate on the uses of heptanoic and pelargonic acids in alkyds since they can be obtained synthetically. Over the years, purified cuts of fatty acids have become available for use in highly critical applications. The monobasic synthetic heptanoic and pelargonic acids are competitive and sometimes superior in these uses.

Oils and their fatty acids can be categorized as drying, semidrying, or nondrying, based on the type and degree of unsaturation. The highr the level of unsaturation, the faster the oil or alkyd containing that oil or its fatty acids will dry. Conjugated unsaturation accelerates drying even further. An extreme example of the high end of the drying scale would be tung oil, which contains 80 to 90% eleostearic acid ester with three conjugated double bonds and very little saturated acid content. Tung oil dries very quickly, even when used alone. Coconut oil is a good example of the low end of the drying scale for natural products. Coconut oil contains only about 10% unsaturated acids, with saturated acids making up the bulk. Coconut oil and coconut fatty acids (55 to 65% C_8, C_{10}, and C_{12}) are considered nondrying and are limited in use to baked-enamel-type applications. Pelargonic and heptanoic acids, being fully saturated, may be substituted for coconut oil/fatty acids in alkyd resins, where they impart desirable characteristics, for example, very light color and good color retention.

The same unsaturation that accelerates drying also promotes discoloration of the coating. During drying, after solvent evaporation, the double bonds polymerize via auto-oxidation and enable cross-linking, even at ambient temperatures. This oxidation also imparts color to the film. A typical use for an alkyd resin containing pelargonic acid as a substitute for coconut oil/acid would be high-grade baking enamel in combination with urea or melamine resins. This combination results in a premium coating products where nonyellowing whites or light colors, such as pastels, are desired. Gloss retention for these alkyds is also excellent. Appliance coatings are a good example of the type of finish that requires these properties. Alkyds based on unsaturated resins will impart plasticity to urea and melamine resins that are otherwise brittle by themselves. Coconut-type alkyds have found use in automotive finishing lacquers due to their desirable qualities, such as good durability.

3.5 Miscellaneous

There are several other uses for synthetic heptanoic and pelargonic acids that will be mentioned but not discussed in detail, including [1,6] lithium and calcium salts of both heptanoic and pelargonic acids to add desirable properties to specialty greases, surfactants, barium and cadmium salts of both acids as stabilizers for poly(vinyl chloride) resins. Heptanoic acid is used in a series of amine condensates which have been found useful in release agents, freeze-thaw stabilizers, pigment dispersing agents, corrosion inhibitors, and sludge dispersants.

4 CHEMISTRY AND PROCESS DESCRIPTION

The key to synthetic production of fatty acids from alpha olefins is the high specificity to linear aldehydes obtainable from rhodium-catalyzed hydroformylation.

$$RCH_2CH{=}CH_2 + CO + H_2 \xrightarrow[\text{heat,pressure}]{\text{Rh/PPh3}} RCH_2CH_2CH_2CHO$$

$$+ RCH_2\underset{\displaystyle CH_3}{\underset{|}{C}}HCHO + RCH{=}CHCH_2 + RCH_2CH_2CH_3 \qquad (1)$$

The aldehydes so produced are then oxidized to carboxylic acids.

$$RCH_2CH_2CH_2CHO \xrightarrow[\text{metal catalyst}]{\frac{1}{2}O_2} RCH_2CH_2CH_2COOH \quad (2)$$

The primary oxo reaction produces branched aldehydes, isomerized olefins, and alkanes as unwanted by-products. Other by-products are formed by secondary reactions involving the aldehydes (aldol condensation, Tischenko ester formation, etc.). The two steps will be discussed separately, beginning with hydroformylation.

4.1 Hydroformylation

The oxo or hydroformylation reaction was discovered by Roelen [12–15] while studying the effects of ethylene under Fischer-Tropsch reaction conditions. He found that both aldehydes and ketones were made in the reaction and coined the word "oxo" (from the German word "oxierung," meaning ketonization) to describe the process. This was an unfortunate choice since ketones are only obtained with ethylene. A better name is hydroformylation because, formally, the elements of formaldehyde are being added across the double bond. Hydroformylation and oxo are used interchangeably in this chapter.

A number of metal carbonyls (e.g., those of Co, Ni, Fe, Ir, Pt, Rh) will catalyze hydroformylation of olefins. Until recently, however, cobalt was the commercial choice. The process operates at high pressure and requires considerable catalyst reprocessing. The selectivity to linear aldehyde (including about 10% alcohol) is about 70 to 75% with a 2 to 4:1 linear-to-branched ratio [13]. By addition of trialkylphosphines, Shell workers have shown that it is possible to increase the linear-to-branched ratio to about 10:1 [15,16]. However, reaction rates are lower. The major products are alcohols instead of aldehydes, and the overall selectivity to linear product remains the same because an increased amount of olefin is hydrogenated to alkane.

The initial studies using rhodium carbonyls as hydroformylation catalysts resulted in rates 100 to 1000 times higher than cobalt carbonyl catalysts but with much lower linear-to-branched aldehyde ratios [17]. Some of the branched aldehyde arises from isomerization of the olefin, while the remainder arises from addition of the carbonyl group to the beta carbon of alpha olefins. Addition of phosphines reduces olefin isomerization [18,19] without necessarily affecting the position of addition (1- or 2-) of the carbonyl group to alpha olefins. The selectivity to linear versus branched aldehydes during rhodium-catalyzed hydroformylation of alpha olefins is determined by both the identity and concentration of the phosphine used.

Pruett and Smith (Union Carbide) demonstrated that both the rate of hydroformylation and selectivity to linear aldehyde increased as the basicity of the phosphine (or phosphite) ligand decreased [20,21]. Unruh, Hughes, and Christenson (Celanese) quantified the effect using ferrocene-based bidentate phosphine ligands [22–24]. However, most of the rhodium-phosphine-catalyzed hydroformylation studies, as well as nearly all commercial operations, have used triphenylphosphine. Therefore, the rest of this discussion will be confined to the rhodium-triphenylphosphine catalyst system.

A number of research groups, including those from Union Oil [25] and Union Carbide [20,21], and Wilkinson's group [26–39], have studied the effects of triphenylphosphine concentration on linear-to-branched aldehyde ratio and the rate of reaction. Figure 4 shows typical curves for these variables using 1-hexene as the feed. As the concentration of triphenylphosphine increases, the rate initially increases until the phosphine-to-rhodium mole ratio is 20:1 to 50:1, and decreases thereafter. The linear-to-branched aldehyde ratio increases monotonically with increasing phos-

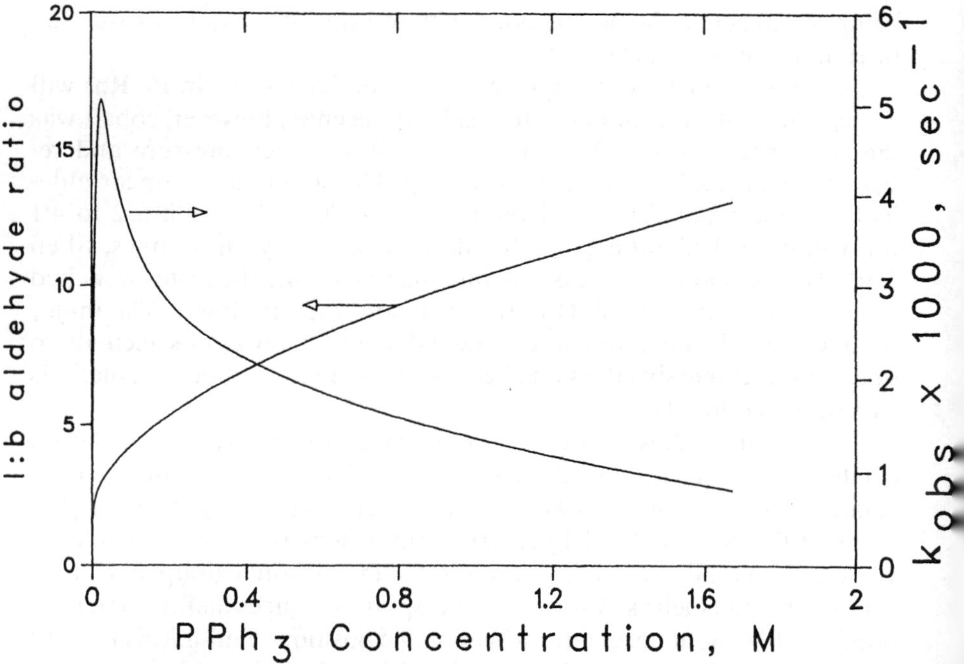

Figure 4 Effects of triphenylphosphine concentration on the linear-to-branched aldehyde ratio and rate of rhodium-catalyzed hydroformylation of 1-hexene.

phine concentration. Figure 5 shows the effects of triphenylphosphine concentration on 2-hexene, 2-methylhexanal, and heptanal efficiencies. The efficiency to heptanal increases rapidly until the triphenylphosphine reaches approximately 0.5 M. After that, only minor increases in efficiencies occur, even though the straight chain-to-branched aldehyde ratio continues to increase almost linearly. The reason is that both 2-hexene and 2-methylhexanal efficiencies decrease until about 0.3 M but, after that, increasing triphenylphospine concentration results in increasing 2-hexene at the expense of 2-methylhexanal.

The mechanism of the reaction explains these results. Rhodium-catalyzed hydroformylation has been studied extensively by Wilkinson's group [26–39]. They found the most active catalyst source to be hydridocarbonyltris(triphenylphosphino)rhodium(I), $HRhCO[P(C_6H_5)_3]_3$ [30]. However, a molecule of triphenylphosphine is presumed to dis-

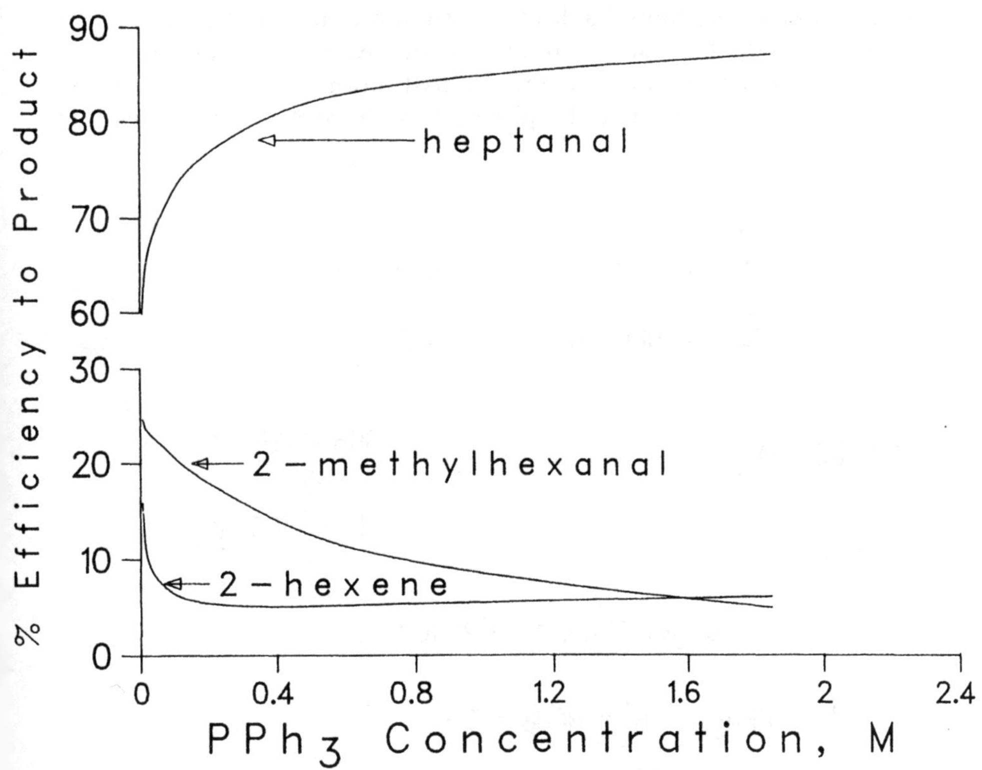

Figure 5 Effects of triphenylphosphine on selectivities to products.

sociate to form the active species [30,38]. Further dissociation could occur as follows:

$$HRhCOL_3 \underset{L}{\overset{CO}{\rightleftharpoons}} HRh(CO)_2L_2 \underset{+L}{\overset{-L}{\rightleftharpoons}} HRh(CO)_2L \quad (3)$$

where L = $P(C_6H_5)_3$.

For each active species [$HRh(CO)_2L_2$ or $HRh(CO)_2L$] there exists a catalytic cycle for hydroformylation, which is shown in Figure 6 using $HRh(CO)_2L_2$ and the formation of linear aldehyde for illustration purposes.

It has been suggested that rhodium with three phosphines bound to it (e.g., $HRhCOL_3$) can also catalyze hydroformylation [22–24,40]. It has also been suggested that CO, rather than phosphine, might dissociate prior to coordination of the olefin [22,39]. Whatever the case, the greater the number of phosphine molecules bound to rhodium, the slower the rate of reaction but the greater the efficiency to linear aldehyde.

Several other points are of interest. 2-Hexene is essentially inert in the process under conditions that lead to optimum values for the efficiency to linear aldehyde, the linear-to-branched aldehyde ratio, and the rate of reaction. At these conditions, the rate of hydroformylation of internal olefins is only 1 to 5% of that of alpha olefins. Since it is quite difficult to

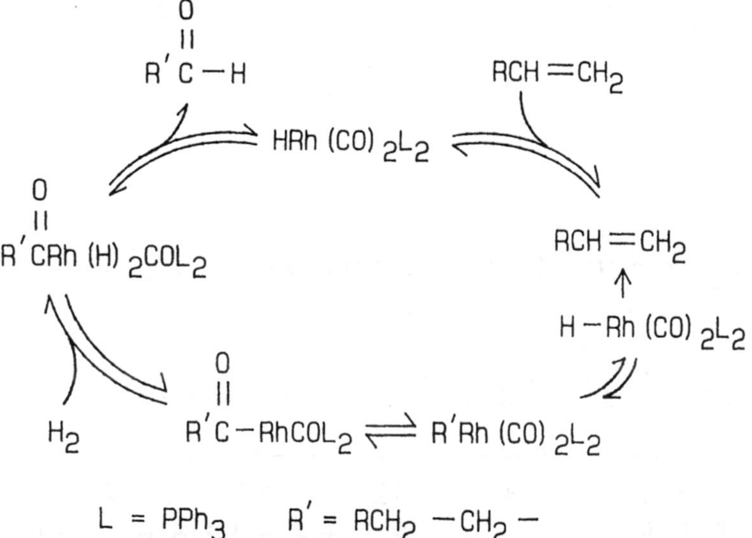

Figure 6 Hydroformylation of alpha olefins by rhodium-triphenylphosphine catalyst system. (Adapted from Ref. 28.)

separate alpha olefins from the inernal olefins, it is most economical to operate so that nearly complete conversion of the alpha olefin is obtained per pass.

Finally, there is not a unique set of conditions that leads to a particular product mix and reaction rate. Instead, there are "families" of conditions that will lead to approximately the same product mix and unit activity. Since the formation of the various active catalyst species is determined by competition between CO and triphenylphosphine for active sites, what is required is that the conditions result in approximately the same mix of active species in solution. Exactly what conditions of temperature, pressure, rhodium concentration, and triphenylphosphine concentration will be chosen will therefore be determined by the economics of individual plants.

4.2 Oxidation of Aldehydes to Acids

Conversion of the aldehydes from hydroformylation to acids is accomplished by the well-known oxidation of aldehydes to acids [41].

$$RCHO + \tfrac{1}{2}O_2 \xrightarrow{\text{catalyst}} RCOOH + \text{by-products} \tag{4}$$

While catalysts are not necessary for this process, the efficiency of the reaction is improved by their use. A number of transition metals may serve as catalyst; the carboxylic acid salts of manganese and copper are particularly effective [42]. The general mechanism of these oxidations is as follows:

Initiation

$$RCHO + \text{initiator} \rightarrow R\overset{\overset{\displaystyle O}{\|}}{C}\cdot + \text{H-initiator} \tag{5}$$

Chain Propagation

$$R\overset{\overset{\displaystyle O}{\|}}{C}\cdot + O_2 \rightarrow R\overset{\overset{\displaystyle O}{\|}}{C}O_2\cdot \tag{6}$$

$$R\overset{\overset{\displaystyle O}{\|}}{C}O_2\cdot + RCHO \longrightarrow R\overset{\overset{\displaystyle O}{\|}}{C}O2H + R\overset{\overset{\displaystyle O}{\|}}{C}\bullet \tag{7}$$

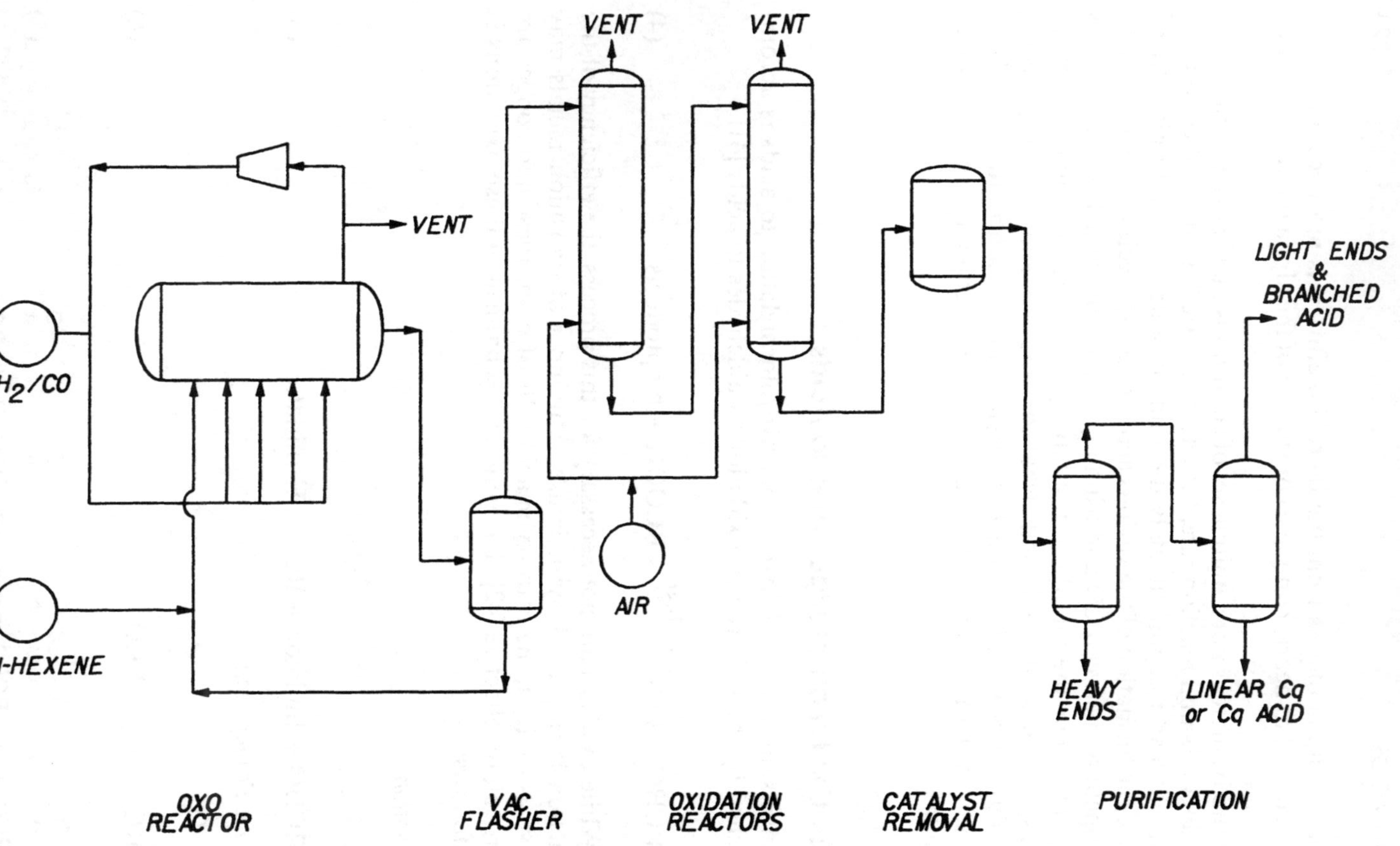

Figure 7 Hydroformylation/oxidation process for synthetic fatty acids.

Product Formation

```
 O                              O..H    OH
 ||                            /    \  /
RCO2H  +  RCHO  ---------->  RC      C   ---------->  2 RCOOH
                               \    / \
                                O-O    R
```

As in all free-radical reactions, a number of by-products are possible, including CO, CO_2, alcohols, lower-molecular-weight aldehydes, and alkanes. To minimize side reactions and still maintain high conversions, it might be appropriate to run the reaction in several stages. This allows maintenance of adequate O_2 concentration at optimum radical flux.

4.3 Process

A schematic of the Celanese fatty acids process is shown in Figure 7. This process received an Honor Award in Chemical Engineering's Kirkpatrick Award competition in 1981 [43].

Synthesis gas is generated in a partial oxidation unit and then sparged into the hydroformylation reactor, while olefin (1-hexene or 1-octene) is fed in as a liquid. High conversion per pass of olefin eliminates the need for olefin recycle, but synthesis gas recycle is required.

Reaction conditions are quite mild—150 psig and 100 to 150°C [43]. Product is removed from the reactor by taking a liquid stream to a vacuum flasher, where aldehydes, light ends, and some heavy ends go overhead. The catalyst solution is recycled to the hydroformylation reactor.

The aldehydes are sent to a two-stage oxidation reactor, where they are air oxidized to acids using a mixed transition metal catalyst at less than 100 psig. The oxygen conversion is kept high to minimize by-product formation. The process is highly efficient and the aldehyde conversion is high enough to obviate recycle.

Crude acid is conducted from the oxidation reactors through a catalyst removal step in order to remove catalyst at the ppm level [43]. This is followed by a two-column purification train. In the first column, heavy ends are removed. The overhead is sent to the finishing column, where light ends and some branched acid are removed. The result is heptanoic or nonanoic acids that have about a 20:1 linear-to-branched acid ratio [43].

REFERENCES

1. *Heptanoic Acid Product Bulletin,* Celanese Chemical Co., Dallas, Tex., 1984.
2. *Synthetic Lubricants III,* Vol. 2, Ozimek Data Corporation, Rye, N.Y., 1981.

3. *Synthetic Lubricants III,* Vol. 1, Ozimek Data Corporation, Rye, N.Y., 1981.
4. Campen, M., *A Chemical Introduction to Synthetic Automotive Engine Oils—Their Sources, Classes, Advantageous Properties, and Fuel Saving, Cost/Performance Benefits, Synthetic Automotive Engine Oils,* Society of Automotive Engineers, Warrendale, Pa., 1981.
5. Fulmer, R. W., Applications of fatty acids in protective coatings, in *Fatty Acids and Their Industrial Applications* (E. S. Pattison, ed.), Marcel Dekker, New York, 1968.
6. *Pelargonic Acid Technical Bulletin,* Celanese Chemical Co., Dallas, Tex., 1984.
7. *High Water Content Synthetic Metalworking Fluids Using Heptanoic Acid,* Technical Bulletin, Celanese Chemical Co., Dallas, Tex., 1984.
8. *Heptanoic Acid Salt as a Lubricity Additive to Commercial High Water Content Fluids for Metalworking,* Technical Bulletin, Celanese Chemical Co., Dallas, Tex., 1984.
9. *Kirk-Othmer Chemical Encyclopedia of Chemical Technology,* 3rd ed., Vol. 2, Wiley, New York, 1978, pp. 18–50.
10. Martens, C. R., Alkyd resins, in *Technology of Paints, Varnishes and Lacquers,* (C. R. Martens, ed.), R. E. Kriegler, Melbourne, Fla., 1974.
11. W. von Fischer (ed.), *Paint and Varnish Technology,* Reinhold, New York, 1948.
12. Hatch, L. F., *Higher Oxo Alcohols,* Enjay Co., New York, 1957, pp. 1–29.
13. Falbe, J., in *Propylene and Its Industrial Derivatives,* (E. G. Hancock, ed.), Wiley, New York, 1973, pp. 333–366.
14. Paulik, F. E., *Catal. Rev, 6,* 49 (1972).
15. Slaugh, L. H., and Mullineaux, R. D., *J. Organomet. Chem., 13:* 469 (1968).
16. Slaugh, L. H., and Mullineaux, R. D., U.S. patent 3,293,569 (Mar. 8, 1966), to Shell Oil Co.
17. Wakamotsu, H., *Nippon Kagaku Zasshi, 85:* 227 (1964).
18. Fell, B., Rupilius, W., and Asinger, F., *Tetrahedron Lett.,* 3261 1968.
19. Asinger, F., Fell, B., and Rupilius, W., *Ind. Eng. Chem. Prod. Res. Dev., 8:* 214 (1969).
20. Pruett, R. L., and Smith, J. A., *J. Org. Chem., 34:* 327 (1969).
21. Pruett, R. L., and Smith, J. A., U.S. patent 3,527,809 (Sept. 8, 1970), to Union Carbide.
22. Unruh, J. D., and Christenson, J. R., *J. Mol. Catal., 14:* 1 (1982).
23. Hughes, O. R., and Unruh, J. D., *J. Mol. Catal., 13:* 71 (1981).
24. Hughes, O. R., and Young, *J. Am. Chem. Soc., 103:* 6636 (1981).
25. Olivier, K. L., and Booth, F. B., *Hydrocarbon Process.,* 112 (1970).
26. Osborn, J. A., Wilkinson, G., and Young J. F., *Chem. Commun.,* 17 (1965).

27. Jardine, F. H., Osborn, J. A., Wilkinson, G., and Young, J. F., *Chem. Ind. (London),* 560 (1965).
28. Osborn, J. A., Jardine, F. H., Young, J. F., and Wilkinson, G., *J. Chem. Soc. A.,* 1711 (1966).
29. Hallman, P. S., Evans, D., Osborn, J. A., and Wilkinson, G., *Chem. Commun.,* 305 (1967).
30. Evans, D., Yagupsky, G., and Wilkinson, G., *J. Chem. Soc. A,* 2660 (1968).
31. Evans, D., Osborn, J.A., and Wilkinson, G., *J. Chem. Soc. A,* 3133 (1968).
32. Baird, M. C., Mague, J. T., Osborn, J. A., and Wilkinson, G., *J. Chem. Soc. A,* 1347 (1967).
33. Baird M. C., Nyman, C. J., and Wilkinson, G., *J. Chem. Soc. A,* 348 (1968).
34. Wilkinson, G., *Bull. Soc. Chim. Fr., 12:* 5055 (1968).
35. Brown, C. K., and Wilkinson, G., *Tetrahedron Lett., 22:* 1725 (1969).
36. Yagupsky, M., Brown, C. K., Yagupsky, G., and Wilkinson, G., *J. Chem. Soc. A,* 937 (1970).
37. Yagupsky, G., Brown, C. K., and Wilkinson, G., *J. Chem. Soc. A,* 1392 (1970).
38. Brown, C. K., and Wilkinson, G., *J. Chem. Soc. A,* 2753 (1970).
39. Wilkinson, G., West German patent 1,939,322 (Feb. 12, 1970), to Johnson, Matthey and Co.
40. Pruett, R. L., *Science, 211:* 11 (1981).
41. Soyers, L., and Seree De Roch, I., in *Comprehensive Chemical Kinetics,* Vol. 16, (C. H. Banford and F. Tipper, eds.), Elsevier, New York, 1980, pp. 89–124.
42. Scott, R. H., Thigpen, H. H., and Wood, F., U.S. patent 4,289,708 (Sept. 15, 1981), to Celanese Corp.
43. *Chem. Eng., 88*(22): 68 (1981).

CHAPTER 12

Lube Oil Additives

WILLIAM Y. LAM and EDMUND F. PEROZZI Ethyl Corporation–Petroleum Additives, St. Louis, Missouri

1 INTRODUCTION

The lubricant additive market has grown from an approximately $300 million business in 1965 (*Chemical Week,* 1965) to $1 billion business in recent times (Booser, 1981). Alpha olefins have long been important in the synthesis of lubricant oil additives. Since additives for motor and gear oils, hydraulic and industrial fluids, and greases often contain molecular functionality which is more polar than the petroleum base stocks into which they are to be dissolved, a relatively nonpolar functionality is needed to provide solubility. Alpha olefins provide a nonpolar "tail" that dissolves in the base stock while providing a carbon-carbon double-bond "head" which allows attachment of the portion of the molecule that gives the additive its special properties. The choice of specific alpha olefins employed in a certain reaction often depends on factors such as economics, availability, and functional properties derived from chain length, branching, and substitution.

A number of types of additives are used in typical lubricating oils. These include antioxidants, metal corrosion inhibitors, dispersants, detergents, pour point depressants, viscosity index improvers, rust inhibitors, antifoam agents, and extreme pressure and antiwear agents. These additives, which comprise about 20% of a typical commercial motor oil formulation, have important functions, but the properties of the base oil cannot be neglected, properties such as viscosity and stability that can limit a lubricant's usefulness (Pritzker, 1945).

A number of compilations of the lubricant additive patent literature have been made (Ramney, 1973, 1978, and 1980; Satriana, 1982). Because of the extensive nature of the literature, we will deal with only a small number of recent patents in which alpha olefin chemistry is exemplified. For general treatment of the subject of lubricant oil additives, see the references by Georgi (1950), Braithwaite (1967), and O'Brien (1984).

We have divided the chapter into sections discussing areas where alpha olefins have had the most activity. These areas are dispersants and viscosity index improvers, sulfonates, antioxidants, and antiwear and extreme pressure agents. In each section, we outline specific classes of reactions that have been utilized.

2 DISPERSANTS AND VISCOSITY INDEX IMPROVERS

Dispersants, detergents, and viscosity index improvers are three classes of lubricant oil additives in which alpha olefins are used extensively. To distinguish between dispersants and detergents, we use the definition of Crail et al. (1963): "Additives capable of dispersing cold sludge are referred to as dispersants, whilst those for handling higher temperature dirt are referred to as detergents." The "dirt" referred to can result from the high-temperature oxidation and mechanical shearing forces present in the internal combustion engine. The dispersants and viscosity index improvers are discussed here and the detergents are covered in the next section, on sulfonates.

Viscosity index (VI) improvers help maintain a desirable viscosity range of the engine oil over a fairly wide temperature range by using relatively long oil-soluble polymers. At lower temperatures, these polymers are considered to be in the form of micelle spheres, giving the oil molecules the ability to move fairly freely. At higher temperature, the polymer uncoils, enabling it to interact more freely with the base oil. This would have a viscosity-increasing effect, counteracting the natural tendency of the hot oil toward lower viscosity. Thus, the overall effect is only a moderate range of viscosities over a broad temperature range.

In modern lubricant additive technology, the dispersant and viscosity index improving characteristics are incorporated into one molecule, often referred to as a dispersant viscosity index improver. For convenience, we have grouped the recent literature on dispersant VI improvers into five general classes: dispersants from (1) reactions utilizing halogen, (2) reactions of homo- or cross-polymers, (3) reactions of oxidized polymers, (4) reactions of vinyl pyridine or vinylpyrrolidone with polymers, and (5) miscellaneous reactions.

2.1 Reactions Utilizing Halgen

Chlorination of polyisobutylene of molecular weight about 950 was achieved by heating a benzene solution of this polymer to 73°C and bubbling chlorine gas through the solution (Plonsker, et al., 1972). During the course of the chlorination, 1 mol of hydrogen chloride was produced for every mole of chlorine consumed. This gave evidence that the chlorination was a free-radical substitution at the hydrogen atoms allylic to the double bonds rather than a 1,2-addition reaction. The chlorinated polymer was treated with nitrilotrisethylamine (*1*) to give a quaternary ammonium salt such as *2*. The chlorine atom of the chloropolymer above reacted with the hydrogens on the primary amine groups of *1* causing hydrogen chloride to be produced and the polymer to bond to the nitrogen atom. The hydrogen chloride bound to the nitrogen atoms was removed simply by washing with strong aqueous base.

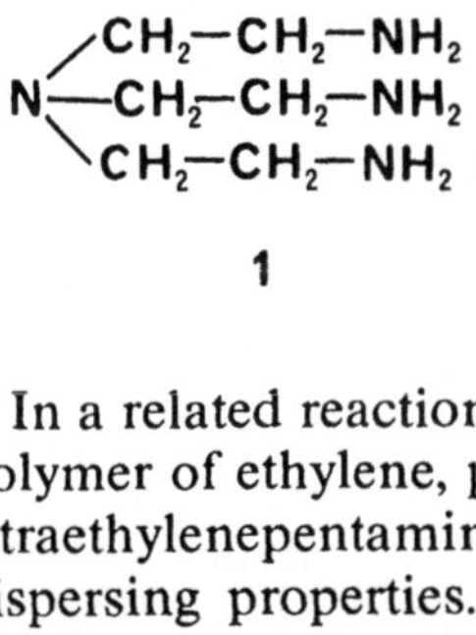

$N(CH_2-CH_2-NH_2)_3$

1

$N\!\left(CH_2-CH_2-\overset{+}{N}H_2\,Cl^- -CH(R)-C(CH_3)=CH_2\right)_3$

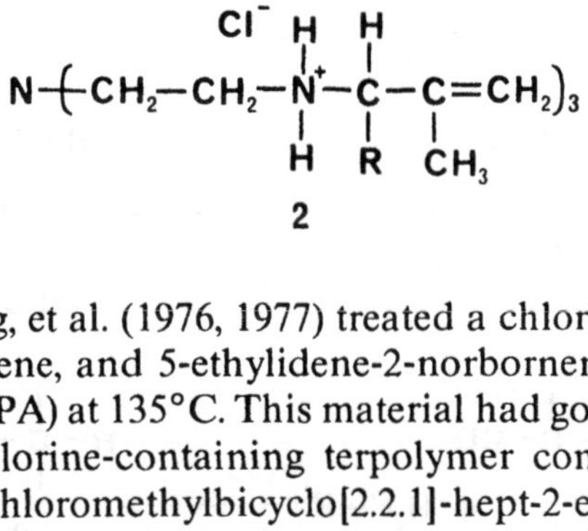

2

In a related reaction, Song, et al. (1976, 1977) treated a chlorinated terpolymer of ethylene, propylene, and 5-ethylidene-2-norbornene (*3*) with tetraethylenepentamine (TEPA) at 135°C. This material had good sludge-dispersing properties. A chlorine-containing terpolymer comprised of ethylene, propylene, and 5-chloromethylbicyclo[2.2.1]-hept-2-ene (*4*) was synthesized with a Ziegler-Natta catalyst of $VOCl_3$ and $Et_3Al_2Cl_3$. Treatment of this chloropolymer with TEPA gave a product with no dispersing ability, demonstrating the importance of polymer type.

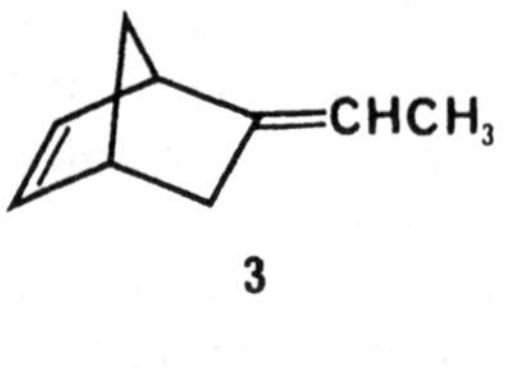

=CHCH$_3$

3

—CH$_2$Cl

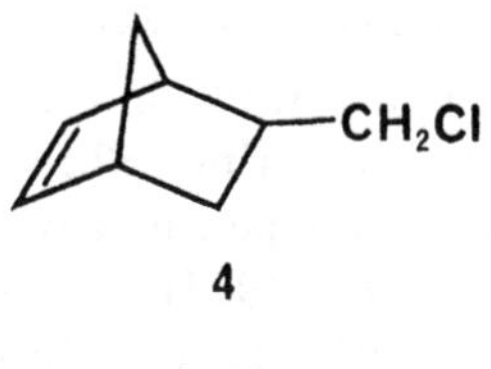

4

Chandler (1974) reported the reaction of a chlorinated polybutene of 950 average molecular weight with acrylic, methacrylic, or crotonic acids at 232°C. The acrylic acid derivative was tested in a cyclic temperature sludge test in which a Ford 6-cylinder engine is alternately cycled at prescribed temperatures for periods of time to simulate stop-and-go driving. The additive gave a rating of 9.85 after 84 hr (10 indicates clean parts), while a control run using a prior art dispersant gave a rating of 5.8.

Lee (1976b) described the reaction of a polyolefin in the presence of acetonitrile and chlorine with an iodine catalyst. The product (*5*) was then reacted with a primary amine to complete the process, giving imidazoline (*6*). A butene or propene polymer of molecular weight about 2000 has been used in this reaction as sequenced below.

$$R-C=C- \;+\; Cl_2 \xrightarrow[I_2]{\text{dark}} R-\overset{Cl^{+}}{C\!-\!C}- \;+\; Cl^- \qquad \textbf{Step 1}$$

$$R-\overset{Cl^{+}}{C\!-\!C}- \;+\; CH_3CN \longrightarrow R-C(-N{=}C^{+}-CH_3)-C-Cl \qquad \textbf{Step 2}$$

$$R-C(-N{=}C^{+}-CH_3)-C-Cl \;+\; Cl^- \longrightarrow R-C(-N{=}C(Cl)-CH_3)-C-Cl \;\; (\mathbf{5}) \qquad \textbf{Step 3}$$

$$R-C(-N{=}C(Cl)-CH_3)-C-Cl \;+\; H_2N-R' \longrightarrow \underset{\mathbf{6}}{R-C-C(-N-R')\text{ ring: } N{=}C(CH_3)} \;+\; 2HCl \qquad \textbf{Step 4}$$

Other materials have been reacted with chloropolymer (*5*). These include piperazine or an aminoalkyl-substituted piperazine (Lee, 1976a) and a primary aliphatic amine followed by maleic anhydride and then an aliphatic amine with at least one primary amino group per maleic anhydride function (Lee, 1975, 1977). Of the additives derived from the chloropolymer (*5*), the most effective (Table 1) in the spot dispersancy test (SDT) was the product of *5* and 1,4-bis(aminopropyl)piperazine (*7*). In this test a measured amount of the additive under test is mixed with a measured

Table 1 Spot Dispersancy and Panel Coker Test Results for Reaction Product of Chloropolymer *5* with Various Amines

Sample, product of *5* and the following compounds	Spot dispersancy test[a]	Panel coker test[b]	Ref.
None (control sample not reacted with *5*)[c]	46	N.D.	Lee (1977)
Tetraethylene pentamine[c]	73.5	6	Lee (1976b)
1,4-Bis(aminopropyl) piperazine[c]	95.5	10	Lee (1976a)
Butylamine, then maleic anhydride, then tetraethylene pentamine[d]	87.5	N.D.	Lee (1975)
Butylamine, then maleic anhydride, then 1,4-bis(aminopropyl) piperazine[d]	84	N.D.	Lee (1975)

[a] Perfect is 100; approximate rating: 90, excellent; 85, very good; 80, pass.
[b] Perfect is 10; N.D., not determined.
[c] Samples were formulated by adding 2% of the additive to an oil formulation designated MS MIL B plus 1% of calcium alkylbenzene sulfonate and 1% calcium alkyl phenate.
[d] Samples were formulated using 4.5% of the additive together with 2% overbased magnesium sulfonate, 1.1% of a zinc dialkyl dithiophosphate, and 0.5% of a pour point depressant in a base oil blend.

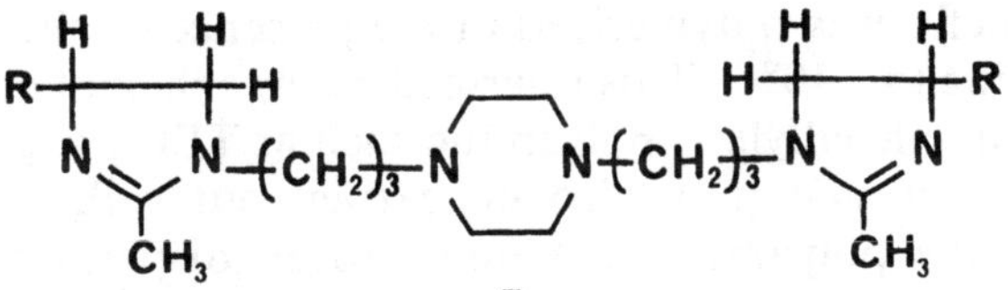

H H H H
R—+——+—H H—+——+—R
N N—$(CH_2)_3$—N N—$(CH_2)_3$—N N
CH_3 CH_3

7

amount of used crankcase oil which was used in a Lincoln sequence V engine test for 394 hr (twice the normal test time). This mixture is heated and stirred at 150°C for 16 hr and a small sample is placed on blotting paper. A control is made from a sample of the Lincoln sequence engine oil, which is heated at 150°C for 16 hr and deposited on blotting paper. The deposits on the blotter paper are measured to obtain the average diameter of the outer oil ring (D_o) and the average diameter of the inner sludge ring (D_a). The ratio of D_a/D_o is a measure of the dispersant property of the additive, with a value of 1.00 (expressed as 100%) being complete dispersancy. See Table 1 for a comparison of SDT results.

Alcoholates have been reacted with polymers derived from ethylene and alpha olefins. For example, a terpolymer of 49% ethylene, 41.3% propylene, and 9.7% 5- ethylidene-2-norbornene (*3*) was chlorinated at 60°C

(Gardiner, 1975). To a solution of this polymer was added the sodium salt of pentaerythritol (*8*) with heating to 180°C to give the final product after removal of sodium chloride.

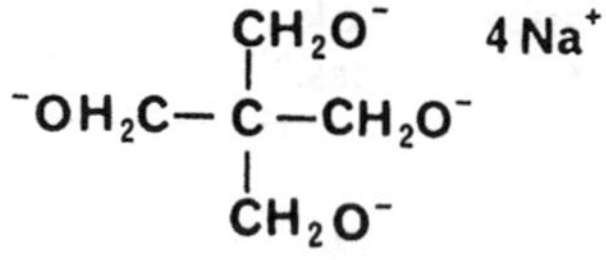

$$\begin{array}{c} \quad CH_2O^- \quad 4\,Na^+ \\ | \\ {}^-OH_2C-C-CH_2O^- \\ | \\ CH_2O^- \end{array}$$

8

Chlorinated polymer is optional in the case of the reaction with maleic anhydride reported by Kiovsky (1979). A solution of ethylene-propylene polymer of 202,000 weight-average molecular weight was heated to 50°C in the presence of chlorine. This product was next heated with maleic anhydride (180 to 200°C) followed by TEPA at 160 to 190°C to give a product that had a SDT result of 67% for a 2% solution of the product in a base oil. This reaction was repeated without the chlorine. A higher temperature of 225°C was used in the reaction with maleic anhydride. This product gave a comparable spot dispersancy test result of 64% for a 2% solution in base oil.

Chlorine can also be introduced together with the maleic anhydride, as demonstrated by Hayashi (1984a,b). A mixture of isobutylene and a commercial C_{16}–C_{18} alpha olefin was polymerized in the presence of aluminum choride while cooling to −10°C. This intermediate was then treated with a commercially available ethylene polyamine such as TEPA to give the final product. The last step may also be carried out with pentaerythritol in place of the polyamine if a nonnitrogen dispersant is desired.

Chlorinated alpha olefins have been used as corrosion inhibitors (Rothert, 1977) in automatic transmission fluids. For example, an alpha olefin of about 20 to 38 carbons was isomerized and chlorinated to 30 to 50% chlorine. When 0.05 to 0.5% of this material was present in the transmission fluid, corrosion of the copper brazing alloy present in the transmission cooling system was prevented or retarded.

2.2 Reactions of Single or Mixed Olefin Polymers

Holler and Youngman (1973) described the addition polymerization of alpha olefins containing a polar group. The materials were produced using a Ziegler-type catalyst consisting of a mixture of $TiCl_3$ and Et_2AlCl (with $EtAlCl_2$ added in some cases). Examples of the polymerization conditions of the polar-substituted alpha olefins are shown in Table 2. Although reaction times were long, generally yields were high. DeVries

Table 2 Homopolymerization of Polar Olefins

		Catalyst (Mmol)		Polymerization conditions		
Olefin	Mmoles	$TiCl_3$	Metal alkyl	Time (days)	Temp. (°C)	Polymer yield (wt %)
$CH_2{=}CH(CH_2)_9 \cdot N(n{-}C_4H_9)_2$	25	0.5	4.0 Et_2AlCl, 25 $EtAlCl_2$	20	20	100
$CH_2{=}CH(CH_2)_8 \cdot P(C_6H_5)_2$	24	4.0	24 Et_2AlCl	7	50	74
$CH_2{=}CH(CH_2)_5 \cdot OC_6H_5$	50	0.5	2.0 Et_2AlCl	14	50	94
$CH_2{=}CH{-}CH_2(p{-}C_6H_4)OCH_3$	50	1.0	8.0 Et_2AlCl	11	50	15
$CH_2CH(CH_2)_8CO_2 \cdot n{-}C_4H_9$	25	1.0	4.0 Et_2AlCl, 25 $EtAlCl_2$	14	50	50

Source: Holler and Youngman (1973).

and DeJovine (1979) have shown that the reaction of an ethylene, propylene, 1,4-hexadiene terpolymer, and *N,N*-dimethylaminoethyl methacrylate at 80 to 100°C produced a grafted polymer with dispersant properties. For instance, this material incorporated into an oil blend has given a sequence VC test result of 8.7 overall sludge rating. The minimum rating for a pass on this cyclic engine test is 8.5.

2.3 Reactions Employing Oxidized Polymers

For certain dispersants it has been desirable to oxidize the product to form a product with reduced molecular weight. Engel and Gardiner (1978b) reported the mechanical-oxidative breakdown of a polymer made by grafting a polyamine such as TEPA onto an ethylene-propylene copolymer. The reaction was carried out at 170 to 185°C under a blanket of air for 3 hr. The extent of reaction is expressed as the thickening efficiency (TE), which is defined as the ratio of the weight percent of a polyisobutylene (sold by Exxon Chemical Company as Paratone N), required to thicken a solvent-extracted neutral oil [viscosity of 150 to 155 Saybolt Universal Seconds (SUS) at 37.8°C] to a viscosity of 12.4 cSt at 98.9°C to the weight percent of a test copolymer required to thicken the oil to the same viscosity at that temperature:

$$\text{TE} = \frac{\begin{array}{l}\text{weight percent of Paratone N needed to}\\ \text{thicken a solvent extracted neutral}\\ \text{oil to 12.4 cSt at 98.9°C}\end{array}}{\begin{array}{l}\text{weight percent of test copolymer needed}\\ \text{to thicken the same neutral oil to}\\ \text{12.4 cSt at 98.9°C}\end{array}}$$

In an example in which an ethylene-propylene copolymer grafted with diethylenetriamine, the TE before mechanical-oxidative mastication was 2.86 and decreased to 1.0 after 3.25 hr and 160 to 186°C when air was introduced. Chludzinski et al. (1978) also reported a reaction of a terpolymer of ethylene, propylene, and 5-ethylidene-2-norbornene with air at 170°C, and West and Culbertson (1977) described a reaction of ethylene-propylene copolymer in the presence of soluble benzene sulfonic acids or their salts.

The compounds prepared by Chludzinski et al. (1978) were evaluated by the sludge inhibition bench (SIB) test. The test is described as follows. A used oil sample from a fleet of taxicabs was centrifuged and the sludge-free supernatant oil containing dissolved sludge precursor served as the test medium. Either 0.4 to 0.5 wt % of dispersant was dissolved into the used, centrifuged oil and added to preweighed centrifuge tubes. After storing for 16 hr at 138°C along with control tubes containing no dispersant, the tubes were centrifuged and the supernatant oil discarded. After pentane washing, a final centrifugation, and drying, the tubes were reweighed. The effectiveness of the dispersant was found by the ratio of the weight of sludge in the tubes containing dispersants to that in the tubes containing no dispersants. A value of 100% would indicate no dispersancy, while lower values indicate dispersancy.

The importance of the proper polyamine in designing dispersants is evident in the results of Chludzinski et al. (1978). According to the SIB test the oxidized terpolymer of ethylene, propylene, and 5-ethylidene-2-norbornene, when treated with ethylenediamine, had a dispersancy of 55% of the blank tube. When diethylenetriamine (DETA) was grafted onto the polymer, the dispersancy increased to 47%. The best improvement (to 39%) was observed using triethylenetetramine (TETA). Even better dispersants were produced when TEPA was grafted onto similar oxidized polymers. For example, two different TEPA- grafted terpolymers gave SIB dispersancies of 6% and 25%. When comparing these results with those above, one must take into account the use of different polymers in the preparations, which may account for some of the differences.

2.4 Reactions Using Copolymers Vinyl Pyridine and Pyrrolidone

Ethylene-propylene copolymer at 2.0% concentration will not disperse 0.4% of asphaltenes at 150°C (Stambaugh and Galluccio, 1979). However, the reaction of a copolymer of 60% ethylene/40% propylene and 2-vinylpyridine (in the presence of *t*-butylperbenzoate catalyst) produced an effective dispersant. When 0.0625% of this graft copolymer is used, all of the asphaltenes were dispersed at 150°C.

Copolymers of C_{10}–C_{20} alpha olefins and *N*-vinyl-2- pyrrolidone (*9*) or *N*-vinyl-3-morpholinone (*10*) have been prepared using di-*t*-butyl peroxide as a radical initiator (General Aniline and Film Corporation, 1968). These polymers are effective dispersants when present in oils at concentrations from about 0.03 to 5 wt %.

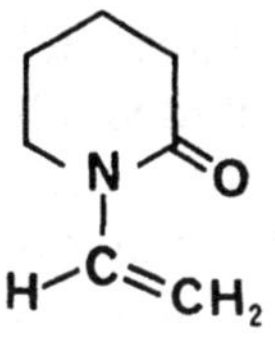

9 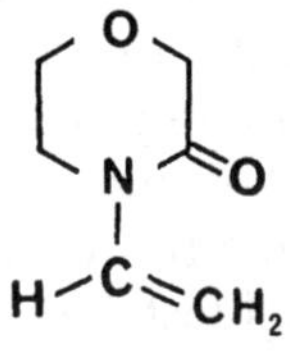 10

2.5 Miscellaneous Reactions

A number of different types of reactions that produce dispersant VI improvers have been investigated. Hayashi (1982) has described the *ene* reaction of maleic anhydride with a terpolymer of ethylene, propylene, and 1,4-hexadiene at 240°C to give the intermediate adduct *11*. This was then heated at 180°C with an amine mixture such as 95% *N*-(3-aminopropyl) morpholine and 5% ethylene polyamine with three to seven amino groups per molecule.

ENE Reaction

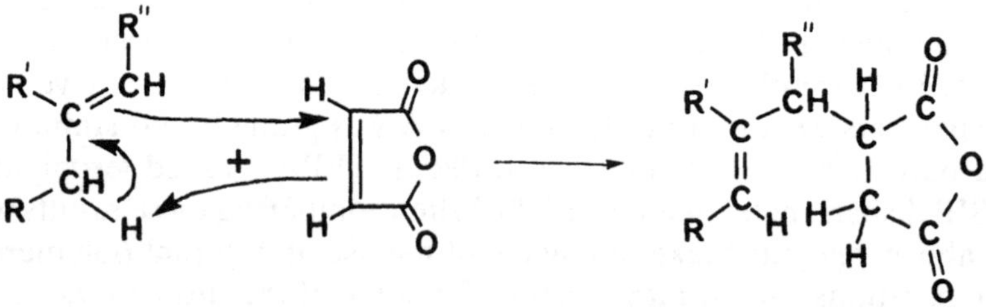

11

The reaction of amines with chloropolymers was discussed earlier. Engel and Gardiner (1978a) have found that amines can be grafted directly onto polymers. For example, an ethylene-propylene copolymer containing 57 wt % ethylene was placed into a Bramley Beken Blade Mixer, which served as a mechanical masticator. A commercial grade of TEPA was added at 177 to 202°C with mastication and heating continued for 3.5 hr to produce the dispersant.

A different type of amine was used in the synthesis by Papay and O'Brien (1985). They reported the reaction of the terpolymer of ethylene, propylene, and 1,4-hexadiene with maleic anhydride in the presence of di-*t*-butyl peroxide at 180°C. This product was then treated with 2-ethylhexyl alcohol to esterify the succinic acid groups. Finally, this esterified product was treated with *N*-[3-(dodecyloxypentadecyloxy)propyl]-1,3-propane diamine (Jet Amine DE 12/15 from Jetco Chemicals, Inc.) at 160°C. The additive was tested in a bench dispersancy test and was found to be effective at dispersing sludge a concentrations as low as 0.125%. The bench dispersancy test was conducted by mixing 1 mL of a prepared sludge solution into 10 mL of 100 neutral oil containing test additives in different concentrations. After standing 16 hr the tubes were rated visually as to levels of precipitate or haze, with the lightest haze or clear tubes showing the best dispersancy.

Another approach for introducing nitrogen was outlined in the work of

DeVries (1977). An olefin or polyisobutylene was treated with methanol and hydrogen cyanide in the presence of hydrogen fluoride to give a formamidate ester (*12*). Treatment of *12* with diethylamine gave an amidine (*13*), which can be used as a dispersant and also as an oil-soluble source of base for neutralization of acids formed in the internal combustion engine.

$$R-\overset{H}{\overset{|}{C}}=CH_2 + CH_3OH + HCN \xrightarrow{HF} R-\overset{H}{\overset{|}{C}}(-CH_3)-N=\underset{H}{\underset{|}{C}}-OCH_3$$

12

$$\mathbf{12} + HN(C_2H_5)_2 \longrightarrow R-\overset{H}{\overset{|}{C}}(-CH_3)-N=\underset{H}{\underset{|}{C}}-N(C_2H_5)_2$$

13

A terpolymer of monoalkene, methyl oleate, and 1,3-butadiene was converted to a dispersant by Malec (1978). The monoalkene contains 19.5 wt % paraffin and 78.6 wt % mono-olefin in the range C_{10}–C_{26} with a large contribution of C_{12}–C_{18}. The mono-olefins were 12.6 mol % linear alpha olefins, 35.7 mol % branched-chain vinylidine olefin, and 51.7 mol % internal olefin. The terpolymer was reacted with TEPA at 243 to 250°C and then coupled with adipic acid at 200 to 210°C. In a low-temperature dispersancy test in which a gasoline engine was operated on a controlled cycle, an oil formulated with this additive gave a rating of 9 (10 = clean).

Brois (1978) has investigated the reaction of an olefin with dimeric thionophosphine sulfides. For example, 2-methyl-1-tridecene was reacted with methylthiophosphine sulfide (*14*) to give product (*15*). Related products were obtained using polyisobutylene as the olefin and *p*-anisylthionophosphine sulfide in place of *14*. These polymeric products were then treated with polyamines to give effective dispersants which also have an antiwear capability, presumably due to phosphorus.

3 SULFONATE DETERGENTS

Modern lubricant detergents are generally derived from metal salts of organic acids that belong to the following classes of compounds: sul-

$CH_3(CH_2)_8CH_2CH_2-C(CH_3)=CH_2$ + $[H_3C(S=)P(-S-)_2P(=S)CH_3]$ (**14**) $\xrightarrow{\Delta}$ $CH_3(CH_2)_8CH_2(H)C=C[CH_2-P(=S)(CH_3)]_2S$ (**15**) + H_2S

fonates, phenates, phosphonates, thiophosphonates, and salicylates. Among these, sulfonates are by far the most commonly manufactured.

The two main olefinic sources for sulfonate synthesis are the oligomers of lower 1-alkenes and alkylated benzenes prepared by Friedel-Crafts reactions of benzene with polyalkenes or higher linear olefins. For example, polyisobutylene with an average molecular weight of 950 was used by Bakker (1979). A'Court et al. (1971) reported the chlorination of the polyolefins and subsequent dehydrochlorination of the adduct at 180°C, giving a low chlorine (<1%), olefinic mixture with an average molecular weight of 1175, from which sulfonates with better performances were prepared. Segessemann (1967) prepared sulfonates from "bodied" olefins, which were C_{15}–C_{20} alpha olefins aerated at 115°C for 20 hr. Oldham (1969) described the treatment of propylene tetramer with silica/alumina at 120 to 150°C and the alkylation of benzene with the treated olefins using anhydrous hydrogen fluoride catalyst. Kerfoot and Krehbiel (1969) alkylated diphenylalkanes with 1-tetradecene using anhydrous aluminum trichloride or hydrogen chloride catalyst. More recently, Osselet and Tirtiaux (1980) reported the alkylation of benzene with a mixture of propylene oligomer and C_{24} linear olefin followed by sulfonation of the alkylated benzene.

Sulfonation of the olefins may be effected by sulfurous acid or its alkali metal salts (Segessemann, 1967), chlorosulfonic acid (A'Court et al., 1971; Bakker, 1979; DeVries, 1979), and sulfur trioxide (Osselet and Tirtiaux, 1980). The resulting sulfonic acids are neutralized with a stoichiometric amount of metal (usually calcium, barium, or magnesium) oxides, or hydroxides to give the neutral or normal sulfonates (A'Court et al., 1971). Neutral calcium polyisobutenyl sulfonates were also prepared from the corresponding sodium salt by a continuous metathesis process (DeVries, 1978, 1979; Bakker, 1979).

Sulfonic acid or its neutral salt may also be treated with a large excess of the metal base and carbon dioxide (carbonation) to form an overbased or basic sulfonate. The carbonation converts the metal base to colloidally dispersed metal carbonate. The amount of overbasing is usually expressed in terms of the total base number (TBN), the number of milligrams of potassium hydroxide equivalent to the amount of acid required to neutralize the basic or alkaline constituents per gram of the sulfonate composition. Overbased calcium sulfonates with 300 TBN or greater have been prepared from various olefinic sources (DeClippeleir and Vanderlinden, 1978; Bakker, 1979; Osselet and Tirtiaux, 1980). Sabol (1979) reported the preparation of overbased magnesium sulfonates with TBN over 400. In addition to reducing deposits in high-temperature internal combustion engines, overbased detergents are also capable of neutralizing acidic and corrosive contaminants and suspending debris formed in oxidized lubricating oils. Although sulfonates are useful, indeed essential lubricating components, especially in engine oils, they perform best when in the presence of antioxidants.

4 ANTIOXIDANTS

An important function of lubricant additives is to provide protection against oxidation of lubricating oils under the conditions of high temperature and mechanical shearing forces encountered in an engine. In general, the antioxidants employed in modern lubricant formulations do not make direct use of alpha olefins to a large extent. However, these olefins bear an incidental relationship to antioxidants in, for example, the formation of alcohols used to make zinc dialkyldithiophosphates. See the references cited in Section 1 for the different antioxidant types. We will examine a few examples of antioxidants in which alpha olefins were used.

Lee and Richardson (1980) investigated the reaction of thio-oxamide $[H_2NC(S)C(S)NH_2]$ with C_{12}–C_{14} alpha olefins and capryl aldehyde. The reaction was carried out at 1 to 9°C in the presence of boron trifluoride etherate prior to warming to room temperature. Oils containing this additive were oxidized by blowing air through a sample at 171°C and catalyzed with 5% of a Ford VC drain oil. Samples were taken periodically and the viscosity was measured. The time to reach a fourfold increase in viscosity was 33 hr for a commercial hindered phenol (with a 30- to 40-hr minimum to pass) whereas the tested antioxidant required 77 to 96.5 hr to achieve the same viscosity increase, indicating an improved performance.

A method of stabilizing base oils to light and thermal degradation is disclosed by Yan and Bridger (1980). For example, a Kuwait oil was stirred with propylene tetramer at 125°C for 23 hr in contact with attapulgus clay as a catalyst. In an oxidation test in which oxygen was bubbled through a sample at 175°C for 20 hr, the reference oil produced 0.23 g of sludge, whereas the oil treated with propylene tetramer produced only 0.012 g of sludge.

Aromatic amines represent an important class of lubricant antioxidants. Wheeler (1976) has disclosed the reaction of diisobutylene (2,4,4-trimethyl-1-pentene) and diphenylamine using aluminum chloride catalyst to get 4-(1,1,3,3- tetramethylbutyl)diphenylamine (*16*). This product was heated with triphenylmethyl chloride in acetic acid together with concentrated hydrochloric acid to give 4-(1,1,3,3-tetramethylbutyl)-4′-triphenylmethyldiphenylamine (*17*). A 2% solution of *17* in a synthetic lubricant gave an improvement in viscosity increase over a control after bubbling air for 72 hr in the presence of copper and iron washers (40.6 SUS versus 43.2 SUS for the control).

$$H_2C{=}C(CH_3){-}CH_2{-}C(CH_3)_2{-}CH_3 + C_6H_5{-}NH{-}C_6H_5 \xrightarrow{AlCl_3} H_3C{-}C(CH_3)_2{-}CH_2{-}C(CH_3)_2{-}C_6H_4{-}NH{-}C_6H_5$$

16

$$\mathbf{16} + (C_6H_5)_3CCl \xrightarrow{H^+} H_3C{-}C(CH_3)_2{-}CH_2{-}C(CH_3)_2{-}C_6H_4{-}NH{-}C_6H_4{-}C(C_6H_5)_3$$

17

5 EXTREME PRESSURE/ANTIWEAR AGENTS

Extreme pressure (EP)/antiwear agents are load-carrying lubricant additives that minimize metal-to-metal contact of moving machine parts that are under boundary lubrication conditions. They function by forming a protective film on the metal surfaces by either a chemical or a physical adsorption mechanism.

EP/antiwear additives are generally organic or organometallic compounds containing one or more of the elements, such as sulfur, phosphorus, or halogen (usually chlorine). Olefinic substrates are commonly reacted with inorganic reagents to make this type of lubricant component. Examples in the recent literature that involve alpha olefins in the preparations of EP/antiwear additives are: (1) reactions utilizing sulfur halides, (2) reactions utilizing elemental sulfur, and (3) reactions with phosphorus reagents.

5.1 Reactions Utilizing Sulfur Halides

Sulfur monochloride (S_2Cl_2) and dichloride (SCl_2) are the most common sulfur halides used to react with alpha olefins. The resulting sulfochlorinated product usually has a high chlorine content and has been reported to be useful as an antiseize and an antiwear additive for cutting fluids (Veretenova et al., 1974).

The reaction products of sulfur chlorides with olefins are often treated with various reagents to produce compounds with low chlorine content. For example, Caspari (1978) reported the reaction of isobutylene with an approximately equimolar amount of sulfur monochloride. The resulting adduct was treated with the monosodium salt of 2,5-dimercapto-1,3,4-thiadiazole (*18*) in a methanolic solution at reflux. The product, containing 55.5 wt % sulfur, was claimed to be an effective EP agent and copper and lead corrosion inhibitor.

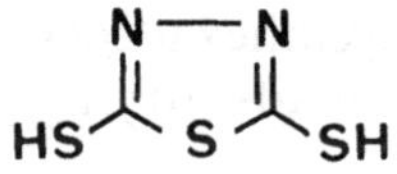

18

Sulfochlorinated olefins can be used to produce organic polysulfides as EP additives for gear oils by reaction with alkali metal sulfides. Myers (1969) described the preparation of sulfurized isobutylene by reacting S_2Cl_2 with isobutylene to form an adduct followed by treatment with sodium sulfide and caustic. A similar product, claimed to have better EP performance, was prepared by Papay and O'Brien (1980) by reacting S_2Cl_2 with isobutylene in the presence of a promoter amount of methanol and treating the resulting adduct with sulfur and sodium sulfide in an aqueous alkanol at a ratio of 0.1 to 0.4 g atom of sulfur per gram mole of sodium sulfide. Braid (1980b) reported that the presence of a catalytic amount of iodine in the sulfur halide/isobutylene reaction increased the olefin-to-sulfur halide ratio and subsequently the yield of the sulfurized product. Bolle and Dabir (1981) prepared EP additives from mixtures of isobutylene and diisobutylene by sulfochlorinating the olefins with S_2Cl_2 and reacting the adducts with sodium sulfide in aqueous methanol.

This type of sulfurized olefin may be further treated with other chemicals to form products with better or improved lubricating properties. Horodysky (1980a,b), for instance, treated sulfurized isobutylene with zinc oxide and 2-propanol at 80 to 85°C to reduce its copper corrosivity, and with dimethyl hydrogen phosphite, $(CH_3O)_2P(O)H$, in the presence of 2,2-azobisisobutyronitrile (*19*) catalyst, to give a product having excellent antioxidant and antiwear properties. A lubricant additive having good copper corrosivity, antioxidant, and antiwear properties was prepared by Horodysky and Braid (1979) by reacting sulfurized isobutylene with *O,O*-dialkylphosphodithioic acid (*20*) at 85 to 90°C with a slow sparge of hy-

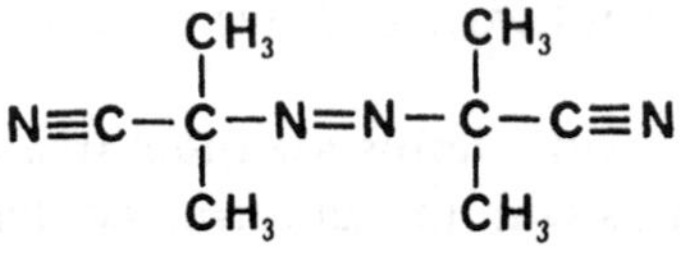

$$N\equiv C-C(CH_3)_2-N=N-C(CH_3)_2-C\equiv N$$

19

$$(R_1O)(R_2O)P(=S)-SH$$

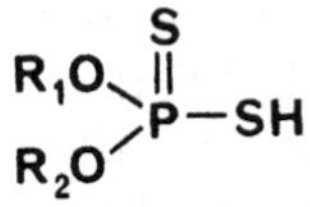

20

drogen sulfide. This reaction product was further treated with molybdenum oxide (Horodysky, 1979) or with an unsaturated compound such as vinyl acetate at 75 to 85°C (Horodysky, 1980c) to give a product having increased antioxidant and antiwear properties. Braid (1980a) claimed that the reaction product of 29 parts of sulfurized isobutylene with 1 part of 1,1,5,5,9,9,13,13-octamethyl-3,4,7,8,11,12,15,16-octathiocyclohexadecane (*21*) in a fully formulated gear oil, passed the federal load-carrying test (Method 6504).

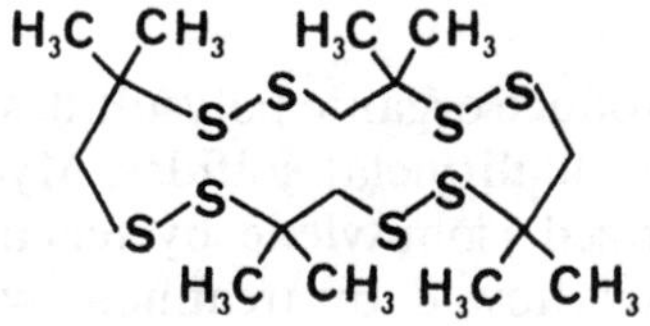

H$_3$C CH$_3$ H$_3$C CH$_3$ S S S S S S S S H$_3$C CH$_3$ H$_3$C CH$_3$

21

5.2 Reactions Utilizing Elemental Sulfur

The reaction of sulfur with an olefinic substrate is another synthetic method to produce chlorine-free organic sulfides. The olefinic substrate is often a mixture of alpha olefins and fatty acid esters or triglycerides. Sanson and Hartman (1978) described the sulfurization reaction of 85:15 (v/v) mixture of lard oil and C_{15}–C_{20} alpha olefins at 190°C, giving an 8.34% sulfur product which can substitute for sulfurized sperm oil. Lee and Boslett (1979) reported the preparation of EP additives containing 11 wt % sulfur by sulfurizing a 50–15:50–85 (w/w) mixture of C_{15}–C_{18} alpha olefins and cottonseed oil at 160 to 180°C. More recently, Recchuite (1984) described a process of sulfurizing an 85:15 (v/v) mixture of lard oil and triisobutylene, producing an 8.6 wt % sulfur material having good EP, antiwear, and antiweld properties. A similar type of additive was prepared by reacting 65.2 parts of C_{20}–C_{28} alpha olefins and 34.8 parts of diethyl fumarate with 2 mol of sulfur at 200°C (Yamada et al., 1975). An antiwear additive containing low levels of chlorine (1 to 3%) was made by reacting C_9–C_{10} or C_{15}–C_{18} alpha olefins with 0.8 mol of sulfur and 0.1 mol of S_2Cl_2 at 160°C. It was also claimed to have good antioxidant activity (Hotten, 1979, 1982).

Lower alkenes can be sulfurized by reaction with sulfur and hydrogen sulfide at elevated temperature and pressure. Davis and Holden (1978) described the process for preparing a sulfurized lubricant additive containing 42.5 wt % sulfur by reacting a 1:1:0.5 (mol) mixture of isobutene, sulfur, and hydrogen sulfide at 182°C in a high-pressure reactor. The pressure was reported to have reached 1350 psig before the gaseous reactants were consumed. Basic catalysts such as butylamine, sodium sulfide, and ammonia have also been used in related reactions to give this type of high-sulfur (48 to 53%), EP, and antioxidant additive (Davis, 1978, 1980).

5.3 Reaction with Phosphorus Reagents

Gordon, et al. (1975) prepared a stable noncorrosive EP additive by reacting a 2:1 (mol/mol) mixture of C_{15}–C_{20} alpha olefins and polyisobutylene of 950 molecular weight with phosphorus pentasulfide (P_2S_5) at 200°C and treating the resulting phosphosulfurized material with C_{14}–C_{15} alcohols at 150°C followed by ethylene oxide at 25°C. Caspari (1976) showed that the product obtained by reacting phosphosulfurized C_{15}–C_{20} alpha olefins with bis(*o*-aniline) disulfide (*22*) had EP/antiwear and antioxidant properties.

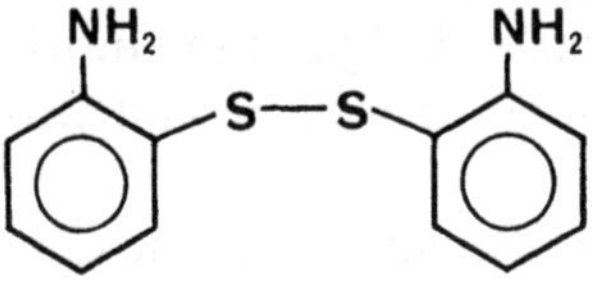

22

Kreutzer (1979) described the preparation of a monoalkylphosphoric acid by reacting a mixture of diisobutylene and oleyl alcohol with sulfur at 150°C and treating the sulfurized material first with phosphorus trichloride and then with water at 50°C. The acid was neutralized with dioctylamine to give the corresponding amine salt that was claimed to be an excellent EP additive.

6 FUTURE TRENDS IN LUBRICANT ADDITIVES

The impetus for improved lubricant additives is founded on a number of fronts. Governmental and regulatory requirements continue to challenge the industry for improved products with lower toxicities. New engine developments, such as the ceramic diesel, are on the horizon presenting opportunities for EP additives that can function at very high operating temperatures. Space technology presents new challenges to the industry. And, of course, there will always be a need for low product costs and ease of production.

Three particular developments are expected to have an impact on the near-term lubricant industry: (1) a move toward synthetic base stocks, (2) the trend toward lower phosphorus content in engine oils, and (3) a desire to reduce chlorine contant in gear additives. Considering the synthetic base stocks such as polyalpha olefins, together with hydrotreated petroleum base stocks, the trend toward these types of lubricants is not expected to abate. These types of materials have no aromatic hydrocarbons or greatly reduced amounts of aromatic hydrocarbons, which are potentially carcinogenic. However, upon removing these solubilizing aromatics, the additives tend to precipitate out of the oil. Alpha olefins are expected to play an important part in solubilizing newly designed additives for these newer base stocks.

Because of the large number of automobiles equipped with catalytic converters and the concern that phosphorus derived from zinc dialkyldithiophosphates (ZDDP) in the crankcase oil and present in engine blowby reduces catalytic efficiency, a need exists for oils with lower phosphorus contents. This is expected to create a need for alternative antioxidant and antiwear/extreme pressure additives, which could have an impact on the use of alpha olefins. In addition, redesign of the lubricant antioxidant could have further repercussion on detergent and dispersant types.

Gear additives are another area of concern. Because of the toxicity of chlorine compounds, their use in gear oils has been greatly reduced. However, a number of processes for making gear additives utilize chlorine or a chlorine-containing reagent at some point in the reaction sequence. Small amounts of chlorine still remain in the final product. The complete removal of the chlorine is therefore expected to become an important priority.

In the future the lubricant additive business will continue growing, with possible new markets opening up in robotics, supersonic transport, space technology, ceramics, and bionics, as well as expanding traditional markets in crankcase and diesel, automatic transmission fluids, and gear, hydraulic, and industrial lubricants. Healthy growth for alpha olefin usage is undoubtedly expected in many of these areas.

ACKNOWLEDGMENTS

We would like to thank Dr. David L. Wooton for assisting us in computerized literature searches, Mr. Willard G. Montgomery for procurring the cited patents, Dr. Thomas P. Stocky for proofreading the manuscript, and Mrs. Merri Von Hatten for typing the drafts. The financial support of this work by the Ethyl Corporation is gratefully acknowledged.

REFERENCES

A'Court, A. J., Manners, D. S., and Morris, A. L. (1971). Improved high temperature detergents, British patent 1,246,545, assigned to Esso Research and Engineering Company.

Bakker, N. (1979). Superbasic sulfonates, U.S. patent 4,137,184, assigned to Chevron Research Company.

Bolle, J., and Dabir, A. (1981). Stable and noncorrosive polysulphides derived from olefins and having different sulphur contents: process for their manufacture and applications thereof, U.S. patent 4,284,520, assigned to Institut National de Recherche Chimique Appliquée.

Booser, E. R. (1981). Lubrication and lubricants, in *Kirk-Othmer Encyclopedia of Chemical Technology* (H. F. Mark, D. F. Othmer, C. G. Overberger, and G. T. Seaborg, eds.), Wiley, New York, p. 490.

Braid, M. (1980a). Sulfurized olefin lubricant additives and compositions, U.S. patent 4,194,980, assigned to Mobil Oil Corporation.

Braid, M. (1980b). Process of preparing sulfurized olefins, U.S. patent 4,240,958, assigned to Mobil Oil Corporation.

Braithwaite, E. R. (1967). *Lubrication and Lubricants,* Elsevier, Amsterdam, pp. 119–165.

Brois, S. J. (1978). Olefin-thionophosphine sulfide reaction products, their derivatives and use thereof as oil and fuel additives, U.S. patent 4,100,187, assigned to Exxon Research & Engineering Co.

Caspari, G. (1976). Bis-aniline disulfide reaction products as multifunctional lubricating oil additives, U.S. patent 3,981,809, assigned to Standard Oil Company.

Caspari, G. (1978). Olefin-dimercaptothiadiazole compositions, U.S. patent 4,097,387, assigned to Standard Oil Company (Indiana).

Chandler, R. E. (1974). Process for forming monocarboxylic acid, U.S. patent 3,786,077, assigned to Esso Research and Engineering Company.

Chemical Week (1965). $300-million market on wheels, *96:* 71.

Chludzinski, G. R., Gardiner, J. B., and Engel, L. J. (1978). Dispersant and V. I. additives for lubricants, British patent 1,511,501, assigned to Exxon Research and Engineering Company.

Crail, I. R. H., Elliott, J. S., Pearman, D. C., and Simmons, H. J. (1963). The effect of polymeric dispersants on engine sludge, *J. Inst. Pet., 49:* 189.

Davis, K. E. (1978). Sulfurized compositions, U.S. patent 4,119,549, assigned to Lubrizol Corporation.

Davis, K. E. (1980). Sulfurized compositions, U.S. patent 4,191,659, assigned to Lubrizol Corporation.

Davis, K. E., and Holden, T. F. (1978). Sulfurized compositions, U.S. patent 4,119,550, assigned to Lubrizol Corporation.

DeClippeleir, G., and Vanderlinden, A. (1978). Overbased calcium sulfonates, U.S. patent 4,086,170, assigned to Labofina SA, Belgium.

DeVries, L. (1977). HCN-olefin adduct using HF, U.S. patent 4,044,039, assigned to Chevron Research Company.

DeVries, L. (1978). Lubricating oil composition containing group I or group II metal or lead sulfonates, U.S. patent 4,116,873, assigned to Chevron Research Company.

DeVries, L. (1979). Succinate dispersant combination, U.S. patent 4,159,958, assigned to Chevron Research Company.

DeVries, D. L., and DeJovine, J. M. (1979). Solid particles containing lubricating oil composition and method for using same, U.S. patent 4,132,656, assigned to Atlantic Richfield Company.

Engel, L. J., and Gardiner, J. B. (1978a). Aminated polymeric additives for fuel and lubricants, U.S. patent 4,068,057, assigned to Exxon Research and Engineering.

Engel, L. J., and Gardiner, J. B. (1978b). Aminated polymeric additives for fuel and lubricants, U.S. patent 4,068,058, assigned to Exxon Research and Engineering.

Gardiner, J. B. (1975). Hydroxylated polymers useful as additives for fuels and lubricants, U.S. patent 3,899,434, assigned to Exxon Research and Engineering Company.

General Aniline and Film Corp. (1968). Simultaneous polymerization and alkylation of heterocyclic N-vinyl monomers, British patent 1,101,163.

Georgi, C. W. (1950). *Motor Oils and Engine Lubrication,* Reinhold, New York, pp. 157–217.

Gordon, C. D., Lowe, W., and Hotten, B. W. (1975). Phosphosulfurized lubricating oil additives, U.S. patent 3,904,535, assigned to Chevron Research Company.

Hayashi, K. (1982). Nitrogen-containing terpolymer-based compositions useful as multi-purpose lubricant additives, U.S. patent 4,357,250, assigned to Lubrizol Corporation.

Hayshi, K. (1984a). Carboxylic acylating agents substituted with olefin polymers of high molecular weight mono-olefins, derivatives thereof, and fuels and lubricants containing same, U.S. patent 4,486,573, assigned to Lubrizol Corporation.

Hayashi, K. (1984b). Carboxylic acylating agents substituted with olefin polymers of high/low molecular weight mono-olefins, derivatives thereof, and fuels and lubricants containing same, U.S. patent 4,489,194, assigned to Lubrizol Corporation.

Holler, H. V., and Youngman, E. A. (1973). Polymerization process and products, U.S. patent 3,761,458, assigned to Shell Oil Company.

Horodysky, A. G. (1979). Metal salts of sulfurized olefin adducts of phosphorodithioic acids, U.S. patent 4,175,043, assigned to Mobil Oil Corporation.

Horodysky, A. G. (1980a). Metal salt treated sulfurized olefins and organic compositions containing same, U.S. patent 4,200,546, assigned to Mobil Oil Corporation.

Horodysky, A. G. (1980b). Sulfurized olefin adducts of dihydrocarbyl phosphites and lubricant compositions containing same, U.S. patent 4,207,195, assigned to Mobil Oil Corporation.

Horodysky, A. G. (1980c). Reaction products of sulfurized olefin adducts of phosphorodithioic acids and organic compositions containing same, U.S. patent 4,212,753, assigned to Mobil Oil Corporation.

Horodysky, A. G., and Braid, M. (1979). Sulfurized olefins adducts of phosphorodithioic acids and organinc compositions containing same, U.S. patent 4,152,275, assigned to Mobil Oil Corporation.

Hotten, B. W. (1979). Sulfur-and chlorine-containing lubricating oil additive, U.S. patent 4,132,659, assigned to Chevron Research Company.

Hotten, B. W. (1982). Process for preparing a sulphurized alkane useful as a lubricating oil, British patent 2,011,392, assigned to Chevron Research Company.

Kerfoot, O. C., and Krehbiel, D. D. (1969). Conversion of diphenylalkane to higher value products, British patent 1,156,110, assigned to Continental Oil Company.

Kiovsky, T. E. (1979). EPR dispersant VI improver, U.S. patent 4,169,063, assigned to Shell Oil Company.

Kreutzer, I. (1979). Sulphurized phosphoric acid ester salts and method of preparation, U.S. patent 4,154,779, assigned to Rhein-Chemie Rheinau GmbH.

Lee, R. J. (1975). Oil-soluble reaction products of (A) a high molecular weight olefin polymer, acrylonitrile, chlorine, an amine and maleic anhydride with (B) an aliphatic amine; and lubricant compositions containing the same, U.S. patent 3,914,203, assigned to Standard Oil Company.

Lee, R. J. (1976a). Oil-soluble reaction products of intermediate (A) from a high molecular weight olefin polymer, acetonitrile, and chlorine with (B) a piperazine, and lubricant compositions containg the same, U.S. patent 3,953,348, assigned to Standard Oil Company.

Lee, R. J. (1976b). Oil-soluble reaction products of (A) a high molecular weight olefin polymer, acetonitrile, and chlorine with (B) an amine, and lubricant compositions containing the same, U.S. patent 3,953,349, assigned to Standard Oil Company.

Lee, R. J. (1977). Oil-soluble reaction products of (A) a high molecular weight olefin polymer, acrylonitrile, chlorine, an amine, and maleic anhydride with (G) an aliphatic amines [*sic*]; and lubricant compositions containing the same, U.S. patent 4,005,021, assigned to Standard Oil Company (Indiana).

Lee, D. A., and Boslett, J. A. (1979). High pressure lubricant additive, U.S. patent 4,149,982, assigned to Elco Corporation.

Lee, R. J., and Richardson, E. E. (1980). Oxidation resistant lubricant composition, U.S. patent 4,210,545, assigned to Standard Oil Company (Indiana).

Malec, R. E. (1978). Lubricant composition containing a dispersant which is a condensation product of a copolymer, polyamine and a polycarboxylic acid, U.S. patent 4,120,803, assigned to Ethyl Corporation.

Myers, H. (1969). Lubricating compounds containing polysulfurized olefin, U.S. patent 3,471,404, assigned to Mobil Oil Corporation.

O'Brien, J. A. (1984). Lubricating oil additives, in *CRC Handbook of Lubrication (Theory and Practice of Tribology)*, Vol. II, *Theory and Design* (E. R. Booser, ed.), CRC Press, Boca Raton, Fla., pp. 301–315.

Oldham, W. J. (1969). Production of heavy alkylate, British patent 1,140,138, assigned to Grance Chemicals, Ltd.

Osselet, A., and Tirtiaux, R. (1980). Alkylates and sulphonic acids and sulphonates produced therefrom, U.S. patent 4,235,810, assigned to Exxon Research & Engineering Company.

Papay, A. G., and O'Brien, J. P. (1980). Lubricant composition containing sulfurized olefin extreme pressure additive, U.S. patent 4,204,969, assigned to Edwin Cooper, Inc.

Papay, A. G., and O'Brien, J. P. (1985). Lubricating oil compositions, Canadian patent 1,184,554, assigned to Edwin Cooper, Inc.

Plonsker, L., Perilstein, W. L., and Malec, R. E. (1972). Fuel and lubricating oil compositions, U.S. patent 3,700,598, assigned to Ethyl Corporation.

Pritzker, G. G. (October 3, 1945). Sulfonic derivatives as lubricating additives, *Nat. Pet. News, 37*(40): R793–800.

Ramney, M. W. (1973). *Lubricant Additives,* Noyes Data Corporation, Park Ridge, N.J.

Ramney, M. W. (1978). *Lubricant Additives Recent Developments,* Noyes Data Corporation, Park Ridge, N.J.

Ramney, M. W. (1980). *Synthetic Oils and Additives for Lubricants Since 1977,* Noyes Data Corporation, Park Ridge, N.J.

Recchuite, A. D. (1984). Sulfurizing lard oil and an olefin, U.S. patent 4,481,140, assigned to Sun Research and Development Company.

Rothert, K. (1977). Corrosion-inhibiting functional fluid, U.S. patent 4,010,107, assigned to Chevron Research Company.

Sabol, A. R. (1979). Overbased magnesium sulfonates, U.S. patent 4,137,186, assigned to Standard Oil Company (Indiana).

Sanson, H. E., III, and Hartman, W. R., Jr. (1978). Cosulfurized olefin and lard oil, Canada patent 1,041,077, assigned to Mayco Oil and Chemical Company, Inc.

Satriana, M. J. (1982). *Synthetic Oils and Lubricant Additives, Advances Since 1979,* Noyes Data Corporation, Park Ridge, N.J.

Segessemann, E. (1967). Process for the preparation of alkane sulfonates, U.S. patent 3,349,122, assigned to Atlas Refinery, Inc.

Song, W. R., Gardiner, J. B., and Engel, L. J. (1976). Aminated polymers useful as additives for fuels and lubricants, British patent 1,457,328, assigned to Exxon Research and Engineering Company.

Song, W. R., Gardiner, J. B., and Engel, L. J. (1977). Process for the preparation of aminated polymers useful as additives for fuels and lubricants, U.S. patent 4,032,700, assigned to Exxon Research & Engineering Company.

Stambaugh, R. L., and Galluccio, R. A. (1979). Polyolefin graft copolymers, U.S. patent 4,146,489, assigned to Rohm and Haas Company.

Veretenova, T. N., Balin, A. I., and Lebedev, E. V. (1974). Sulfur-and-chlorine containing additives for cutting fluids and technological lubricants, *Chem. Abstr., 82*(22): 142402z.

West, C. T., and Culbertson, G. S. (1977). Oxidation of polymers in presence of benzenesulfonic acid or salt thereof, U.S. patent 4,011,380, assigned to Standard Oil Company (Indiana).

Wheeler, E. L. (1976). Lubricant compositions containing N-substituted naphthylamines as antioxidants, U.S. patent 3,944,492, assigned to Uniroyal, Inc.

Yamada, T., Tanaka, Y., and Kuroda, K. (1975). Mixed sulfides of unsaturated esters and olefins as extreme-pressure additives, *Chem. Abstr., 84*(8): 47018m.

Yan, T.-Y., and Bridger, R. F. (1980). Method of stabilizing lube oils, U.S. patent 4,181,597, assigned to Mobil Oil Corporation.

CHAPTER 13

Synthetic Lubricants

RONALD L. SHUBKIN Ethyl Corporation, Baton Rouge, Louisiana

1 INTRODUCTION

Synthetic functional fluids manufactured from linear alpha olefins are an increasingly important class of commercial materials. A functional fluid is a fluid that is used in the performance of some kind of work or operation. It may be a chemically pure compound or it may be a mixture of compounds. Typically, it consists of one or more base fluids and minor amounts of performane-enhancement additives. The applications for functional fluids include

Engine lubrication	Hydraulic fluids
Compressor oils	Heat transfer media
Electrical insulation	Grease bases
Gear oils	Circulating oils
Brake fluids	Turbine oils
Cutting fluids	

Historically, the base stocks for functional fluids have been derived from petroleum. Synthetics, on the other hand, are products manufactured by chemical synthesis. The development of synthetic functional fluids can be traced back over 50 years [1]. The initial impetus was the military need for fluids with high-performance characteristics. German shortages of petroleum during World War II also spurred development of synthetics. Until the mid-1970s, however, synthetics were used only in those applications where the required performance characteristics could not be met by petroleum-based products (i.e., mineral oils). The large cost differential virtually prohibited the use of synthetics in any application adequately served by mineral oils.

The oil embargo of 1974, and the subsequent escalation of petroleum prices, brought a new and urgent need for the conservation of oil reserves and the development of alternative raw materials. By early 1987, the pressures on the world petroleum supply had eased considerably, but the lessons of the late 1970s had not been lost. The emphasis on cost-effectiveness led to the development of machines and systems that required fluid performance characteristics that were becoming increasingly difficult to satisfy with mineral oil products. The pattern of growth for synthetic functional fluids was firmly established.

A variety of synthetic functional fluids have been developed to meet the demanding performance characteristics required by different specialized applications. Commercially, the most important class of functional fluids manufactured from alpha olefins are the saturated olefin oligomers. These materials are variously referred to as polyalphaolefins (PAOs) or synthesized hydrocarbons fluids (SHFs or SHCs). Gulf Oil popularized the term PAO, and Mobil Oil popularized the terms SHF and SHC. These companies have been leaders in the development of saturated olefin oligomers. All three terms are used generically.

Two other synthetic functional fluids are manufactured from alpha olefins—linear dialkylbenzenes and polyol esters. Dialkylbenzenes (DABs) are manufactured by the alkylation of benzene, either with alpha olefins or with chloroparaffins. They can also be made by the telomerization of xylene with ethylene. This product temporarily had a fairly large market during construction of the Alyeska pipeline in Alaska. Conoco has promoted its Polar Start DN600 for cold climate automotive crankcase applications. DABs are higher priced than either polyoefins or diesters and possess no physical or chemical advantages.

The other class of synthetic functional fluids that may be manufactured from alpha olefins are the polyol esters. The acid precursor of the ester can be manufactured from alpha olefins via an oxo route. This application is discussed in Chapter 11. Neither DABs nor polyol esters are considered further in this chapter.

2 Chemistry

Cationic catalysts such as $AlCl_3$ were reported for the polymerization of olefins derived from thermal cracking of wax by Sullivan et al. [2] as early as 1931. In 1951, a patent was issued to Montgomery et al. of Gulf Oil Company describing the use of $AlCl_3$ for the oligomerization of 1-octene [3]. Free-radical or peroxide initiators for alpha olefin oligomerization were patented by Norwood of Socony-Mobil in 1960 [4]. Coordination complex catalysts such as the ethylaluminum sesquichloride/titanium tetrachloride system were disclosed in a patent issued to Southern et al. (Shell Research) in 1961 [5]. The fluids produced by these various catalyst systems contained oligomers with a wide spectrum of molecular weights, ranging from dimers through hexamers and higher. The composition and internal structure of these fluids resulted in relatively poor viscosity/temperature characteristics.

The discovery of a series of unique catalyst systems containing boron trifluoride and a protic cocatalyst (e.g., RCO_2H, ROH, or H_2O) by Brennan at Mobil Oil [6] and by Shubkin at Ethyl Corp. [7,8] enabled the production of oligomers with pronounced peaking of the trimer (Fig. 1). In addition, hydrogenation of the predominantly trimeric product of the oligomerization of C_6–C_{14} alpha olefins produced unexpectedly good temperature/viscosity characteristics. It was later shown by Shubkin et al. [9] that approximately two-thirds of the oligomeric product molecules have one more branch in the carbon skeleton than would be predicted by the generally accepted mechanism for cationically catalyzed oligomerization. For instance, normal oligomerization would produce a trimer with two branches and therefore four terminal methyl groups. Using proton NMR, these workers found that a distilled sample of trimer contained an average of 4.7 methyl groups per molecule. The extra branch was also found in distilled samples of dimer, tetramer, and pentamer. The extra branching is responsible for the low pour point and the viscosity characteristics of the product. Three different mechanisms have been proposed to rationalize the degree of branching and the large number of isomers found in the oligomer product [9–11]. From a practical point of view the mechanism does not matter. The unique combination of molecular weight distribution and isomeric configuration makes the products from the BF_3-catalyzed oligomerization of linear alpha olefins exceptionally useful as high-performance functional fluids.

3 PROPERTIES OF SATURATED OLEFIN OLIGOMERS

Selection of a functional fluid for a particular application involves a careful evaluation of the relative strengths and weaknesses of the various materials available. In Table 1 a qualitative comparison of the physical and chemical properties is given for the most important functional fluids. It serves as a useful first screen in the selection of candidates for a particular application. An examination of the table gives a good indication of

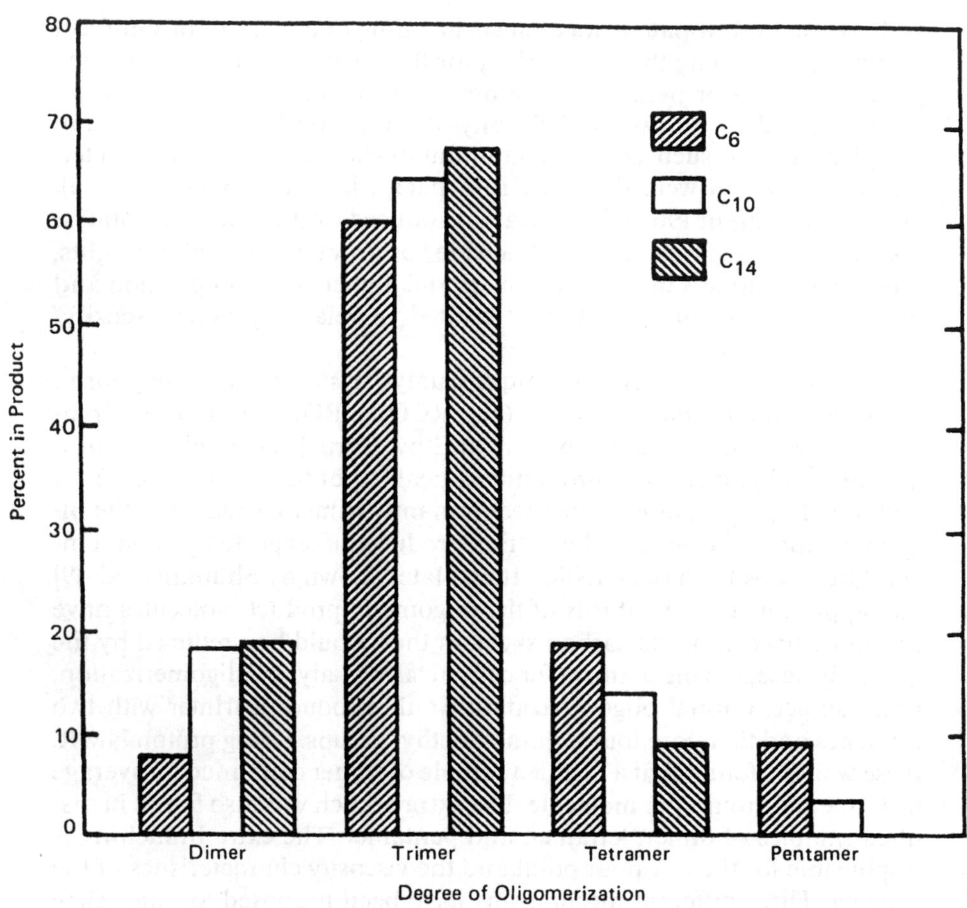

Figure 1 Product distribution from BF_3-catalyzed oligomerization, three different starting alpha olefins. (From Ref. 9.)

why saturated olefin oligomers are enjoying steady growth as a replacement for mineral oils in high-performance applications. They are the only product rated good or excellent in all categories. They outperform mineral oils in viscosity/temperature relationships, low-temperature fluidity, high-temperature oxidation stability, and low volatility. Very importantly, they are compatible with mineral oils and can therefore be used as replacement fluids in hardware designed for mineral oils. They can also be used in blends with mineral oils to enhance certain properties. Significantly better hydrolytic stability and antirust activity give saturated olefin oligomers an important advantage over ester products for many ap-

Table 1 Properties of Synthetic Fluids[a]

Properties	Mineral oil	Saturated olefin oligomer	Dialkylated benzene	Dibasic ester	Polyol ester	Polyglycol	Phosphate ester	Silicone fluid
Viscosity/temperature characteristics	F	G	F	VG	G	VG	P	E
Low-temperature performance	P	G	G	G	G	G	F	G
High-temperature oxidation stability (with inhibitor)	F	VG	G	G	E	G	F	G
Compatibility with mineral oils	E	E	E	G	F	P	F	P
Low volatility	F	E	G	E	E	G	G	G
Compatibility with most paints and finishes	E	E	E	VG	G	G	P	VG
Hydrolytic stability	E	E	E	F	F	VG	F	G
Antirust (with inhibitor)	E	E	E	F	F	G	F	G
Additive solubility	E	G	E	VG	VG	F	G	P
Seal-swell performance; BUNA rubber	E	G	E	G	F	E	F	E

Source: Ref. 12.

[a]Letter signifies performance level: P, poor; F, fair; G, good; VG, very good; E, excellent. A rating of E indicates that the fluid does not cause swelling of the elastomer.

Table 2 Physical Properties of Decene Oligomer and Mineral Oil

	Oligomer	Mineral oil
Viscosity (cSt)		
100°C	5.7	5.2
40°C	29.0	29.5
−17.8°C	1010	[a]
−40°C	7790	[a]
Viscosity index	140	102
Pour point (°C)	−54	−18[b]
Flash point (°C)	235	218
Distillation		
Percent overhead at 400°C	2	20

[a]Too viscous to measure.
[b]Using pour point depressants.

plications. The only areas of weakness for saturated olefin oligomers are in additive solubility and seal-swell performance. Both of these deficiencies are easily overcome with the proper choice of additives or by the use of blends. For example, one very popular synthetic oil is a combination of about 70% saturated olefin oligomer, 20% ester, and 10% additives. In Table 2 a direct comparison of the physical properties is shown for a saturated olefin oligomer and for a mineral oil of about the same viscosity at ambient temperature.

In addition to superior physical properties, saturated olefin oligomers have outstanding oxidative stability when compared to mineral oils. In Figure 2 the results of a test are shown in which air is bubbled through the lubricant in a glass tube at 163°C in the presence of iron, copper, lead, and aluminum catalysts. Viscosity increase is recorded as a function of time throughout the test. Both oils contained 0.5% of a phenolic oxidation inhibitor. As shown, the viscosity of the saturated olefin oligomer remained relatively unchanged after 80 hr, while the mineral oil viscosity began to increase significantly after 20 hr, indicating that oxidation was occurring [13].

4 END-USE APPLICATIONS

The major applications for saturated olefin oligomers are engine lubricants, hydraulic fluids, gear oils, compressor pump and turbine oils, and grease bases.

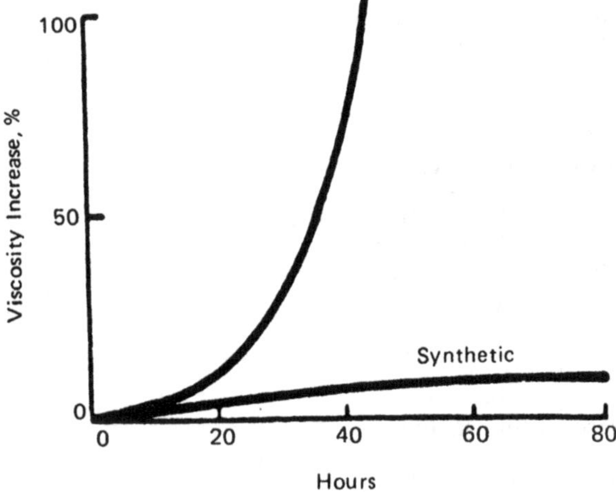

Figure 2 Saturated olefin oligomer oxidation stability.

4.1 Engine Lubricants

Saturated olefin oligomers have excellent properties for both gasoline and diesel engine lubrication. Mobil Oil introduced Mobil 1 in 1975 as the first premium-priced, all synthetic, automotive crankcase oil. Mobil 1 contains 70% decene oligomer, 20% synthetic esters, and 10% additives. The esters are a necessary part of the base-fluid package in order to increase additive solubility and seal swell. Delvac 1 is a similar Mobil product formulated for diesel engines and marketed since 1977.

The movement that began in the mid-1970s toward more fuel-efficient automobiles led to the development of smaller engines operating at higher temperatures. This places additional burdens on the crankcase lubricant. Low-viscosity motor oils alleviate some of the problems by making cold starting easier and by providing reduced friction to improve fuel economy. The former is a particular problem with smaller engines having fewer cylinders because they have to be cranked at higher speeds. In Table 3 a complilation of test data acquired in Europe shows the fuel savings resulting from the use of synthetic oligomer-based engine oils compared to mineral oil reference lubricants.

The Society of Automotive Engineers has published a collection of 26 papers dealing with synthetic automotive engine oils [14]. In Appendix B of this volume, the editors have summarized the "eight superior performance features of synthetic engine oils." Their conclusions are based on a compilation of the data in the various papers. The features include:

Table 3 Synthetic 10W-30 Engine Oil Fuel Savings (SAE 15W-40 and 15W-50 Reference Mineral Oil)

	Percent of test in given driving cycle		
Driving cycle	I	II	III
ECE 15 urban cold	100	40	40
ECE 15 urban hot	—	60	—
90 km/h	—	—	60

	Percent fuel savings		
Car	I	II	III
A	3.5	4.1	2.7
B	1.7	1.4	1.0
C	7.7	5.8	5.7
D	2.9	2.4	2.4
E	3.7	3.4	3.3
F	5.0	4.1	4.2
G	5.8	3.8	3.9
H	8.1	6.7	6.0
I	3.1	3.5	2.7

Source: Ref. 13.

1. *Improved engine cleanliness.* This is based on a test using four taxicabs and a SAE 5W-20 olefin oligomer-based oil. Oil drains at 12,000 miles for 60,000 miles were followed by a 40,000-mile "no drain" period.
2. *Improved fuel economy.* The results of 10 different test programs involving a total of 182 vehicles showed a weighted average fuel savings of 4.2%.
3. *Improved oil economy.* In 10 different tests on oil consumption, the percent improvement in miles per quart ranges from 0% (for a military arctic lubricant) to 156%. The average improvement was 55.9%.
4. *Excellent cold starting (low-temperature fluidity).* See Table 4.
5. *Outstanding performance in extended oil drain field service.* See the data in Table 5, abstracted from a test conducted on parkway police cruisers.
6. *High-temperature oxidation resistance.* Viscosity increase was measured

Table 4 Cold Temperature Starting

	SAE 5W-20 mineral oil	SAE 5W-20 synthetic oil
Pour point (°F)	−37	−65
Kinematic viscosity at −20°F (cSt)	Too viscous	3812
400 CID V-8 engine cranking tests		
Start (°F)	−29	−39
No start (°F)	−34	−43

Table 5 Parkway Police Cruisers
455 CID V-8, 1976 model, Operated at speeds of 55 to 100 mph
Test length: 100,000 miles

	Oil A, 10W-40 "SE" mineral	SHC oil syn 1,[a] "SE-CC," SAE 5W-20
Test vehicles	3	12
Oil and filter changes	5000 mi	25,000 mi
Total sludge rating (10 = clean)	9.3	9.4
Total varnish rating (10 = clean)	5.1	7.0
Piston skirt varnish	5.5	6.6
Oil screen (% clogging)	0	0
Final drain (% insolubles)	0.21	0.23
Oil consumption (mi/qt)	1586	3677

[a] Oil Syn 1 passed CAT 1-H, sequence II-C and double sequence III-D, double V-C and double L-38. These are engine tests used to qualify for severe service.

in a 2-L Renault after 64 hr of operation with a sump temperature of 302°F. The synthetic oil showed a 10% increase and the mineral oil showed a 135% increase. Both were SAE 10W-50 oils.

7. *Outstanding single- and double-length SAE-ASTM-API "SE" performance tests.* The results of all these standard tests are presented in Appendix B of this volume [14]. The synthetics meet or exceed all requirements.

The outstanding properties of an all-synthetic automotive engine oil are quite impressive in the eyes of an automotive aficionado. But the average motorist is more interested in getting the job done in the most efficient, cost-effective manner. One approach to utilizing the advantageous prop-

Table 6 Properties of Fully Formulated 5W-30 Oils: All Mineral Versus Partial-Synthetic

	Viscosity at:		
	100°C (cSt)	−18°C (cP)	Volatility index
All mineral (110 SUS)	11.5	1850	27.4
Partial-synthetic (110 SUS + 15% oligomer D)	11.1	1150	27
All mineral (125 SUS)	10.2	2600	21.4
Partial synthetic (125 SUS + 35% oligomer D)	11.0	1125	18

erties of saturated olefin oligomers without costly "overkill" is to use blends of oligomers and mineral oils [15]. In this manner miniumum performance requirements can be achieved using the most economical blend of the two. In Table 6 the effects on both the viscosity and volatility that can be achieved in a fully formulated 5W-30 oil by the choice of the mineral oil and by the proportion of oligomer included in the blend are illustrated.

But the question of cost-effectiveness remains. In Table 7 an approximate calculation is given of the savings that the average motorist can expect using either a partial- or full-synthetic saturated olefin oligomer motor oil in the crankcase.

4.2 Hydraulic Fluids

The market for synthetic hydraulic fluids may be divided into two sectors. The first is the military, where the impetus has come form the need to reduce fire hazards in military combat equipment (air, land, and sea). The market for saturated oligomers barely existed until the late 1970s. Three military specifications adopted at that time led to the rapid displacement of mineral oils for certain applications. MIL-H-83282 was adopted for U.S. Navy and U.S. Air Force aircraft. MIL-H-46170 and MIL-H-46167 were adopted for U.S. Army tank hydraulic systems. In Table 8 the base stock requirements are shown for MIL-H-83282. Saturated decene oligomer meets all of these requirements. In Table 9 the specifications for the finished fluid are shown.

Table 7 Saturated Olefin Oligomer Versus Conventional SF/CC F.E. Engine Oils
Annual basis, "do-it-yourself"
Distance: 25,000 km
Oil drain intervals: 12,500 km (9 L/yr)
Gasoline cost: $0.30/L

	Conventional	Partial-synthetic	Full-synthetic
SAE viscosity	10W/40	5W/30	5W/20
Fuel			
Economy (km/L)	8	8.2(+2.5%)	8.4(+5%)
Use (L/25,000 km)	3125	3049	2976
Cost ($)	937.50	914.70	892.80
Saving ($)		22.80	44.70
Oil			
Cost ($/L)	1.00	2.00	4.00
Cost of changes and 2 $5 filters ($)	19.00	28.00	46.00
Consumption (L/yr)	6	5	4
Cost of oil consumed ($)	6.00	10.00	16.00
Total cost of oil ($)	25.00	38.00	62.00
Additional cost ($)		13.00	37.00
Savings			
Same oil drains ($)		9.80	7.70
One drain for full-synthetic (5 L oil, 1 filter)		9.80	28.70
Other benefits			
Easier starting, especially in cold weather			
Less wear, less maintenance, longer engine life			

The compatibility of saturated olefin oligomers with mineral oil systems led to a rapid changeover to synthetic hydraulic fluids by the military. By 1980, the market had reached 1.0 million gallons per year and, by 1983, it was 1.2 million gallons. The market is expected to flatten at this level and then begin to decrease toward the end of the decade. This is because the U.S. Air Force has decided to use chlorotrifluoroethylene fluids (CTFE) for the hydraulic systems in the next generation of aircraft. CTFEs are many times more expensive but they are nonflammable. They are not compatible with mineral oil systems, however, and cannot be used

Table 8 MIL-H-83282: Properties of Synthetic Hydrocarbon Base Stock

Property	Value
Viscosity in centistokes at 37.4°C, (100°F) (min.)	16.5
Viscosity in centistokes at 98.0°C, (210°F) (min.)	3.5
Viscosity in centistokes at −40°C, (−40°F) (max.)	2800
Flash point (min.)	202.4°C (400°F)
Fire point (min.)	243.7°C (475°F)
Evaporation [wt % (max.)]	14.0
Acid or base number (mas.)	0.10
Specific gravity at 15.6°C/15.6°C (60°F/60°F)	Report
Color, Saybolt (min.)	Report
Pour point [°F (max.)][a]	−65

[a] No pour point depressant materials or viscosity index improvers may be used.

Table 9 MIL-H-83282: Properties of the Finished Fluid

Property	Value
Viscosity in centistokes at 37.4°C (100°F) (min.)	16.5
Viscosity in centistokes at 98.0°C (210°F) (min.)	3.5
Viscosity in centistokes at −40°C (−40°F) (max.)	3000 ± 60
Flash point (min.)	202.4°C (400°F)
Fire point (min.)	243.7°C (475°F)
Autoignition temperature (min.)	339.9°C (650°F)
Acid or base number (max.)	0.10
Bulk modulus (isothermal secant, 0 to 10,000 psi) at 37.4°C (100°F) [psi (min.)]	200,000
Pour point [°F (max.)][a]	−65

[a] No pour point depressant materials nor viscosity index improvers may be used.

as replacement fluids in the current generation of military aircraft. Unlike the military, the civilian aircraft market chose to use phosphate ester hydraulic fluids. These have proven quite satisfactory, and the hardware is not compatible with mineral oils or oligomers. Add to this the natural reluctance on the part of fleet operators to use more than one type of fluid in their entire fleet and the lack of acceptance of saturated olefin oligomers in the civilian market becomes understandable.

The second market for synthetic hydraulic fluids is for industrial applications [16]. This is a relatively new market, but it appears to have good potential. The first commercial product, Mobil SHC524, was introduced in 1980. The market grew to 25,000 gal per year by 1983 and is expected to grow to 300,000 gal by 1993.

The thermal and oxidative stabilities of saturated olefin oligomers make them ideal for "sealed-for-life" or "closed-loop" applications. These characteristics, combined with a wide operating temperature range (−20 to 250°F), make them particularly attractive for use in hostile environments. Some current and developmental applications are:

Sealed-for-life hydraulic systems in mobile equipment for construction and mining, especially in dusty or dirty atmospheres and where low temperature is a factor

The lumber (logging) industry, especially where the off-the-road equipment is operated in low temperatures and far from the base of operations

Cargo ships that use hydraulic lifts for loading and unloading

Airport ground equipment employing hydraulic systems that must operate in cold climates

Other mobile hydraulic equipment that is operated in cold climates and stored out-of-doors or in unheated garages (e.g. electric utility "cherry pickers," aboveground mining equipment, and road construction equipment)

4.3 Gear Oils

Saturated olefin oligomers make it possible to formulate efficient low-viscosity gear oils, providing a further means for improving vehicle fuel economy. To maintain oil film strenght at high temperatures, polymeric viscosity improvers have been used. These can lose effectiveness as a result of mechanical shearing. The use of saturated olefin oligomers has proven to be a cost-effective and technically superior solution.

Axle efficiency tests show that partial-synthetic gear oils containing 30% of a 4.0-cSt saturated oligomer give higher load transmission efficiency than that of fluids containing only mineral oil base stocks, which may or

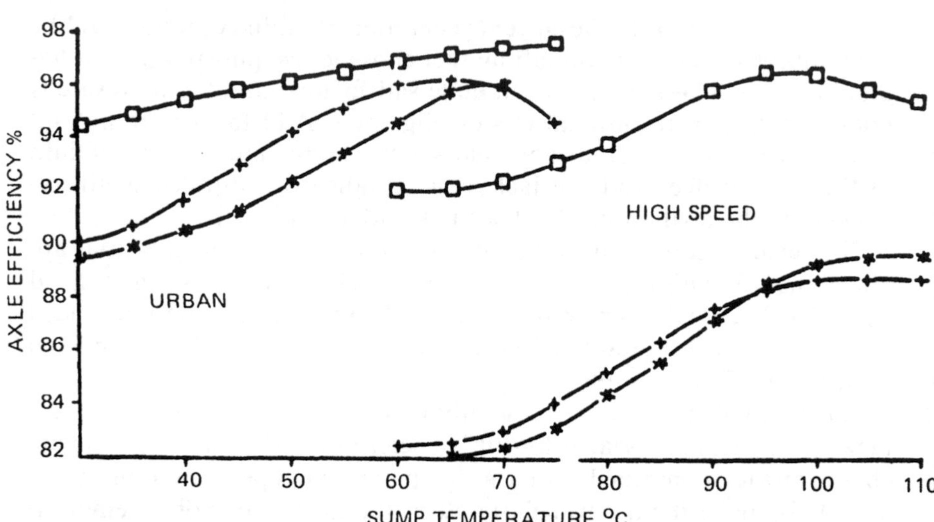

Figure 3 SAE 75W80 partial-synthetic versus SAE 90 mineral gear oil. +, SAE 75W mineral; *, SAE 90 mineral; □, SAE 75W80 partial-synthetic. (From Ref. 17.)

may not contain a viscosity improver. In Figure 3 the results are shown for one such test. The graph compares a SAE 75W80 partial-synthetic oil with a widely used SAE 90 mineral oil. It clearly shows the improved efficiency obtained with the partial-synthetic oil. Under urban conditions, the improvement over the SAE 90 grade was 3 to 5%. Under highway conditions, the improvement was an impressive 6 to 10%. The SAE 75W mineral oil, which is also compared in Figure 4, is generally better than the SAE 90 mineral oil, but is still significantly less efficient than the SAE 75W80 partial-synthetic oil.

Benefits in gearbox efficiency for saturated olefin oligomers relative to mineral oil fluids are well documented. In Figure 4 the relative gains are shown that were found in three laboratory tests, three manufacturer tests, and four different user tests. In Table 10 the savings are shown that were realized when the gear oil for one mineral grinding plant was switched from mineral oil to saturated olefin oligomer base.

4.4 Compressor, Pump, and Turbine Oils

The many desirable properties of saturated olefin oligomers are leading to the rapid growth in their use in industrial applications, such as oils for compressors, pumps, and turbines. The properties that are especially important in these applications are low pour points, low volatility, good

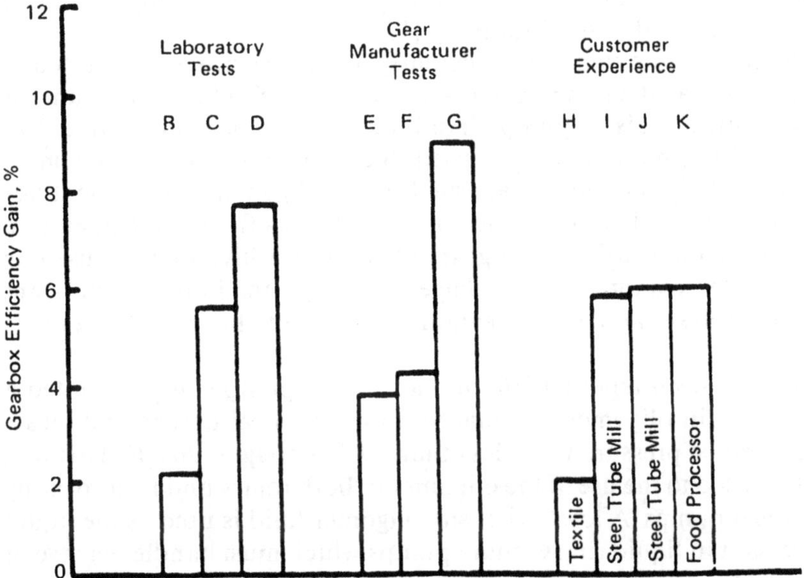

Figure 4 Average efficiency benefits demonstrated for SHF lubricants (relative to mineral-based gear oil). (From Ref. 13.)

Table 10 Approximate Annual Savings with Saturated Olefin Oligomer Gear Oil, Mineral Grinding Plant, Three Raymond Mills (4800 Tons/Day Each)

	Mineral oil	Saturated olefin oligomer synthetic oil
Overhauls and oil changes	8	4
Oil use (gal)		
For changes	120	60
For makeup	600	300
Total	720	360
Oil price ($/gal)	3	13
Oil cost ($)	2,160	4,680
Labor at $15/hr, 8 hr/overhaul	960	480
New bearings, bushings, packaging, etc. ($)	4,000	1,000 (one bearing/2 years)
Lost production ($)	8,000	4,000
Total cost ($)	15,120	10,160
Savings per mill ($)		4,960
Savings, three mills ($)		14,880

hardware compatibility, thermal stability, hydrolytic stability, chemical inertness, and good natural lubricity [16].

High-temperature stability, particularly in the absence of air, has led to the use of saturated olefin oligomers as the sealing fluid and lubricant for the mechanical seals of pumps handling polystyrene process liquid at 450°F. A nitrogen atmosphere is used to exclude air from the system.

The hydrolytic stability of saturated olefin oligomers has made it advantageous to use them to replace polyol esters as the high-temperature lubricant for the bearings and gears of rotary positive blowers used as steam booster compressors. Leakage of live steam into the lubricant causes hydrolysis and thus premature failure of the polyol ester oils [16].

It has also been reported [16] that a vacuum pump company tested a saturated olefin oligomer formulation in mechanical pumps and measured the vapor pressure to be less than 0.01 mmHg at 200°F. This test, and others, led to the use of the oligomer in both rotary and reciprocating mechanical pumps. A lower-viscosity oligomer fluid is used as the liquid compressant in liquid ring vacuum pumps which must handle reactive or corrosive chemical vapors.

The low pour points of saturated olefin oligomers make them attractive for use in refrigeration compressors. Applications include both ammonia and fluorocarbon compressors. Other applications where the low pour point is a major attraction include a variety of compressors and pumps that must operate in very cold climates.

Research in the mid-1970s led to the selection of saturated olefin oligomers as the fluids of choice for an extended drain oil-flooded rotary screw air compressor [16]. The objective of the research was to develop a lubricant/coolant for 100-psi plant air operations to be used in both new and existing equipment which would be suitable for a 10,000-hr drain interval. The results of an accelerated test are shown in Table 11. The test was accelerated by operating the compressor under full load and maintaining a 230°F discharge temperature as compared to the 180°F normal fluid operation. As a result of these tests, the saturated olefin oligomer designated PAO L was chosen for field evaluations. These were successful and the fluid was commercialized. Today, thousands of machines are running on this fluid. Recommended operating procedures call for a yearly (8000 to 10,000 hr) oil change, but some machines are reported to have run up to 15,000 hr with no apparent problems [16].

Saturated olefin oligomers, unlike mineral oils, are manufactured by a carefully controlled chemical process. No by-products contaminate the clear, colorless fluid that is produced. They have been evaluated as baby oils and found to have superior performance for this application. Unfortunately, anticipated reluctance by the genral public to use a "synthetic"

Table 11 Accelerated Test: 25 HP (18.6 kW) Rotary Screw Air Compressor, 230°F (110°C) Discharge Temperature

Lubricant	Viscosity at 100°F (37.8°C) (cSt)			Time to lubricant failure[a] (hr)
	Initial	1000 hr	2000 hr	
Formulated turbine oil A	32.8	43.2	—	330
Formulated turbine oil B	30.9	—	—	520
Automatic transmission fluid A	43.2	—	—	280
Automatic transmission fluid B	40.8	—	—	280
Adipate ester	28.4	35.2	35.6[b]	600
Adipate phthalate diester	61.6	67.9	—[c]	920
PAO J	32.1	36.0	—	1,020
PAO K	43.3	47.2	47.5[d]	3,600
PAO L	28.0	30.0	31.7	2,660

[a] Lubricant failure in all cases was determined when an increase in viscosity of 6 cSt was observed. Field experience indicates that an increase of 6 cSt in viscosity could "ordinarily be expected to give (compressor) operational problems."
[b] Fluid viscosity very erratic after 960 hr.
[c] Fluid precipitated fine particles and caused separator element failure at 1560 hr. Viscosity very erratic thereafter.
[d] Makeup added at 1320 hr.

Table 12 Estimate of Screw Compressor Lubricant Economics

	Mineral oil	Saturated olefin oligomer
Lube service life (months)	2	12
Annual lube requirement (gal)[a]	132	23
Lube price ($/gal)	3	12
Lube annual cost ($)	396	276
Cost of oil change, 2 hr/change, labor at $10/hr	120	20
Cost of filter replacement ($)	120	30
Maintenance and repairs	?	Less
Operating cost	?	Less
Total ($)	636	326
Savings ($)	—	310
Savings (%)	—	49

[a] Lubricant capacity 22 gal; includes 5% makeup for long-life oligomer lube.

Table 13 Lubricant Life: Laboratory Test versus Field Compressors

Lubricant	Lubricant life (hr) 230°F (110°C) Test compressors	250°F (212°C) Air oven test	180°F (82°C) Field compressors	Field compressors recommended lube change (internal hours)
Automatic transmission fluid A, widely used	280	750	2,000	1,000
Non-food-grade olefin, oligomer base lubricant	2,660	6,200	15,000+	8,000–10,000 (one year)
Food-grade olefin, oligomer base lubricant	—	2,680	6,000+	3,000–4,000

on their babies has prevented commercialization for that purpose. Application as a "food-grade" lubricant for rotary screw compressors is not as sensitive. Fluids formulated from specially purified saturated olefin oligomers and FDA approved additives have been found to have greatly extended lifetimes compared to the automatic transmission fluid that is commonly used [19,20]. In Table 13 the results are shown of comparative testing under both accelerated laboratory and actual field conditions. The food-grade oligomer does not perform as well as the non-food-grade oligomer, but it is still significantly better than the mineral oil-based transmission fluid. The development of new additives is expected to improve the lifetime of the food-grade oligomers still further.

4.5 Grease Bases

Synthetic greases prepared from saturated olefin oligomer and a nonsoap, clay-type thickener offer many advantages over conventional petroleum-based greases [21]. Among the advantages are:

Good low-temperature performance
Improved oxidation resistance
Longer life at high temperature

Very long life at normal operating temperature
Hydrolytic stability
Protection from rust
Protection from fretting corrosion
Longer regreasing intervals
Reduced energy consumption
Compatible with mineral oil systems
Do not contain problem components found in mineral oil greases

The U.S. military has established three grease specifications that are based on saturated olefin oligomers. These are MIL-G-81322C and MIL-G-24508A, both of which are wide-temperature greases, and MIL-G-81827, a high-performance grease.

In Table 14 a synthetic grease prepared with saturated olefin oligomer is compared to a typical automotive grease based on lithium 12-hydroxystearate. The advantages for the synthetic far outweigh the 200 to 300% higher price.

Table 14 Synthetic Versus Typical Automotive Grease

Property	Oligomer-based synthetic grease	Lithium 12-hydroxystearate grease
Dropping point (°F)	574	372
Penetration point		
60 strokes	325	330
100,000 strokes	352	348
Evaporation, 22 hr, 350°F (%)	4.9	—
Water washout, loss at 175°F (%)	0.0	6.0
Timken EP test, O.K. load (lb)	45.0	55.0
4-Ball EP test. L.W.I.	56.4	45.5
Weldpoint (kg)	400.0	250.0
High-temperature performance, FTMS 791-333, pope spindle, 10,000 rpm, 350°F, life (hr)	724.0	56.0
Fafnir friction oxidation, weight loss (mg)	0.9	8.8
Low-temperature torque, −80°F		
Starting torque (g · cm)	6343.0	(Too stiff)
Running torque (g · cm)	885.0	(Too stiff)

5 PROJECTED DEMAND

The outlook for synthetic lubricant base stock has begun to brighten [22]. Worldwide growth was about 5% per year for the first half of the 1980s, reaching a market value of approximately $500 million in 1986. Growth for the rest of the decade and the first half of the 1990s is projected at 10 to 15% per year, reaching $1 billion by 1996 [22]. Two elements figure prominently in these forecasts. In-plant industrial experience has shown that synthetic lubricants are cost-effective compared to conventional lubricants because they allow the use of higher temperatures and machine speeds with longer drain intervals and reduced maintenance costs. The second element is the trend toward tougher engine oil specifications, particularly in Europe. A trend toward partial and full synthetics has allowed marketers to meet manufacturer's engine oil specifications.

6 CONCLUSION

Saturated olefin oligomers prepared from linear alpha olefins have become an important member of the class of materials known as synthetic functional fluids. In this chapter we have attempted to present the reader with an overview of some of the many applications for which saturated olefin oligomers are proving to be a high-performance, cost-effective alternative to conventional mineral oil–based materials. In some applications the future for oligomers is assured because of the high-performance demands of those applications. In other areas the growth of saturated olefin oligomers is more directly tied to the cost of crude oil. As the cost of petroleum-based products increases, oligomers become more cost-competitive. When the prices of mineral oils and fuels are low, less efficient operation can be tolerated. Nevertheless, most industry observers predict strong long-term growth for saturated olefin oligomers in an ever-widening scope of performance applications.

REFERENCES

1. Gunderson, R. C., and Hart, A. W., *Synthetic Lubricants,* Reinhold, New York, 1962.
2. Sullivan, F. W., Jr., Vorhees, V., Neeley, A. W., and Shankland, R. V., *Ind. Eng. Chem., 23:* 604 (1931).
3. Montgomery, C. W., Gilbert, W. I., and Kline, R. E., U.S. patent 2,559,984 (1951), to Gulf Oil Co.
4. Garwood, W. E., U.S. patent 2,937,129 (1960), to Socony Mobil.
5. Southern, D., Milne, C. B., Moseley, J. C., Beynon, K. I., and Evans, T. G., British patent 873,064 (1961), to Shell Research.

6. Brennan, J. A., U.S. patent 3,382,291 (1968), to Mobil Oil.
7. Shubkin, R. L., U.S. patent 3,763,244 (1973), to Ethyl Corp.
8. Shubkin, R. L., U.S. patent 3,780,128 (1973), to Ethyl Corp.
9. Shubkin, R. L., Baylerian, M. S., and Maler, A. R., *Olefin Oligomers: Structures and Mechanism of Formation,* Synthetic Lubricant Symposium, 178th National Meeting, American Chemical Society, Washington, D.C., Sept. 9–14, 1979; also, *Ind. Eng. Chem. Prod. Res. Dev., 19:* 15–19 (1980).
10. Onopchenko, A., Cupples, B. L., and Kresge, A. N., *Ind. Eng. Chem. Prod. Res. Dev., 22:* 182–191 (1983).
11. Drascoll, G. L., and Linkletter, S. J. G., *Synthesis of Synthetic Hydrocarbons via Alpha Olefins,* Air Force Wright Aeronautical Laboratories report designation AFWAL-TR-85-4066, May 1985.
12. Manley, L. W., and Jublot, R. M., *New Developments in Synthetic Lubricants,* paper presented at the World Petroleum Congress, Bucharest, Rumania, Sept. 11, 1979.
13. Law, D. A., Lohuis, J. R., Jr., Breau, J. Y., Harlow, A. J., and Rochette, M., *J. Synth. Lubr., 1*(1): 6–33 (1984).
14. Patter, R. I., Campen, M., and Lowther, H. V., *Synthetic Automotive Engine Oils,* Progress in Technology Series 22, Society of Automotive Engineers, Warrendale, Pa. 1981.
15. Papay, A. G., Rifkin, E. B., Shubkin, R. L., Jackisch, P. F., and Dawson, R. B., *Advanced Fuel Economy Engine Oils,* Paper 790947 presented at the SAE Fuels and Lubricant Meeting, Houston, Oct. 1–4, 1979.
16. Miller, J. W., Synthetic lubricants and their industrial applications, *J. Synth. Lubr., 1*(2): 136–152 (1984).
17. *Hitech PAO Systems Applications Guide,* Ethyl Corporation, Baton Rouge, La.
18. Doperalski, E. J., and List, K. R., *Search for a Practical Extended Life Coolant/ Lubricant for a Rotary Screw Compressor,* AICHE 86th National Meeting, Apr. 1979.
19. Galli, R. D., Cupples, B. C., and Rutherford, R. O., *A New Synthetic Food Grade White Oil,* 36th ASLE Annual Meeting, May 1981.
20. Miller, J. W., *New Synthetic Food Grade Rotary Screw Compressor Lubricant,* 38th ASLE Annual Meeting, Apr. 1983.
21. Synthetic greases take on the tough jobs, *Plant Eng.,* 87–88 (Jan. 26, 1984).
22. Synthetic lubes seen on brink of dramatic turn for the better, *Chem. Mark. Rep.,* 4 (Apr. 21, 1986).

CHAPTER 14

Mercaptans

R. VIJAY SRINIVAS and GLENN T. CARROLL Pennwalt Corporation, King of Prussia, Pennsylvania

1 INTRODUCTION

Mercaptans and alkyl sulfides are the sulfur analogs of alcohols and ethers, respectively. They can be characterized by their extremely unpleasant odor. These compounds play an important role in biological systems as well as in the application of chemistry to everyday life. Some of the alkyl sulfides are found in several plant and animal oils, and are minor components of petroleum distillates, shale oil, and coal tar [1].

2 PREPARATION

2.1 Alkyl Mercaptans and Sulfides

There are several methods for the preparation of mercaptans and sulfides. These are cataloged in many standard organic chemistry textbooks. It is our intention here to emphasize one of the most facile methods for their preparation—by the direct addition of hydrogen sulfide and mercaptans to alpha olefins. In the presence of strong acids, electrophilic addition of H_2S and mercaptans to olefins takes place, yielding Markovnikov adducts. Since divalent sulfur compounds are stronger nucleophiles than alcohol or water, the addition of mercaptans to an intermediate carbonium ion takes place quite readily [2]. This type of reaction yields secondary mercaptans and secondary thioethers from alpha olefins. Since 1965, the ready availability of alpha olefins has opened the possibility of utilizing several of their basic reactions known for a long time. Friedel-Crafts catalysts such as anhydrous aluminium chloride, fluoboric acid, mixtures of hydrogen fluoride and boron trifluoride, and their hydrocarbon complexes have been used for the addition of H_2S to olefins—specifically decene [3]. Silica along with alumina has been used to make high-molecular-weight mercaptans which have found use as rubber modifiers [4]. Elemental sulfur along with the bases ammonia or alkylamine [5] or rubber vulcanizing agents such as mercaptobenzothiazole, thiurams, and dithiocarbamates [6] have been used as catalysts to make mercaptans and sulfides from olefins and hydrogen sulfides. Acid cation exchangers (e.g., wet sulfonated styrene-divinylbenzene copolymers in the acid form or more advantageously in combination with sulfonated phenol-formaldehyde resins) have been used for making mercaptans and sulfides from olefins [7]. Acidified clays such as montmorillonite [8], cation-exchanged zeolites [9], and aluminum mercaptides [10] have resulted in catalyzing the Markovnikov addition of H_2S and mercaptans to olefins to give mercaptans and sulfides, respectively. When used, trifluoromethanesulfonic acid not only catalyzes the addition of H_2S to olefins but also polymerizes them to give higher-molecular-weight mercaptans than those corresponding to the initial olefin used [11].

It has been well established that in the presence of free-radical initiators [12], H_2S and mercaptans add to olefins in an anti-Markovnikov fashion by a mechanism similar to the well-known inverse addition of hydrogen bromide to olefins. Several free-radical initiators [13] and ultraviolet light (UV) [14] catalyze this reaction. Olefins exposed to air even for short periods contain enough peroxides to initiate this reaction, while radical inhibitors (e.g., hydroquinone) retard it. A general mechanism may be written, involving a secondary free-radical intermediate that undergoes chain transfer to give a primary mercaptan or sulfide from the addition of H_2S or mercaptan, respectively, to an alpha olefin.

$$\mathrm{RSH} \xrightarrow{\text{Initiation}} \mathrm{RS}^{\bullet} \tag{1}$$

$$\mathrm{CH_2{=}CH{-}R^1} + \mathrm{RS}^{\bullet} \longrightarrow \mathrm{R^1{-}\dot{C}H{-}CH_2{-}SR} \tag{2}$$

R^1 (SR) + RSH ⟶ R^1 (SR) + RS•

R^1 (SR) + R^1 ⟶//⟶ RS R^1 R^1

The direction of addition of the mercaptan can be predicted by assuming that the more stable free-radical intermediate is formed. Since the primary sulfur compound is the major product, it is apparent that chain transfer with the mercaptan or H_2S [Eq. (3), Scheme 1] is faster than polymerization [Eq. (4)]. In the reaction of H_2S with olefins, the mercaptan is formed first, which then reacts with unreacted olefin to form the sulfide. Depending on the structure of the olefin and mercaptan, the rates will differ, but for a simple straight-chain mercaptans and alpha olefins, the rate of addition of mercaptan to the olefin is faster than that of H_2S to the olefin.

It is beyond the scope of this chapter to cite all the work that has been reported on the inverse addition of H_2S and mercaptans to olefins. However, we will attempt to concentrate on the important developments in the area of promoters/inhibitors for peroxide and ultraviolet light-catalyzed processes. Several different types of promoters may be added to the alpha olefins to augment the catalytic activity of peroxides. Examples of suitable peroxides [15] are those of unsaturated organic compounds, ethers, ketones, aldehydes, carboxylic acids and esters, aromatic peroxides (benzoyl peroxide and tetralin peroxide), and the terpene peroxides such as ascaridole. Several organic acyclic azo compounds of the formula

$$R_1-N=N-R_2$$

where R_1 and R_2 are aliphatic or cycloaliphatic and at least one is tertiary, have been used as initiators [16]. These are used at temperatures high enough to decompose them to free radicals and yet low enough to be able to control their half-life, so that a sustained rate of reaction is achieved for a reasonable length of time. Therefore, the optimum temperature for the addition of H_2S or mercaptans to alpha olefins depends not only on the structure of the starting compounds, but also on the half-life of the initiator used. Promoters have been used in these radical-initiated reactions to enable working at lower temperatures and pressures than otherwise possible, and to increase the yield of the desired products. The metal salts of strong inorganic acids which under the conditions of the reaction do not react with the thiol group; simple and complex salts of iron, chromium, manganese, molybdenum, vanadium, thorium, uranium, cerium, lanthanum, beryllium, magnesium, aluminum, and osmium are some examples. Specifically, these promotors were used to convert 1-cetene and 1-hexene to their corresponding primary mercaptans and primary sulfides by their reaction with H_2S [17]. In a related study, acyclic azo compounds along with metal salt promotors were used to produce mercaptans and sulfides from alpha olefins and the addition of a bisthiolester greatly increased the ratio of mercaptan to sulfide [18]. Pentavalent phosphorus compounds containing at least one phosphorus-to-sulfur bond have been shown to have excellent promoting activity. These compounds are illustrated by the organic thiophosphates, thiophosphonates, and thiophosphinates, containing from 1 to 4 sulfur atoms adjacent to the phosphorus atom. For instance, examples in which trimethyl, triethyl, tripropyl, tributyl and triphenyl thiophosphates and thiophosphonates were used as promoters are known [19]. Anti-Markovnikov addition has been observed in the presence of the mercaptides of sodium, potassium, magnesium, and calcium [10].

The observation by Vaughan and Rust (1945) that UV radiation greatly enhances the rate of addition of H_2S and mercaptans to olefins [14] was an important one. With the availability of newly designed continuous-operation photochemical reactors [20], the production of several straight-chain mercaptans and primary sulfides has been made economical and efficient. Catalysis of the H_2S addition to alpha olefins by other types of high-energy radiation [21] such as ^{60}Co radiation [22] to make mercaptans is known. The use of several different kinds of promoters has increased the applicability of this UV-catalyzed addition of H_2S and mercaptans to double bonds to a wide range of alkene and mercaptan substrates with varied reactivities. Acetophenone derivatives such as 2,2-diethoxyacetophenone and benzoin ethyl ether [23] and aromatic thiols such as benzenethiol and naphthalene thiol [24] have been shown to be promoters. A groupVA compound such as phosphoric acid and its alkyl esters, trialkyl phosphites coupled in general with ketones or thioketones, were found to be excellent promotors for the photo addition of H_2S or mercaptans to double bonds [25]. Straight-chain alpha olefins such as 1-decene, 1-dodecene, and 1-tetradecene have been converted to the corresponding primary mercaptans by the UV-catalyzed addition of H_2S [26]. Electron-deficient halogenated olefins react with H_2S or mercaptans in the presence of UV light and epoxy compounds or anion-exchange resins as promotors to yield mercaptans or sulfides [27]. Several reactors have been designed to produce mercaptans and/or sulfides, the selectivity being increased by keeping the conversion level between 7 and 22% [28].

There are several indirect ways to prepare mercaptans and sulfides: for example, by the addition of a thiolcarboxylic acid across a double bond in the presence of benzpinacol as initiator followed by alkaline hydrolysis, to give mercaptans [29], and by the addition of a sulfenyl halide across a double bond followed by treatment with a base such as an alkali-metal mercaptide, to give sulfides [30].

2.2 Derivatives from Mercaptans

Mercaptans can be converted to various functional groups containing sulfur and therefore are building blocks for a variety of useful intermediates. It would be prohibitive as well as redundant to list all the reactions that effect these conversions. This information has been well reviewed in numerous books. A summary of frequently used transformations is shown in Figure 1.

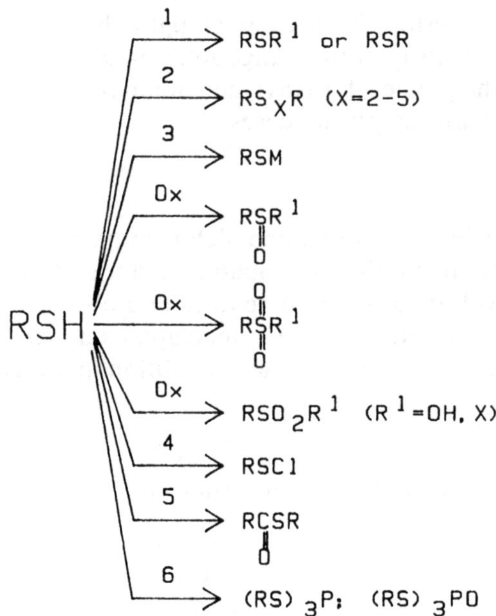

Figure 1 Sulfur derivatives from mercaptans. 1, Alcohol/catalyst *or* alpha olefin/UV light; 2, mild oxidation or sulfur/catalyst; 3, reaction with metal, metal salts, and so on; 4, chlorine or mild oxidation followed by chlorine treatment; 5, acid chlorides; 6, phosphorus trichloride; phosphorus oxychloride or phosphorus trichloride followed by oxidation. (From Ref. 31)

3 PHYSICAL PROPERTIES

The properties of mercaptans and alcohols are quite different, although they appear to be similar in nature. The differences in the bond dissociation energies, acidities, and boiling points for ethyl mercaptan and ethyl alcohol, as shown in Table 1, help explain why the chemistry of these two classes of compounds differs in so many ways.

The bond dissociation energy of the S–H bond is over 10 kcal/mol lower than the corresponding O–H bond. The ease of free-radical hydrogen abstraction from a mercaptan supports this fact and permits these compounds to be included in preparative free-radical chemistry.

The hydrogen bonding in thiols has been confirmed by NMR spectroscopy to be lower than in alcohols [33]. The lower boiling points observed for mercaptans also support this issue.

Mercaptans also exhibit a lower pK_a than their oxygen analogs. While alkoxides are stronger bases, mercaptides are better nucleophiles, due to the increased polarizability of the electrons on sulfur. Additionally, the mercaptide ion is less solvated than the alkoxide ion because of the lower hydrogen bonding observed in mercaptans.

Finally, Table 2 lists some properties for various straight-chain aliphatic mercaptans (C_2-C_{12}). The boiling points range from 34 to 277°C, while their solubility in water is the greatest for ethyl mercaptan (6.76 g/L) and decreases as the aliphatic chain length increases.

4 APPLICATIONS

Mercaptans in general and those derived from alpha olefins in particular have wide applications, ranging from the production of agricultural chemicals and polymers to miscellaneous use in specific end-user consumer products. The existing U.S. market for C_3-C_{12} mercaptans derived from C_3-C_{12} alpha olefins is approximately 15 to 20 million pounds per year.

Table 1 Comparison of Some Properties of Ethyl Mercaptan and Ethyl Alcohol

Property	Ethyl mercaptan	Ethyl alcohol
Bond dissociation energy (kcal/mol)	92	104
Boiling point (°C)	37	78
Acidity (pK_a)	10.6	18

Source: Ref. 32.

Table 2 Some Properties of Aliphatic Mercaptans

Alkyl mercaptan	Melting point (°C)	Boiling point (°C)	Flash[a] point	$D_4^{[25]}$	$n_D^{[20]}$	Solubility in water at 20°C (g/L)
Ethyl	−147.3	34.7	<−18	0.83147	1.427	6.76
Propyl	−113.3	67.5	−18	0.83572	1.4351	1.96
Butyl	−115.9	98	−7	0.83651	1.4401	0.57
Amyl	−75.7	126.5		0.8375	1.444	0.164
Hexyl	−81.0	151.5	32	0.83826	1.4473	0.047
Heptyl	−43.4	176.2		0.83891	1.4498	0.0138
Octyl	−49.2	199.1	52	0.83956	1.4519	0.004
Nonyl	−20.1	220.2		0.84015	1.4537	0.00115
Decyl	−26	240.6		0.8405	1.4572	
Undecyl	−3	259.5		0.8411	1.4585	
Dodecyl	−8	277.3		0.8416	1.4597	

Source: Ref. 34.
[a]The flash points, closed capped, were measured by the Pennwalt Corp.

4.1 Agricultural Chemicals

One of the largest markets for mercaptans is in the manufacture of agricultural chemicals. They have probably been used because they couple high biological activity with good biodegradability. Although the market here is relatively mature, its growth will come from the development of new and sophisticated agricultural chemicals and their increased use abroad. Propyl mercaptan is used in the manufacture of several herbicides, for example, Tillam or Pebulate, a preplant selective herbicide for control of both grassy and broadleaf weeds, specifically in sugar beets, tobacco, and tomatoes; Vernam [35] or Vernolate for control of a variety of weeds, such as crabgrass, barnyard grass, and foxtail; and Mocap or Ethoprop, a nematicide-soil insecticide for control of plant parasitic nematodes that attack plant roots and certain soil-inhabiting insects. Merphos, used in bottom defoliation of cotton to reduce and prevent losses due to boll rot organisms, incorporates butyl mercaptan into its structure. Mercaptans (C_1–C_8) are used in the preparation of 1-alkylthio substituted amines, which are intermediates in the manufacture of ureas [36]. These are pre/postemergence herbicides for total or selective weed control. Alkylthio or sulfone groups (C_1–C_4) are incorporated into substituted diphenyl ethers, to result in compounds that possess both preemergence herbicidal and plant growth regulating activity [37]. Up to C_{10} alkylthio, alkylsulfinyl, and alkylsulfonyl groups have been incorporated into substituted hexahydroindenes to yield compounds that have acaricidal or herbicidal properties [38]. Substituted alkylthioethyl sulfones, and alkylthiovinyl sulfones, where the alkyl group contains up to six carbons, have been shown to have algacidal and fungicidal properties [39]. Immidazoles substituted by alkylthio groups (C_1–C_{12}) show fungicidal, bactericidal, and antiprotozoal activity [40]. Pyridones substituted by phenyl and alkylthio groups are claimed to possess herbicidal activity [41].

Another major agricultural area where mercaptans and their derivatives find use is as pesticides or insecticides. Unsaturated thioethers made from C_1–C_{10} mercaptans and containing an amide group or a cyano group have been claimed to have insecticidal properties [42]. Thioether-substituted benzodioxan derivatives and alkylenedioxy benzenes are known to be insecticides and acaricides [43] and up to C_{10} thioalkyl-substituted unsaturated aliphatic ketones and thioether-substituted benzaldehydes have also been shown to have insecticidal properties [44]. Ether derivatives of phenylthiobenzenes, particularly phenoxyphenyl alkylthio, alkenylthio, alkynylthio, and benzylthio alkoxyethers and their corresponding sulfinyl or sulfonyl derivatives, have been claimed to be useful as insecticides and miticides [45]. In the compounds listed above, the alkyl group attached to the sulfur had up to six carbon atoms. Halogenated esters incorporating alkylthio, alkanesulfinylmethyl, or alkanesulfonylmethyl substituents where the alkyl group has up to four or in some cases up to six carbon atoms are effective as insecticides and miticides [46]. Different compositions of pesticidal dithiophosphoric acid esters have been prepared, some by the facile addition of mercaptans to substituted olefins [47]. Novel compositions that are thiolesters of alkadienoic acids such as phenylheptadienoic acid and trienoic acids such as 3,7,11-trimethyl-dodeca-2,4,11-trienoic acid are useful as insecticides [48]. Alkythiohydroxamates of the type

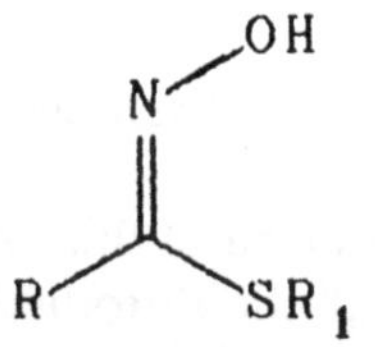

$$R-C(=N-OH)-SR_1$$

are intermediates in the manufacture of Methomyl or Lannate-type insecticides. These are prepared by the chlorination of hydroximines followed by treatment with alkyl mercaptans in the presence of *N*-methylpyrrolidine as solvent [49]. There are various insecticide, herbicide, and fungicide compositions-of-matter patents, wherein alkyl mercaptans have been used directly, resulting in a sulfide or upon oxidation, a sulfone or sulfonyl group. For example, 4-alkylthio-, alkylsulfinyl-, and alkylsulfonyl-substituted butanones are known to be intermediates for fungicides [50], and substituted cyclopropyl sulfones using C_1–C_8 alkyl groups attached to the sulfur atom have shown insecticidal properties [51]. 2-Arylpropyl thioether derivatives [52], a polysulfide obtained from an allylic sulfide [53], and alkylthio 1,2,4-thiazoles [54] are intermediates to novel pesticidal compositions.

Mercaptans, sulfides, sulfoxides, and sulfones are used as plant growth protectors and stimulators. Alkylthiomethylphenols [55], prepared by the reaction of hydroxymethylphenols and C_1–C_{12} mercaptans, and *N,N,N*-trisubstituted-*N*-halomethylthioureas and ureas [56] are useful as plant protection agents. Nitroarylalkylsulfone derivatives and sulfur substituted cyclohexen-1-ones are plant growth stimulants and growth regulators, respectively [57], while substituted dialkyl sulfides have been used as antidotes to thiocarbamate-type herbicides [58].

4.2 Detergents

Polyoxyethlene mercaptans, obtained from the reaction between an alkyl mercaptan and ethylene oxide, have been demonstrated to be good nonionic surface-active agents. Farben reported the first synthesis of these detergents in the 1930s [59]. The majority of the work in this area has been focused on aliphatic mercaptans.

These polyoxyethylene mercaptans are prepared most often by reacting a long-chain alkyl mercaptan with ethylene oxide at approximately 100 to 150°C using a base catalyst [60]. For example, the hydroxides of barium [61], strontium [61], and sodium, and sodium methoxide and mercaptide, have all been employed as catalysts.

$$RSH + n\ \overset{O}{\triangle} \xrightarrow{\text{Base}} RS(CH_2CH_2O)_{n-1}CH_2CH_2OH$$

The mechanism and kinetics for this reaction have been discussed by several groups of workers [60,62].

Other methods have also been utilized for the preparation of this class of compounds. Dodecyl mercaptan, when reacted with β,β′-dichlorodiethyl ether and polyethylene glycol in the presence of sodium hydroxide, gave the mercaptopolyether, 1 [60].

$$C_{12}H_{25}SH + (ClCH_2CH_2)_2O + HO(C_2H_4O)_6H \rightarrow$$

$$C_{12}H_{25}S(CH_2CH_2O)_8H + 2\,NaCl + 2H_2O$$

In more recent years, this class of detergents has become more complex, probably owing to the increased availability of inexpensive and sophisticated starting materials (e.g., alpha olefins). For example, alkylmercaptoethoxylates have been prepared by adding 2-mercaptoethanol to a long-chain alpha olefin, followed by ethoxylation with ethylene oxide [63].

$$R\text{-}CH{=}CH_2 + HSCH_2CH_2OH \rightarrow RCH_2CH_2SCH_2CH_2OH$$

$$\downarrow \overset{O}{\triangle}$$

$$RCH_2CH_2SCH_2CH_2O(CH_2CH_2O)_nCH_2CH_2O$$

Hydroxy-substituted sulfoxides and thioethers have also been prepared in a similar manner and are claimed not only to be good detergents, but also to have antimicrobial activity [64]. The thioethers, obtained by the addition of a hydroxy alkylmercaptan (e.g., 2-mercaptoethanol) to a long-chain aliphatic epoxide (derived from an alpha olefin) using a base catalyst, gave the sulfoxides upon oxidation. Other routes are also available for producing compounds of this type. The reaction of a long-chain

$$\text{R-epoxide} + HS(CR_2)_nOH \longrightarrow \text{R-CH(OH)-CH}_2\text{-}S(CR_2)_nOH \xrightarrow{[Ox]} R'\overset{O}{\overset{\|}{S}}(CR_2)_nOH$$

aliphatic thiol and epichlorohydrin gave 3-alkylthio-2-hydroxy-1-chloropropanes which upon reaction with base generated glycidyl thioethers.

Analogously, alkyl mercaptans have been condensed with glycidol, producing polyhydroxythioethers, which then gave the corresponding sulfoxide surfactants following oxidation [65]. Glycidol was prepared by the dehydrochlorination of glycerol monochlorohydrin using sodium or potassium hydroxide. Previously, the glycidol had been purified by distilla-

$$HO\text{-CH}_2\text{-CH(OH)-CH}_2\text{-}Cl \xrightarrow{NaOH} \text{epoxide-}CH_2OH \xrightarrow{RSH} RS\text{-CH}_2\text{-CH(OH)-CH}_2\text{-}OH$$

tion; however, a process is claimed which utilizes the crude material for condensation [66].

Finally, the salts and esters of long-chain aliphatic thioethercarboxylic acids act as wetting agents, emulsifiers, and thickeners for aqueous products [67]. Their syntheses consisted of reacting glycolic acid and a long-chain aliphatic mercaptan followed by neutralization or esterification. Similarly, mercaptans (C_6–C_{16}) have been added to acetylenemonocarboxylic acid, producing the β-sulfenylacrylic acid [68]. The corresponding sulfoxides, obtained by oxidation using hydrogen peroxide, are reported to have antimicrobial activity in addition to being surface-active agents.

4.3 Lubricants

Lubricants constitute a variety of formulated products such as engines oils, gear lubricants, automatic transmission fluids, hydraulic fluids, metalworking fluids, process oils, and greases. The total volume in the United States for lubricants was about 3 billion gallons in 1975, and at that time a 19% increase per year was predicted [69]. Over 98% of the lubricant volume is based on petroleum fractions, while the balance is synthetic lubricants based on petrochemicals, such as polyalkylene glycols, polyolefins, and other chemicals, such as phosphate esters. Petroleum lubricants are blends of refined oils or base stocks. Over 25 major base stocks are produced for lubricant formulations. These base stocks are classified by the type of crude oil from which they are produced—naphthenic or paraffinic—and the degree to which some fractions such as aromatics, wax, and asphalt have been removed. Viscosity index (VI), pour point, and aromatic content are some properties of lubricants which determine their quality. The higher the VI, the more stable is the base oil viscosity over a wide range of temperatures, and the higher the quality of the base stock. The pour point of an oil is the temperature at which wax crystallizes out of it, thereby interfering with the fluid flow. Naphthenic base stocks are preferred for use in low-temperature applications since they contain lower proportions of crystalline waxes. The high aromatic content of naphthenic oils affects compatibility with rubber seals by lowering the oxidation resistance, and increases the solubility of the lubricant for compounded products such as greases and cutting oils.

Additives are used in about two-thirds of the total volume of lubricants produced, to enhance and control certain specific properties that meet end-user requirements. Oil additives are comprised of a wide variety of chemical compounds. These are generally dissolved in the base stock in concentrations usually below 1% and sometimes as high as 10%. Phenol derivatives, thiophosphates, and sulfonates are the most commonly used lubricant additives. These additives provide specific functions, such as oxidation and corrosion inhibition, for paraffinic oils. The naphthenic stocks which are used for process oils tend to be less flexible as far as additives are concerned. Synthetic base stocks are more expensive than some of the additives themselves, and therefore the trend is to maximize additive use in synthetic lubricants.

There are many organosulfur compounds which have been used as lubricant additives. The sulfide, disulfide, and polysulfide linkages, incorporated into novel compositions-of-matter, serve various functions, such as corrosion inhibition, oxidation resistance, high-temperature stability, and static prevention. Certain 3-halo-1,2,4-thiadiazole disulfides and the bis-disulfides are shown to be useful as corrosion inhibitors [70]. In general, these compounds are prepared by treating hydrazine with carbon disulfide in the presence of a base, followed by acidification and reaction with alkyl mercaptan and sulfenyl chloride to give 1,3,4-thiadiazylbis-disulfides. On the other hand, if cyanamide and chloride are used instead of hydrazine and sulfenyl chlorine, we obtain the 1,2,4-thiadiazolyl-sulfenyl chlorides, which readily react with mercaptans to give the disulfides [71]. The compounds mentioned above may be used in a variety of lubricating compositions based on oils of varying viscosity, including natural oils, synthetic oils, and mixtures of the two. Substituted polysulfides are also used as high-pressure lubricant additives. Polyoxyalkylene compounds and mercaptans react to give polysulfides, which are high-pressure noncorrosive additives in lubricant stabilizers for PVC and hardeners for epoxy resins. These compounds reduce the frictional wear between moving parts and improve their load-bearing properties. Other compositions of sulfide polymers of polyoxyalkylene compounds find use as metalworking lubricants, antistatic agents, and render oxidative resistance to lubricants [72]. Specifically, polyethylene glycol *t*-dodecylthioether is used in lubricant pad coatings to reduce the tendency for the pad to wick water without impairing its ability to wick lubricants [73]. Similarly, extreme-pressure lubricant additives are obtained when olefins containing 3 to 30 carbons are reacted either with mercaptans or S_2Cl_2 followed by elemental sulfur at elevated temperatures. Specifically, substituted dialkyl tetrasulfides are claimed to be extreme-pressure lubricant additives and a possible replacement for sulfurized sperm oil [74].

Sulfurized olefins are obtained when olefins (C_3–C_{30}) are reacted at elevated temperatures and pressures with elemental sulfur and H_2S, in both the absence and presence of catalysts [75]. These are claimed to be lubricants and lubricant additives. By reacting these sulfurized olefins with a group IIA or IIB metal salt, anticorrosive lubricant additives are obtained [76]. Sulfurized olefins have been aminated using polyamines, followed by treatment with a molybdenum-containing compound at approximately 50 to 300°C to yield a composition, which is claimed to be a multifunctional lube additive, providing friction reduction, oxidation resistance, and dispersancy [77]. Other lubricant compositions include a trihydroxyhydrocarbyl sulfide [78], a product obtained by reacting a propargyl alcohol with an alkyl or alkenyl mercaptan which finds use as a corrosion inhibitor in cutting oils [79], a multifunctional ester used as a power transmission fluid additive, formed by reacting a straight-chain (up to C_{30}) alkenyl, alkyl, thioalkyl, or thioalkenyl succinic acid with an equimolar amount of an ethoxylated or propoxylated alkyl mercaptan (C_8–C_{20}) containing one to six ethylene or propylene oxide units [80]. β-Substituted epithio compounds and their addition products are used as extreme-pressure and antiwear stabilizing additives for both mineral and synthetic lubricants [81]. These compounds form no residues upon burning, have good corrosion-inhibiting properties, and are effective in small concentrations (e.g., 0.1 to 3%). Sulfurized alkyl phenols react with olefins in the presence of an alkaline earth metal salt to give compounds that are useful as lubricant additives [82]. Substituted naphthylmethylsulfonium salts are useful as corrosion inhibitors [83]. Several phosphorus-containing compositions [84], where the olefin is reacted with H_2S, S_8, or P_2S_5, in the presence of phosphoric acid and MoO_3 [85], are known to be lubricants and lubricant additives. A sulfur compound produced by the reaction of trithiolane (obtained from a norbornyl compound reacted with sulfur) and an alkyl mercaptan (C_1–C_{12}) is claimed to be a lubricant additive [86] and thioformidate esters obtained from the reaction of nonconjugated olefins with HCN, HF, and alkyl mercaptan are used as lubricant oil dispersant additives [87]. Higher mercaptans (C_{14}–C_{20}) used in concentrations up to 5000 ppm sulfur in refrigerator oil compositions prevented the seizure of moving parts [88].

4.4 Pharmaceuticals

Organosulfur compounds have been incorporated into various different classes of molecules, which have pharmaceutical utility as anti-inflammatory agents, analgesics, antiarrhythmic agents, antibiotics, antiobesity agents, and so on. Inflammation, pain, and fever have been treated in warmblooded animals by administering specific doses of derivatives of ethynyl benzene compounds, where the active substituent is either alkylsulfinyl or alkylsulfonyl groups containing C_1–C_7 carbons [89]. Novel compositions of 3-ketoandrostenes, disubstituted in the 17-position with a R_1S- and a R_2S- group, where R_1 and R_2 are different alkyl, cycloalkyl, or aryl groups, are known to be useful as anti-inflammatory agents and used in topical administration for skin disorders or asthma [90]. In some cases, lower alkyl mercaptans are used as synthetic transformation agents in the preparation of substituted pyrrole-acetic and arylacetic acid derivatives, which are intermediates for a variety of anti-inflammatory, antipyretic, antispasmodic, and analgesic agents [91]. In one case, substituted acetonitriles are converted to substituted acetic acids via the thiolesters, obtained by treating the nitriles with mercaptans. Hydrogen sulfide and lower alkyl mercaptans are used in a process to prepare 3-hydroxycephalosporin antibiotics [92]. Rifamycin S and SV have been substituted in the 3-position with an alkylthio group (C_1–C_5) to give novel antibiotic compositions [93]. α-Aminoalkyl-4-hydroxy-3-alkylsulfonylmethyl phenyl ketones and similarly substituted benzyl alcohols are useful as bronchodilators [94]. Alkyl diaryl sulfonium salts have various therapeutic uses depending on the size of the alkyl group [95]. When the alkyl group is C_1–C_7 and C_{10}–C_{24} or cyclopropyl, these compounds are hypoglycemic and antiobesity agents, respectively. Polyether derivatives containing alkylthio groups having up to 12 carbon atoms have medicinal properties as antilipemic agents and dietetics [96]. Stimulation of the immune response of a mammal has been effected by administering a specific amount of a compound, which is a derivative of a di- or polysulfide containing terminal ether linkages [97]. These are also used as adjuvants for vaccines. Substituted thiazolidine carboxylic acids are claimed to have some analgesic and mucolytic activity [98]. Alkylthiobenzaldehydes (C_{13}–C_{20}) show pharmaceutical use as hypolipidemic agents [99]. Lower alkylthio (C_1–C_5), alkylsulfinyl- or alkylsulfonyl-substituted indene phosphoric acids are anti-inflammatory agents [100]. 1-Alkenylthioethers show central nervous system activity and find use as tranquilizers [101]. Substituted amino-ethyl-thiol derivatives with C_1–C_{10} alkylthio groups are useful in the treatment of cardiovascular disorders, arthritis, hypertension, and cancer [102]. 5-Alkynylaminoalkenylthioalkyl-1H-imidazoles are useful as histamine H-2 receptor antagonists for treating peptic ulcers [103]. In the presence of cationic surfactants, alkyl mercaptans (C_1–C_{20}) react with propargyl alcohol using an alkaline earth metal oxide as catalyst to give novel alkylthioalkenols which have antibiotic activity [104]. Mercaptans react with acetoxyamino acid derivatives to give α-mercaptoamino acids which are intermediates for several antibiotic preparations (e.g., 3,6-dimercapto-piperazine-3,5-dione [105]). Mercaptans, which are easily converted to an alkali mercaptide, are used in substituting corticosteroids in the 20-position. The resulting alkylthio groups may easily be oxidized to the sulfoxides [106].

4.5 Polymers

Polymer applications utilizing mercaptans derived from alpha olefins hold a little more than one-third of the existing mercaptan market. This area has great potential and has an expected growth rate of 4 to 6% per year.

The uses of mercaptans in polymers fall into three major categories: chain transfer agents, additives such as stabilizers against heat or ultraviolet light, and monomers which incorporate an alkylmercapto group into their structure. Chain transfer agents probably hold the greatest share of the market.

Chain transfer agents, often referred to as polymerization modifiers, are used to control the molecular weight and thus many of the polymer's properties [107]. The proposed mechanism of this process is illustrated in Scheme 2 [108]. Just as with any other free-radical process, free-radical polymerization involves three steps: initiation, propagation, and termination. During the initiation step, free redicals are produced, usually by the addition of some initiator such as an organic peroxide. Subsequent attack of these radicals on the monomer begins polymerization. During the propagation and termination steps the growth of the polymer continues or stops, respectively.

Scheme 2

Initiation $\quad I \longrightarrow 2\,R\bullet$

Propagation	M + R•	⟶	P_1•
	P_1• + M	⟶	P_2•
	P_i• + M	⟶	$P_{(i+1)}$•
Termination	P_i• + P_j•	⟶	"Dead" Polymer
Chain Transfer	P_i• + RSH	⟶	P_iH + RS•
	RS• + M	⟶	P_1•
	P_1• + M	⟶	P_2•

The addition of a chain transfer agent such as an alkyl mercaptan controls the molecular weight by donating a hydrogen atom to the polymer chain, thereby terminating it [107]. The resulting mercaptide radical can react with free monomer to initiate a new polymer chain, resulting in a lower molecular weight for the product polymer than might otherwise be expected.

In choosing a mercaptan for use as a chain transfer agent, a number of factors need to be considered. First is the chain transfer constant (C_x), defined as the ratio of the transfer rate constant to the propagation rate constant. The degree of polymerization is related to this constant through an equation developed by Mayo [108]. Of the known chain transfer agents, mercaptans have some of the highest chain transfer constants. For steric reasons, C_x decreases in the order primary > secondary > tertiary. Table 3 lists some selected chain transfer agents for a variety of polymerizations. Solubility considerations, which dictate the choice of alkyl substituent, are also important in cases involving emulsion polymerizations.

Mercaptans (or derivatives) are incorporated into a large variety of compounds which are added to a polymerization recipe to enhance some property of the polymer. In most cases, these additives are used as stabilizers against heat, ultraviolet radiation, and/or oxidation.

Bis(alkylsulfonylvinyl)anilines and benzenes are claimed to act as stabilizers for polymers by preventing their degradation by ultraviolet light [109]. Similarly, other aromatic compounds, such as hindered hydroxy-

Table 3 Chain Transfer Constants (C_x) for Alkyl Mercaptans

n-Alkyl mercaptan	Monomer	Temperature (°C)	C_x
Ethyl	Styrene	50	17.1
Butyl	Ethylene	130	5.8
	Methyl methacrylate	60	0.67
	Styrene	60	22 ± 3
	Vinyl acetate	60	48 ± 14
Amyl	Methyl methacrylate	50	17.1
	Styrene	62.5	17.8
Octyl	Butadiene	50	16
	Styrene	50	19
Decyl	Butadiene	50	18.2
Dodecyl	Styrene	60	18.7 ± 1
Tetradecyl	Styrene	50	19

Source: Ref. 107.

phenylalkanoates of thioether isopropanols [110] and vicinal glycols [111], alkylidene bisphenols containing sulfur [112], thioalkyl phenols [113], and a mixture containing a substituted naphthindole, a metal halide, an ethoxylated alkyl mercaptan, a carboxylic acid, an organic phosphite, and an organic polyamine [114], have been claimed as stabilizers in producing heat and light resistant polyolefins. Polyoxysulfoxide and sulfones have been claimed to be good antistatic agents for plastics [115].

Mercaptans have also been used to modify a polymer's properties or to prepare a new class of polymers by effectively building the mercaptan (or derivative) into the monomer. The mercaptan may be part of the starting monomer or reacted with each monomeric unit after polymerization has occurred. A few selected examples are shown below. An overview of this area is presented in a review by Goethal (1977) [116].

As discussed above, mercaptans and their derivatives serve as additives for the stabilization of polymers against oxidation. They have also been used to prepare resins which are antioxidants bearing free-radical-inhibiting and hydroperoxide decomposition properties. Their preparation involves the condensation of a phenol and a long-chain mercaptoaldehyde [117]. Gel polymers of water-soluble or water-dispersible vinyl compounds (e.g., ethyl acrylate) have been prepared by employing in the recipe: (1) ferric or cerric ions in aqueous medium, (2) a mercaptan, and (3) a sulfur compound capable of producing S_xO_y ions such as SO_2^{2-}

and SO_3^{2-}. The polymers are useful as coatings for metal, wood fiber, and so on, with the added advantage of being capable of adhering to rusty surfaces [118]. Thermoplastics have also been prepared which incorporate a mercaptan in the polymer as a mercaptal. Mercaptoles are produced by reacting two equivalents of a mercaptan with a ketone and are the sulfur analogs of a ketal. If a dithiol of greater than five carbons is used, polymers with a structure like *2* are obtained [119].

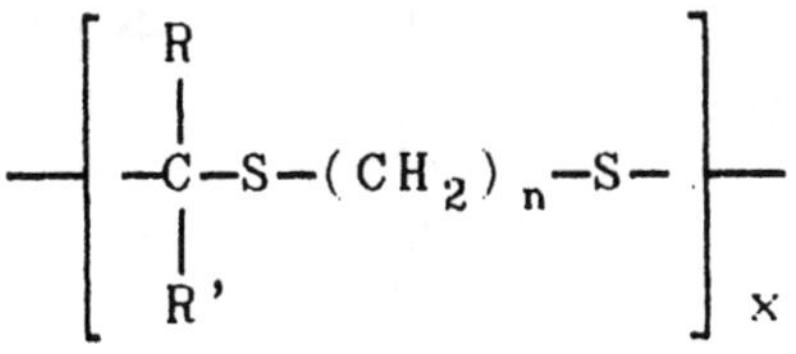

$$\left[-\overset{\overset{\displaystyle R}{|}}{\underset{\underset{\displaystyle R'}{|}}{C}}-S-(CH_2)_n-S- \right]_x$$

2

Mercaptans have been reacted with poly(dichlorophosphazine) polymer, prepared from phosphorus pentachloride and ammonium chloride, to produce polyphosphazine polymers and copolymers having the structure *3*, where X_1 and X_2 are independently alkylthio, alkoxy, or al-

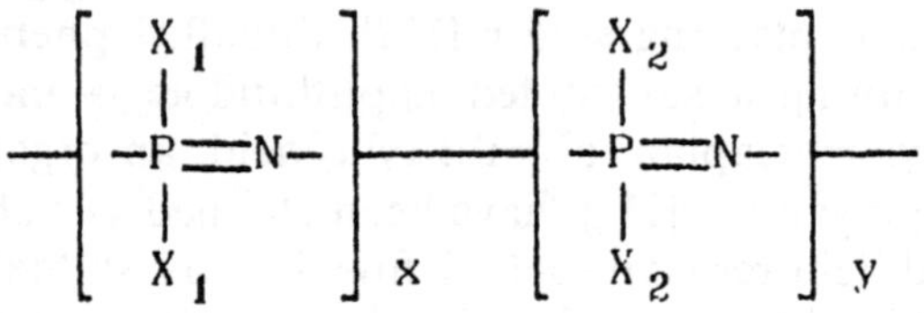

$$\left[-\overset{\overset{\displaystyle X_1}{|}}{\underset{\underset{\displaystyle X_1}{|}}{P}}\equiv N- \right]_x \left[-\overset{\overset{\displaystyle X_2}{|}}{\underset{\underset{\displaystyle X_2}{|}}{P}}\equiv N- \right]_y$$

3

kylamino. These polymers are claimed to be useful as protective films, moldings, coatings, and in foam applications [120].

4.6 Miscellaneous

Mercaptans, sulfides, and other sulfur-based compounds also find use in many end-user products such as food flavorings, perfumes, gas odorants [121], and cosmetics. Therefore, the preparation of organic intermediates, used in the manufacture of these products, warrants a brief discussion. In addition, mercaptans are used to prepare or modify catalysts used for the preparation of various organic intermediates and for the hydrotreating of petroleum feedstock. Another specific use for mercaptans (C_{12}–C_{16}) is as concentrating agents for metallic mineral ores by froth floatation [122].

Some useful reactions of organosulfur compounds are presented below even though the products produced do not have a specific application. Organic compounds may be oxydimethylated by reacting said compound with a novel ether having the formula

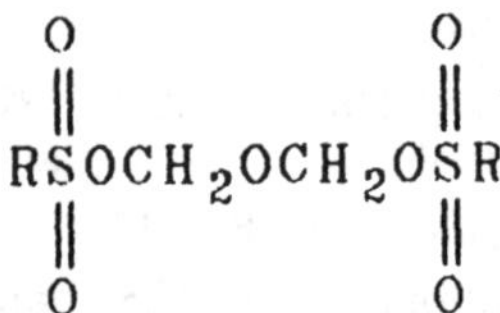

wherein R is an alkyl group having from one to six carbons [123]. Alkylthiomethylphenols were prepared by reacting a dialkylaminomethylphenol with thiocarboxylic acid S-ester [124]. 1-Alkythiobutenyne derivatives were obtained by reacting diacetylene with an alkyl mercaptan in liquid ammonia at -50 to -30°C or in a aqueous ammonia solution at −40 to +20°C [125].

Mercaptans, sulfides, and polysulfides are good pre-sulfiding agents for catalysts used in the hydrotreatment of petroleum feedstock. Although the lower alkyl derivatives are employed most often, butyl mercaptan has also been used [126].

REFERENCES

1. Reid, E. E., *Organic Chemistry of Bivalent Sulfur,* Vol.1, Chemical Publishing Company, New York, 1958, Chap. 1.
2. Ohno, A., and Oae, S., Thiols, in *Organic Chemistry of Sulfur* (S. Oae, ed.), Plenum Press, New York, 1977.
3. Bell, R. T., and Thacker, C. M., U.S. patent 2,498,872. (1950); Bell, R. T., U.S. patent 2,531,602 (1950).
4. Schulze, W. A., U.S. patent 2,502,596 (1950).
5. Jones, S. O., and Reid, E., *J. Am. Chem. Soc., 60:* 2452 (1938); Louthan, R.P., U.S. Patent 3,221,056, (1965); Doss, R. C., U.S. Patent 3,419,614, (1968); Warner, P. F., U.S. patent 3,114,776 (1963).
6. Lang, A., and Vannel, P., U.S. patent 3,333,008 (1967).
7. Ipatieff, V. N., Pines, H., and Friedman, B. S., *J. Am. Chem. Soc., 60:*273 (1938); Macho, V., Czech. patent 185,469 (1980); Arretz, E., Mirassou, A., Landoussy, C., and Auge, P., Fr. Demande FR 2,531,426 (1984).
8. Kubicek, D. H., Belg. Patent 886,261 (1981).
9. Onyestyak, G., Kallo, D., Papp, J., and Detrekoy, E., Hung. Teljes HU29,972 (1984); Fried, H. E., Eur. patent 122,654 (1984).
10. Hahn, W., Ger. patent 1,110,631 (1961).
11. Fields, E. K., U.S. patent 4,347,384 (1982).

12. Kharasch, M. S., Read, A. T., and Mayo, F. R., *Chem. Ind.,* 752 (1938); Posner, H., *Ber. Dtsch. Chem., 38:*646 (1905); Ashworth, F., and Burkhardt, G. N., *J. Chem Soc,* 1791 (1928); Grattan, D. W., Locke, J. M., and Wallis, S. R., *J. Chem. Soc., Perkin Trans., 1:* 2264 (1973).
13. Mayo, F. R., and Walling, C., *Chem Rev. 27:*387 (1940).
14. Evans E. A., Vaughan, W. E., and Rust, F. F., U.S patent 2,376,675 (1945);' U.S. patent 2,411,961 (1946); Brit. patent 567,524 (1946).
15. Ishida, S., and Yoshida, T., U.S. patent 3,459,809 (1969). See Ref. 13.
16. Martin, D. J., U.S. patent 2,551,813 (1951); Stratton, G. B., U.S. patent 3,376,348 (1968).
17. Hoeffelman, J. M., and Berkenbosch, R., U.S. patent 2,352,435 (1944).
18. Kite, G. F., U.S. patent 3,397,243 (1968).
19. Martin, D., U.S. patent 3,780,113 (1973); Elleto, R. J., and Martin, D. J., DE 2,027,206 (1970).
20. Bierker, G. L., and Kivnick, A., U.S. patent 4,043,886 (1977).
21. Kochanny, G. L., Jr., U.S. patent 3,682,604 (1972).
22. Carley-Macauly, K. W., Hills, P. R., and Spindler, M. W., U.S. patent 3,661,745 (1972); Fr. Demande FR 1,582,488 (1969).
23. Dimmig, D. A., U.S. patent 4,140,604 (1979).
24. Louthan, R. P., U.S. patent 3,085,955 (1963).
25. Olivier, J., Souloumiac, G., and Suberlucq, J., U.S. patent 4,233,128 (1980); U.S. patent 4,443,310 (1984); Louthan, R. P., U.S. patent 3,050,452 (1962); Ray, C. A., Jr., U.S. patent 3,652,680 (1963).
26. Seris, J. L., Suberlucq, J., Thuy, T. H., and Leirouici, C., Fr. Demande FR 2,051,887 (1971); Buchholz, B., U.S. patent 3,652,680 (1972).
27. Louthan, R. P., U.S. patent 3,084,116 (1963).
28. Edwards, J. R., U.S. patent 3,412,001 (1968).
29. Bush, R. W., and Morgan, C. R., U.S. patent 4,117,017 (1978).
30. Allen, P. T., and Horodysky, A. G., U.S. patent 3,994,539 (1976).
31. Reid, E. E., *Organic Chemistry of Bivalent Sulfur,* Vols. 2 and 3, Chemical Publishing Company, New York, 1958; March, J., *Advanced Organic Chemistry,* 3rd ed.,Wiley, New York, 1985; House, H. O., *Modern Synthetic Reactions,* Benjamin, The Phillipines, 1972; Capozzi, G. and Modena, G., *The Chemistry of the Thiol Group,* Part 2, (S. Patai, ed.) Wiley, New York, 1974, pp. 785–839; Gilbert, E. E., *Sulfonation and Related Reactions,* Interscience, New York, 1965, pp. 217–239, 202–214; see Ref. 2; Block, E., *Reactions of Organosulfur Compounds,* Academic Press, New York, 1978, p. 16; Kuhle, E., *Synthesis,* 561, (1970); Kuhle, E., *Synthesis,* 563,617 (1971); Scheithauer, S., and Mayer, R., (1979) *Topics in Sulfur Chemistry,* Vol. 4, (A. Senning, ed.), Georg Thieme, Stuttgart, pp. 1–373; Walter, J., and Bode, K. D., *Agnew. Chem. Int. Ed. Engl. 6:* 281–293 (1967). Sandler, S. R., and Karo, W., *Organic Chemistry—A Series of Monographs—Organic Functinal Group Preparations,* Vol. 12-I, 2nd ed., (H. H. Wasserman, ed.) Academic Press, New York, 1983, pp. 478–524.

32. Benson, S. W., *Chem. Rev., 78:*23 (1978). See Ref. 2.
33. Rousselot, M. M., *Compt. Rend.*, 232:2428 (1951); Miller, S. I., and Marcus, S. H., *J. Am. Chem. Soc., 88:*3719 (1969).
34. *Beilstein Organische Chemie* (1930–49), Ed. III, Bd. 1, pp. 1761–1762, 1774, 1789–1790. See Ref. 1
35. Allesandrin, C. G., U.S. patent 4,119,659 (1978).
36. McGuinness, J. A., and Bell, A. R., Eur. patent 93610 (1983).
37. Bayer, H. O., Swithenbank, C., and Yih, R. Y., U.S. patent 4,422,868 (1983); Swithenbank, C., U.S. patents 4,419,122, 4,419,123 (1983), 4,358,308 (1982).
38. Wheeler, T. M., U.S. patent 4,338,122 (1982).
39. Magee, P. S., U.S. patents 4,321,080 (1982), 4,196,152 (1980), 3,984,481, 3,988,376 (1976).
40. Unger, S. H. and Walker, K. A. M., U.S. patent 4,036,973 (1977); Walker, K. A. M., and Marx, M., U.S. patents 4,036,974, 4,036,975 (1977).
41. Abdulla, R., U.S. patent 4,127,581 (1978).
42. Henrick, C. A., and Siddall, J. B., U.S. patent 3,865,852 (1975); Schaub, F., and Schelling, H.-P., U.S. patents 3,925,429, 3,920,711 (1975), 3,932,485 (1976).
43. Schaub, F., and Schelling, H.-P., U.S. patents 4,025,641 (1977), 3,978,097 (1976).
44. Schaub, F., and Schelling, H.-P., U.S. patents 3,935,271, 3,978,134 (1976); Paul, J. H., U.S. patent 4,218,468 (1980).
45. Farooq, S., and Karrer, F., U.S. patents 3,998,891, 3,998,890 (1976); Baum, J. W., U.S. patent 3,891,714 (1975); Roman, S. A., U.S. patent 4,219,562 (1980); Berkelhammer, G., and Kameshwaran, V., U.S. patent 4,199,595 (1980).
46. Punja, N., U.S. patents 4,429,153 (1984), 4,370,346 (1983).
47. Oswald, A., and Schmit, G., U.S. patent 3,904,711 (1975); Oswald, A., and Valint, P., U.S. patent 3,904,710 (1975); Oswald, A., U.S. patent 3,880,735 (1975); Arold, H., and Diehr, H.-J., U.S. patent 4,082,822 (1978).
48. Hendrick, C., and Siddall, J., U.S. patents 3,882,156, 3,897,473, 3,906,020 (1975); Hendrick, C., U.S. patent 3,882,157 (1975).
49. Blackwell, J. T., U.S. patent 4,327,033 (1982).
50. Elbe, H.-L., and Kramer, W., U.S. patent 4,371,708 (1983).
51. Fayter, R. G., Jr., and Hall, A. L., U.S. patent 4,469,891 (1984).
52. Gohbara, M., Inoue, T., Ishii, T., Kodaka, K., Nakatani, K., Numata, S., Tachibana, H., Toyama, T., and Udagawa, T., U.S. patent 4,397,864 (1983).
53. Griesbaum, K., Hall, D. N., and Oswald, A, U.S. patent 3,859,360 (1975).
54. Dawes, D., and Thummel, R., U.S. patent 4,055,572 (1977).
55. Flege, H., and Wedemeyer, K., U.S. patent 4,358,616 (1982). Wedemeyer, K., and Flege, H., U.S. patent 4,304,940 (1981).

56. Mathew, C. T., and Ulmer, H. E., U.S. patent 3,931,331 (1976).

57. Fankhauser, E., and Sturm, E., U.S. patent 4,459,152 (1984); Bohnert, T. J., and Dunbar, J. E., U.S. patent 3,943,176 (1976).

58. Arneklev, D. R., and Baker, D. R., U.S. patent 4,030,911 (1977).

59. Schuette, H., and Scholler, C., U.S. patent 2,129,709 (1938); Schuette, H., Scholler, C., and Wittwer, M., U.S. patent 2,205,021 (1940).

60. Lemaire, H., in *Nonionic Surfactants.* (M. J. Schick, ed.), Marcel Dekker, New York, 1966.

61. Kang, Y., Nield, G. L., and Washecheck, P. H., Eur. patent 46947 (1982).

62. Shvets, V. F., and Lykov, Yu. V., *Kinet. Katal., 12:* 347, 883 (1971); *Tr. Mosk. Khim.-Tekhnol. Inst., 66:* 33 (1970).

63. Onopchenko, A., and Schultz, J., U.S. patents 4,102,932 (1978), 4,009,211 (1977).

64. Lamberti, V., and Lemaire, H., U.S. patent 3,988,377 (1976) and references cited within.

65. Sebag, H., and Vanlerberge, G., U.S. patent 3,984,480 (1976).

66. Sebag, H., and Vanlerberge, G., U.S. patent 4,105,580 (1978).

67. Suzuki, S., U.S. patent 4,172,211 (1979).

68. Jpn. patents J80033715, J80033716, J80030707 (1980).

69. Mawn, P. E., Outlook for lube oil additives, *Impact* (L760704), July 29, 1976.

70. Ripple, D. E., U.S. patents 3,821,236 (1974), 3,904,537 (1975).

71. Thaler, W. A., and McDivitt, J. R., *J. Org. Chem., 36:* 14 (1971).

72. Michaelis, K. P., and Wirth, H. O., U.S. patents 4,246,127 (1981), 4,147,666 (1979); Chakrabarti, P. M., Tracy, D. J., and Wood, L. S., U.S. patents 4,259,474, 4,255,561 (1981).

73. Wintringham, H. J., U.S. patent 4,327,144 (1982).

74. O'Brien, J. P., and Papay, A. G., U.S. patent 4,204,969 (1980); Jayne, G. J. J., and Woods, D. R., U.S. patent 4,188,297 (1980); Gast, L. E., Kenney, H. E., and Schwab, A. W., U.S. patent 4,218,332 (1980); Hotten, B. W., U.S. patent 4,132,659 (1979); Jayne, G. J., O'Brien, J. P., and Woods, D. R., U.S. patent 4,147,640 (1979).

75. Davis, K. E., U.S. patents 4,119,549 (1978); 4,191,659 (1980); Davis, K. E., and Holden, T. F., U.S. patents 4,119,550 (1978); 4,344,854 (1982); Braid, M., U.S. patent 4,240,958 (1980); Horodysky, A. G., and Landis, P. S., U.S. patent 4,225,488 (1980).

76. Horodysky, A. G., U.S. patent 4,200,546 (1980).

77. Spence, J. R., and West, C. T., U.S. patent 4,362,633 (1982).

78. Horodysky, A. G., and Kaminski, J. M., U.S. patent 4,486,322 (1984).

79. Anon., British patent GB 1,483,991-A (1977).
80. Gutierrez, A., Brois, S. J., Ryer, J., and Deen, H. E., U.S. patent 4,411,808 (1983).
81. Michaelis, P., U.S. patents 4,217,233 (1980), 4,260,503 (1981).
82. Liston, T. V., and Lowe, W., U.S. patent 4,228,022 (1980).
83. Frenier, W. W., and Settineri, W. J., U.S. patent 3,969,414 (1976).
84. Worrel, C. J., U.S. patent 3,487,131 (1969); Arakelian, A. N., Hopkins, T. R., and Rhodes, A., U.S. patent 3,211,649 (1965); Fields, E. K., U.S. patent 2,753,306 (1956).
85. Schroeck, C. W., U.S. patent 4,289,635 (1981).
86. Ashew, H., U.S. patent 4,012,331 (1977).
87. DeVries, L., U.S. patent 4,044,039 (1977).
88. Anon., Jpn. patent JP 8381722 (1983).
89. Diamond, J., U.S. patent 4,105,786 (1978).
90. Varma, R., U.S. patent 4,361,559 (1982).
91. Carson, J., U.S. patent 4,213,905 (1980); Fujimoto, T., Kondo, K., Suda, M., and Tunemoto, D., U.S. patent 4,268,442 (1981).
92. Koppel, G., U.S. patent 4,008,230 (1977); Wright, I., U.S. patent 3,941,781 (1976).
93. Celmer, W., U.S. patent 3,923,791 (1975).
94. Kaiser, C., and Ross, S. T., U.S. patents 3,976,694 (1976); 3,917,704 (1975).
95. Kathawala, F. G., U.S. patents 4,251,521, 4,246,259 (1981); 4,233,292 (1980).
96. Krause, H. P., Linke, S. Mardin, M., and Sitt, R., U.S. patent 4,330,677 (1982).
97. Hiestand, P., and Strasser, M., U.S. patent 4,407,825 (1983).
98. Chodkiewicz, M., U.S. patent 4,032, 534 (1977).
99. Blohm, T. R., Grisar, J. M., and Parker, R. A., U.S. patent 4,187,319 (1980).
100. Jones, H., and Shen, T.-Y., U.S. patent 3,860 636 (1975).
101. Hirsch, A. F., U.S. patent 4,086,278 (1978).
102. Snyers, M., Gillet, C., Roba, J., Niebes, P., Lambelin, G., and Lenaers, A., British patent GB 2,083,036-A (1982).
103. Algieri, A. A., and Crenshaw, R. R., U.S. patents 4,200,760, 4,221,737 (1980).
104. Anon., Jpn. patent J 52118410-A (1977).
105. Anon., Jpn. patent J 53007619-A (1978).
106. Hersler, E., and Walker, J., U.S. patents 4,404,142, 4,401,599 (1983).
107. *Pennwalt Corporation Publication S-222B;* Brandrup, J., and Immergut, E. H., *Polymer Handbook,* 2nd ed., Wiley-Interscience, New York, (1975).

108. Mayo, F. R. *J. Am. Chem. Soc., 65:* 2324 (1943); Pryor, W. A., *Mechanisms of Sulfur Reactions,* McGraw-Hill, New York, 1962, pp. 82–88.

109. Richmond, H., U.S. patent 3,917,714 (1975).

110. Dexter, M., and Steinberg, D. H., U.S. patents 4,071,497 (1978); 3,954,839 (1976).

111. Dexter, M., and Steinberg, D. H., U.S. patent 3,987,086 (1976).

112. Rosenberger, S., U.S. patent 4,055,539 (1977).

113. Lind, H., U.S. patent 4,021,468 (1977).

114. Snel, M. A., U.S. patent 4,369,273 (1983).

115. Hoffer, K., Jr., and Hoffer, K., Sr., U.S. patent 4,042,632 (1977).

116. Goethal, E. J., *Topics in Sulfur Chemistry,* Vol. 3, (A. Senning, P. S. Magee, ed.), Georg Thieme, Stuttgart, 1977.

117. Langsley, G. W., U.S. patent 3,985,710 (1976).

118. Nishi, N., and Takashi, S., U.S. patent 4,107,156 (1978).

119. Marvel, C. S., Shen, E. H. H., and Chambers, R. R., *J. Am. Chem. Soc., 72:*2106 (1950); anon., Neth. patent 6,510,593 (1965); Schlack, P., Ger. patent 919,667 (1956); Lennke, G., and McReynolds, K., U. S. patents 3,997,614, 3,997,613, 3,997,612 (1976).

120. Antkowiak, T., Cheng, T., and Schultz, D., U.S. patent 4,258,173 (1981); Halosa, A., and Hergenrother, W., U.S. patent 4,258, 171 (1981) and reference cited within.

121. Chrisholm, J. A., Kneibes, D. V., and Stubbs, R. C.; Study of the properties of the numerous odorants and assessment of their effectiveness in various environmental conditions to alert people to the presence of natural gas, *Department of Transportation Report DOT/OPSO-75/08* (1975); Nevers A. D., U.S. patent 3,826,631 (1974).

122. Wiechers, A., U.S. patent 4,211,644 (1980).

123. Burness, D. M., Maggiulli, C. A., and Perkins, W. C., U.S. patent 4,100,200 (1978).

124. Arold, H. U.S. patent 4,091,037 (1978).

125. Volkov, A. N., Volkova, K. A., and Levanova, E. P. Soviet Union patent SU 806679 (1981).

126. Hallie, H., *Oil Gas J.*, 70 (Dec. 20, 1982).

CHAPTER 15
Applications in the Fiber Industries

JOE D. SAUER Ethyl Corporation, Baton Rouge, Louisiana

1 INTRODUCTION

Several specialty industries have performance requirements that are very similar for a variety of process chemicals and chemical intermediates. These industries include those dealing with fibers or fibrous surfaces, such as leather processing, textile manufacture and fabrication,and paper production. These industries have a common requirement, the need to modify a surface of a fibrous or complex substrate to impart a desirable property such as appearance, feel, or water repellency. These modifications may be achieved through the use of various specialty chemicals, many of which can be derived via one or another routes from alpha olefin feedstocks. As a consequence, consideration of these specialty applications are combined in a single chapter.

Currently, large quantities of alpha olefins do not enter these markets in any single application. Rather, each of these process industries consumes a variety of alpha olefin derivatives, and the required volumes for olefin and olefin derivatives are increasing.

It is not practical to deal in great depth with descriptions of the manufacturing processes for each of the industries covered in this chapter. The various technologies are thoroughly described in appropriate reviews of the specific subjects. Examples are cited in this chapter covering applications dealing with olefin/olefin derivatives within the larger areas.

2 LEATHER PROCESSING

The leather industry consumes small amounts of pure alpha olefins in their normal operations. Several steps of the conventional leather operation may involve the use of surfactants that can be derived from olefin raw materials. These surfactants include linear alkylbenzene sulfonates; alpha olefin sulfonates; a variety of alcohol-based surfactants that are ultimately derived from olefin via the oxo process, such as alcohol sulfates, alcohol ether sulfates, and alcohol ethoxylates; and finally, fatty-amine-based surfactants prepared by derivatization of amines synthesized from alpha olefins.

2.1 Posttanning Fat Liquoring

While some of the uses of surfactants undoubtedly are general in nature and essentially involve a "laundering" or cleaning step, some are specific and unique to the leather industry. For example, during an intermediate step in leather manufacture, the tanned hides are subjected to a process called "fat liquoring." This procedure is carried out to lubricate the leather and make it more supple and flexible. Normally, during this step, the hides are soaked in an emulsion of water, oils, and surfactant. While each processor relies on formulations developed by empirical methods, the goal of each is to fill the pores of the leather with the emulsified oils. Once in the pores, the emulsion "breaks" and the oils and surfactant are deposited while the solvent, mostly water, is carried away. One of the principal functions of the fat-liquor oils is to decrease the cohesiveness of the leather during subsequent drying steps.

The emulsifier package utilized is customarily based on an anionic surfactant for practical, physical reasons. The pores of the substrate are usually electrically positive during this phase of leather manufacture, because of both the chemical makeup of the substrate leather and the pH of the particular process preceding the fat-liquor application. Modifications to the operation, or specific intent to produce leather of special qualities, may result in the leather having a nearly neutral or even negative charge at the surface. If this is the case, nonionic or cationic surfactants would be utilized in place of the anionic type. The result (i.e. the appearance of the finished leather) would be virtually the same as that described for the standard fat-liquor process. The positive charge electrostatically directs the anionic surfactant-rich emulsion to the pore surface, where the charges are neutralized and the surfactant "precipitates." As mentioned earlier, this effect "breaks" the emulsion, and leaves the free oils, along with the complexed surfactant, in the interior of the leather. In this manner, the leather is essentially "plasticized" or made flexible and permanently lubricated.

Traditionally, the free oils used in this procedure have included naturally occuring materials, such as animal, vegetable, or mineral oils, and the surfactants utilized have been prepared from natural sources as well. An excellent, natural product—derived surfactant that has been used is sulfonated sperm oil.

Sperm oil is comprised of monoesters derived mainly from C_{14}, C_{16}, and C_{18} monocarboxylic acids with C_{16} and C_{18} linear alcohols. The carboxylic acid portion of the oil is approximately 75% monounsaturated, while the alcohol portion is about 60% monounsaturated. The resulting monoolefinic chains are responsible for the desirable physical stability of this oil as well as the useful physical properties (melting point, polarity, etc.). Its chemical composition sets this natural oil apart from most other naturally occuring liquid fats, which are typically triglycerides and often contain multiple unsaturation.

Sperm oil, however, is now in very limited supply due to the almost worldwide prohibition on whaling or use of whale-derived products. A very good substitute, fortunately, can be obtained by using heavy olefins as a feed for sulfonation, sulfation, and/or sulfitation. The resulting surfactants provide desirable end properties (often referred to as "hand") to the finished leather goods which are nearly indistinguishable from those obtained when using the more traditional fat-liquoring agents. In addition to the sulfonated, sulfated, or sulfited olefin-type surfactants, alcohol sulfates and alcohol ether sulfates can be utilized in a similar fashion for some of these applications. The olefin-based surfactants typically have carbon chains in the region C_{12}–C_{26}; the alcohol-derived surfactants, on the other hand, more likely contain chains from C_{16}–C_{30}. The exact molecular weights may indeed vary considerably according to the particular formulation utilized and the type of finished leathers required. Obviously, different materials will be utilized in the production of fine glove leathers than when the finished product is to be used in the manufacture of "uppers" for shoe leather or for belting . While small amounts of other surfactants are also used, they are generally not as important as the classes already mentioned.

Some leather producers also incorporate small amounts of fat-liquor-type emulsions into other steps of leather production, including the tanning step. Reportedly, this is done to decrease the total requirement for fat-liquor oils in the entire process. The actual, specific fat-liquoring step can also be done simultaneously with the dyeing process, and is occasionally carried out in this manner.

2.2 Finishing Steps

Some leather, especially that used in production of shoe uppers, is rehydrated and mechanically softened prior to subsequent finishing. This step involves filling the leather with polymeric materials. Examples of polymers utilized include pyroxylin lacquers, polyacrylate emulsions and polyacrylonitriles. These specialty finishes provide increased scuff resistance, enhanced luster, and improved water resistance without drastically affecting vapor permeability of the leather.

2.3 Preservatives in Leather Manufacture

Biocidal additives are another fairly important group of olefin derivatives user in the leather industry. In the preparation of leather goods, many operations involve damp or wet leather in various degrees of finish. These materials are extremely susceptible to attack by mildew, mold, and various bacterial growths. Relatively small amounts of alkyldimethlbenzylammonium chloride can be added to the solution involved in the various processing steps to control the unwanted organisms. Typically, use levels of less than 1% are adequate for control purposes. The quaternary compounds are derived from alkyldimethylamine, which can be prepared directly from alpha olefin or, indirectly, from linear alcohols (see Chapter 11). Other preservatives utilized include phenolics, as well as chlorinated phenols.

Of the areas treated in this chapter, the leather industry is probably the oldest in continuous operation. Although leather and furs have been processed for thousands of years, leather technology still contains much complex, not often completely understood art. In part, this is due to the fact that the final goods, leathers and furs, have value and sales primarily dependent on their beauty and feel, and only secondarily on their chemical and measurable physical properties.

3 PAPER PRODUCTION

Writing paper, as a medium for record keeping and transmission of information, has been a significant factor in the development of both culture and civilization. Over the past several decades, the importance of paper as an archival and communication medium has decreased as other applications have grown more rapidly in volume requirements. In 1969, for example, the U.S. paper capacity was estimated to be 46,000,000 tons per year. Of that amount, only about 2,700,000 tons were accounted for by writing and fine papers, with 2,200,000 tons newsprint and 6,500,000 tons by printing papers. The combined output of paper for data collection/ transmission, therefore, amounted to slightly less than 30% of the total production figure. The remaining portion of paper production obviously is consumed in a variety of end uses not even dreamed of a few decades ago. Paper now has utility as a packaging medium, in disposable clothing, in construction materials, and so forth. If the changes in end uses of paper have been significant, it should be expected that major changes in the art/ science of paper manufacturing also have been significant. One change in the technology has been in the increasing number of specialty chemical adjuvants used by the industry.

Today's paper technology is flexible and its chemical and physico-chemical aspects are tremendously important in determining the final characteristics of the paper being produced. A brief outline of the general paper process is given to help understand the actual and potential role of specialty chemicals in this industry.

Raw materials supplied to the paper mill include (1) fibers, usually cellulosic wood-pulp fiber, although cotton, polymerics, glass and mineral fibers, grasses and reeds, agricultural residues, and reclaimed paper-products are also utilized in different paper types; (2) fillers, such as clays, titanium dioxide, and calcium carbonate; (3) sizing agents, to modify water resistance/repellency of the paper, traditionally rosin/alum, but also waxes, polymerics, and cellulosic-reactive sizes such as alkylketene dimer and alkenylsuccinic anhydrides; and (4) miscellaneous additives, including dyestuffs, wet-strength agents, defoamers, biocides, and deflocculating agents. The continuous paper machines in use today convert a dilute aqueous slurry of the foregoing components into a dry web or sheet of paper at rates varying from a few feet to ½ mile per minute. The process may vary depending on the type of paper machine being used. However, the dilute papermaking slurry is fed onto a moving support, which allows the paper sheet to form and the water to drain from the sheet in a uniform manner. As the wet sheet travels through the machine, slight vacuum is applied and more water is removed until the sheet is approximately 30 to 35% solids. The paper web then progresses into a dryer section and the remaining water is evaporated as the sheet passes over steam-heated rollers. This process is used in all continuous operations; the Fourdrinier machine utilizes a moving wire screen for the initial support, while the cylinder machine has a rotary cylindrical filter in the wet end of the operation.

3.1 General Sizing Processes

In modern usage, the term "sizing" is applied to the process in which a chemical additive provides paper and paperboard with resistance to liquid wetting, penetration, and absorption. On a practical basis, the liquid usually involved is an aqueous fluid such as ink, milk, or water. Cellulose in itself is a hydrophilic material. Thus the pulp fiber surfaces have a high specific energy. As a consequence, water readily wets these surfaces. Sizing agents are used to modify the surfaces selectivity and alter their energy characteristics. The simple diagram if Figure 1 illustrates the manner in which the modification occurs. The sizing agent is normally an amphipathic molecule. The polar (hydrophilic) end of the sizing agent can be attracted to the paper surface while the nonpolar (hydrophobic) end is directed away from the surface and becomes a new boundary layer that no longer has a large affinity for polar materials such as water.

Rosin/Alum Sizes

The traditional sizing agent for most paper manufacturing has been rosin/alum. While rosin is a very complicated natural product–derived material, it consists essentially of the sodium soaps of a mixture of high-molecular-weight organic acids. The carboxylic acid function provides a polar group, while the bulk of the molecule is hydrocarbon and nonpolar (Fig. 2).

The role of papermakers' alum in this system is first to be electrostatically attracted to the rosin particle. Second, the positively charged rosin/alum complex is electrostatically attracted to the negatively charged paper fiber surface and is oriented in the desirable manner described earlier. The actual species present has not been completely defined, and the chemistry involved is apparantly complex. The usual order of addition, which can be reversed, is rosin first, followed by alum. Both ingredients are added as aqueous emulsions to the already prepared fiber suspension.

For proper sizing, the pH of the system is maintained in a distinctly acid region (<4.5) and the paper is typically filled with various compositions of clay/titanium dioxide. Typical use levels of the rosin size is 5 to 10 lb/ton finished paper. Estimates of rosin sizing production (sodium soap) indicated that in 1973 alone, 160,000 tons were consumed by the paper industry in the manufacture of 61,900,000 tons of paper and paperboard.

Cellulose-Reactive Sizes; Alkylketene Dimer

The first synthetic, cellulose-reactive size utilized was alkylketene dimer (AKD), formed from long-chain fatty acids, having the general structure

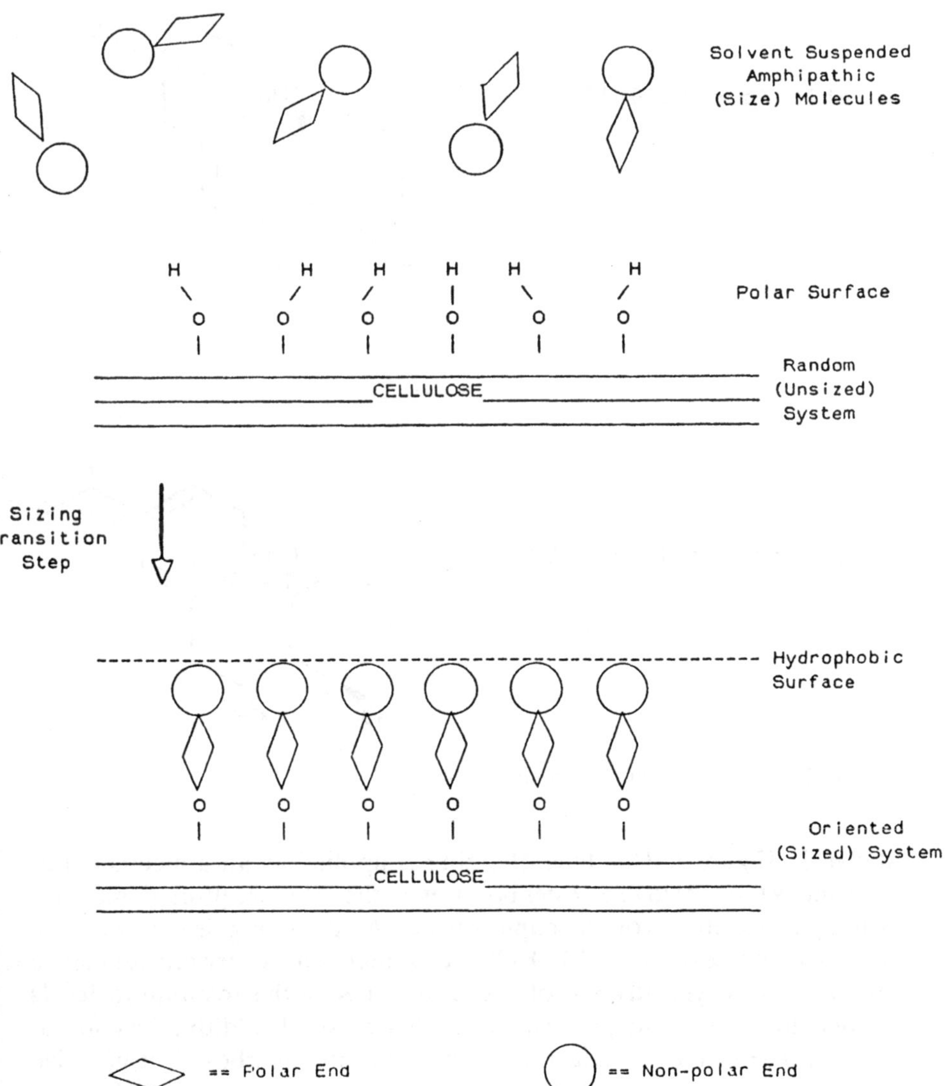

Figure 1 Schematic sizing diagram.

Abietic Acid [514-10-3]

Levopimaric Acid [79-54-9]

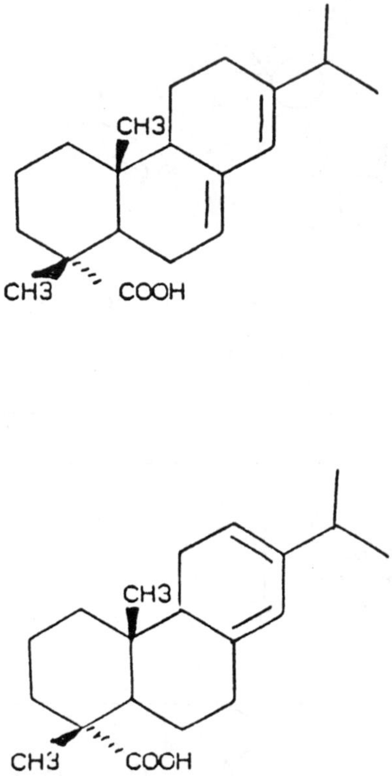

Figure 2 Typical structure of rosin acids.

shown in Figure 3. This type of sizing is applied as an aqueous emulsion and is thought to form ester bonds with the cellulose through reaction with hydroxyl groups on the paper surface. Actual sizing develops during the drying process (around 100°C). AKD treatment is very efficient since end-use levels typically are of the order of <1/10 the rosin/alum levels needed for equivalent performance. This class of additive has found broad usage in the manufacture of milk-carton paperboard; it has also become established in Europe. U.S. utilization has been slower to develop, since the paper machines in this country are faster and the prolonged sizing times needed for AKD introduces response-time difficulties judged unacceptable by many of the mill operators.

It has been reported, however, that new, second-generation sizes based on alkylketene dimer have faster curing times and may eliminate some of this objection. It remains to be seen whether this turn would lead to a dimer system that would be more hydrolytically unstable and would present new processing difficulties to be overcome.

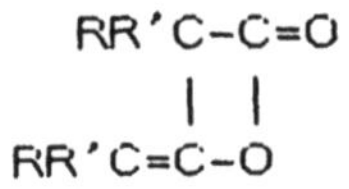

```
    RR'C-C=O
        | |
  RR'C=C-O
```

Typically R = H; R' = Alkyl, C15-C20 range

Figure 3 Alkylketene dimer structures.

Cellulose-Reactive Sizes; Alkenylsuccinic Anhydride

Alkenylsuccinic anhydrides (ASA) comprise a newer class of cellulose reactive sizing. These materials are derived from linear alpha olefins by the "ene reaction" (Figure 4). The first step in the process involves isomerization of an alpha olefin to an internal olefin. The latter is then heated with maleic anhydride to form the branched alkenylsuccinic anhydride in high yield. The isomerized olefin leads to a liquid ASA, which is preferred over the solid, waxy product generally obtained from alpha olefins. The liquid ASA is much easier to handle than the solid product. In paper production practice, the ASA is introduced into the paper process as a dilute aqueous emulsion. Since ASA is very reactive with respect to hydrolysis of the anhydride group, this step is carried out only moments prior to actual addition to the paper slurry. While this hydrolytic instability is a factor in the use of ASA and requires in-plant mixing and emulsification rather than shipment of stable aqueous slurries, the very reactivity of the ASA is a major advantage in other respects. ASA sizing develops "on the machine" or as the paper is being produced. Process changes can be made and monitored in a "real-time" mode with the ASA technology. Thus manufacturing to specifications becomes a much simpler task than in the case of alkylketene dimer systems, where sizing actually only fully develops after the paper is removed from the machine.

Practical advantages of ASA when compared to rosin/alum sizing are many, First, ASA is, like alkylketene dimer, a highly efficient size and end use levels of 1/10 that required of the rosin system are possible. Additionally, it is possible, and in fact desirable, to process the paper at neutral or alkaline pH conditions with ASA. Thus the paper has much improved aging properties compared with that prepared under acidic conditions. Another benefit can be realized as a result of the process pH change; that is the potential of switching from clay/titanium dioxide filler systems to calcium carbonate fillers. The carbonate fillers typically are much less expensive, are much brighter (and subsequently require less whiteners), and

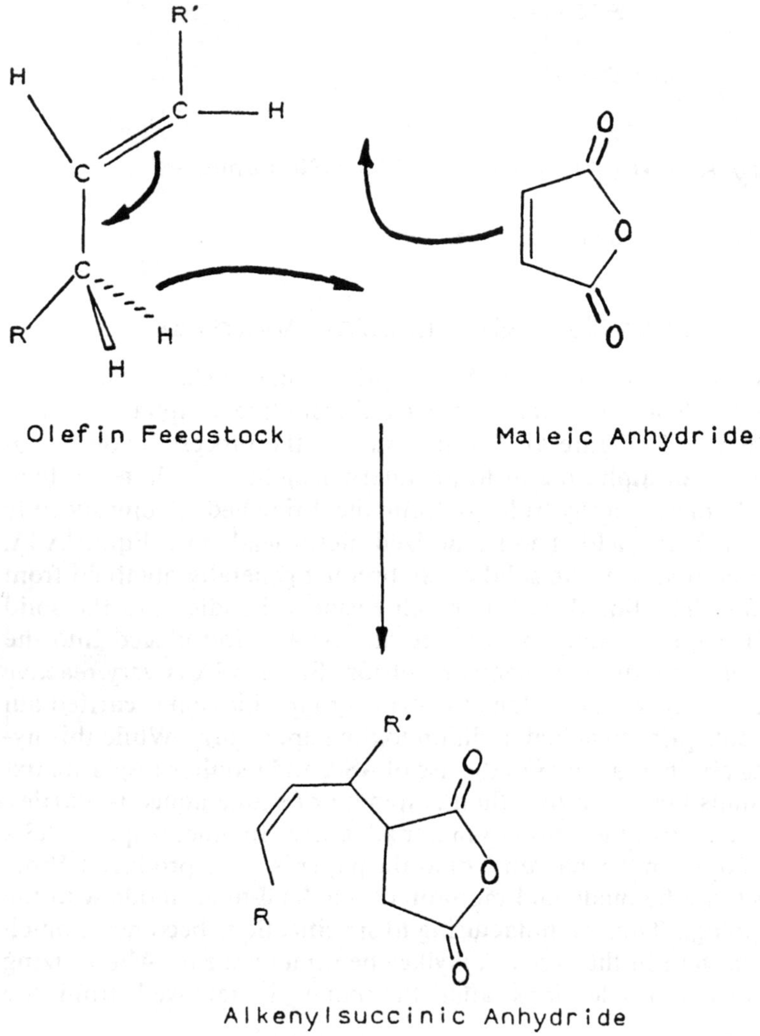

Olefin Feedstock

Maleic Anhydride

Alkenylsuccinic Anhydride

Figure 4 Typical products of the ene reaction.

result in paper having much higher strengths than those typically observed in papers produced with the clay systems. The paper manufacturer may switch to lower grades of fiber and/or lower fiber content (higher filler content) and still maintain end product specifications. These factors all provide for decreased raw material costs. Higher filler loadings also allow higher mill speeds, and consequently lower manufacturing costs. This is largely due to increased wet strengths and improved dewatering in the ASA-sized, carbonate-filled papers. Although it is difficult to provide exact economics for these changes, the incentives are reported to be significant.

Another significant factor is the environmental impact of paper production. In the mills practicing traditional paper manufacture with rosin/alum size, "white water" pollution is significant and very difficult to minimize. In the mills converted to ASA sizing and operating at neutral to alkaline conditions, it has been demonstrated that water pollution can be minimized to the point of practical elimination as a concern. Some mill conversions to alkaline sizing have been based on this consideration alone.

There are some technical difficulties that must be overcome when ASA sizing is utilized. First, and most obvious, the existing rosin/alum process must be changed and the mill must go through a conversion process, which can be expensive and requires managerial commitment. Times for conversion can be six weeks or longer. A total changeover of the water treatment system must be carried out to accommodate the new pH requirements. Special attention must be given to items such as foam control, biocidal control, sizing retention aids, and so forth. The complexity of this technological change is such that specialty chemical companies have developed teams of technical service personnel to aid in the changeover.

A second major obstacle has been that the finished ASA-sized papers display slightly different properties than rosin/alum-sized paper. Paper users have had to approve and qualify the stocks for their applications. Ink manufacturers also have had to modify their products to achieve acceptable results. Ultimately, a new mix of paper products will be available and the specialty chemical market, until such changes have been completed, will continue to undergo varying degrees of change in products.

3.2 Other Specialty Chemicals for Paper Manufacture

A great many specialty chemicals are prepared and sold to the paper manufacturing industry. Many are general surfactants and have been discussed in other chapters. Others are of a more specialized nature and will be touched on briefly here. Many of these products are utilized in water treatment. Examples include biocides, defoamers, retention aids, and flocculation aids. Often, the individual chemical "active" is based on a molecular species derived either directly or indirectly from olefins. For more detailed information on these areas, refer especially to Chapter 11, since many of the useful products are derived from fatty amines (and in turn from alpha olefins).

4 TEXTILE PRODUCTION

Textile manufacturing, like leather and paper manufacturing, is ubiquitous in today's society. These operations are interwoven with our life-styles to an extent that would render it difficult to consider day-to-day activities without their respective products. Items provided by the textile industry include protection from the environment, shelter, comfort, and decoration (apparel); drapes, carpets, bedding (household items): and a multitude of mechanical aids, such as belting for machinery, fabric for tires, and insulation materials for construction.

World production in 1980 for assorted textiles (both natural and man-made) was estimated at >30 million metric tons. Approximately 14 million metric tons represent natural fibers such as cotton, wool, and silk; the balance consisted of cellulosics and synthetic polymers such as rayon, polyester, nylon, acrylics, polyolefins, and so forth. Olefins obviously play an important role in the manufacture of many of these synthetic materials, as covered in other chapters in this book.

In this chapter we consider several separate stages of textile production, including fiber production, fabric manufacture, and fabric finishing steps. Within each of these stages, several operations occur. It is not possible to generalize from one fabric type to another when describing processing, since different fiber compositions often require different procedures. For the purpose of this chapter, we consider only a few of the actual steps.

Both the fiber production and fabric manufacture stages require much handling and processing of filamentous materials. The spinning, drawing, weaving, carding, and miscellaneous other steps all have one thing in common. They all expend energy on a fiber surface, and as a consequence tend to develop static charges, dust (in copious amounts) and altered fiber states. Many chemicals are utilized to minimize the side effects of handling, and many of these chemicals are derived from alpha olefins. Examples include heavy alcohols as dust control agents, quaternary ammonium compounds to eliminate static charge buildup and alkenylsuccinic anhydrides as sizing agents to alter the water affinity of the fiber surface. Synthesis of these materials has been covered in other chapters and will not be reviewed here.

The final stage of textile manufacture, fabric finishing, again is extremely complex and is highly dependent on the nature of the fabric being processed. Typical steps include dyeing or printing the fabric, adding a finish to eliminate wrinkling, introducing flame retardancy, imparting soiling resistance, and so forth. In short, the primary purpose of chemical finishing is to confer special functional properties to the fabric.

4.1 Specialty Chemicals for Textile Manufacture

A major opportunity for use of alpha olefin derivatives exists within the realm of specialty chemicals for the textile industry. Many of the derivatives utilized, as with leather, are surfactants such as alpha olefin sulfonates, alcohol ethoxylates, and alcohol ether sulfates (from olefin-derived alcohols). These surfactants are useful in many operations, including fiber preparation, fiber swelling, and degreasing.

Another general class of materials having broad utilization in this industry is cationic surfactants such as fatty amine quaternary ammonium halides. These quaternary ammonium compounds serve in a variety of applications, including static control agents during fiber/fabric handling; dye levelers to control the quality of dyeing; antimicrobial agents to control mildew, mold, or other biogens that adversely affect fabrics in damp, warm climate; odor control agents; softening agents; and wrinkle-suppression agents. Typically, these "quats" consist of materials such as those illustrated in Figure 5. While usage of these materials in the United States accounts for a relatively small volume of alpha olefin (<30 million pounds per year), the area is believed to be growing rapidly in volume. As energy considerations become more and more important and traditional specialty chemicals are shown to possess objectionable safety/environmental qualities, the usage of the alpha olefin-derived fatty amine derivatives will increase. More information on these materials is given in Chapter 11.

Textile production, as a general topic, is extremely broad. At least 10 different commercially important textile fibers are in use today, and applica-

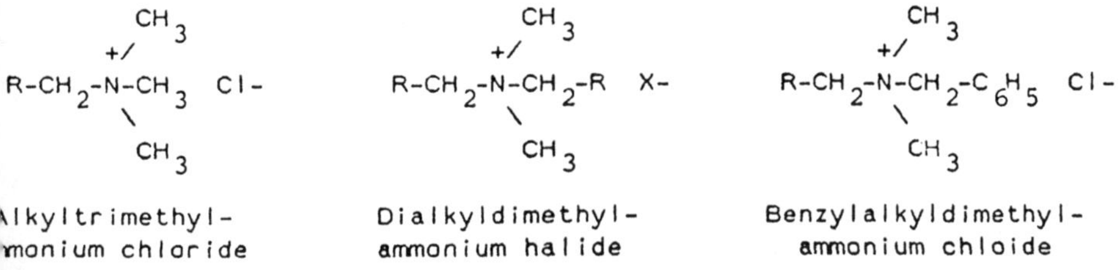

$R\text{-}CH_2\text{-}N^{+}(CH_3)_2\text{-}CH_3 \quad Cl^-$

lkyltrimethyl-
monium chloride

$R\text{-}CH_2\text{-}N^{+}(CH_3)_2\text{-}CH_2\text{-}R \quad X^-$

Dialkyldimethyl-
ammonium halide

$R\text{-}CH_2\text{-}N^{+}(CH_3)_2\text{-}CH_2\text{-}C_6H_5 \quad Cl^-$

Benzylalkyldimethyl-
ammonium chloide

Figure 5 Examples of biocidal quaternary ammonium compounds.

tion of specialty chemicals in their processing can enhance a wide selection of properties already mentioned. The specific manufacturing steps and techniques are based on processes that were developed to accommodate the unique geometries/physical properties/surface characteristics of the individual fibers involved. As a consequence, only a brief treatment can be given here.

CHAPTER 16

Miscellaneous Uses

GEORGE R. LAPPIN and JOE D. SAUER Ethyl Corporation, Baton Rouge, Louisiana

1 INTRODUCTION

Some of the lower-volume uses for alpha olefins are noted in this chapter. The 1987–1988 annual volume estimate of 25 to 30 million pounds of alpha olefins required for this category is approximate. The following five areas may be considered to be the most important, within the miscellaneous set, at this time: (1) epoxides, (2) chlorinated olefins, (3) waxes, (4) fuel additives, and (5) drag flow reducers. The following notes for these particular applications areas are included for perspective.

2 EPOXIDES

As discussed in Chapter 2, alpha olefins may be converted to epoxides by several oxidation methods. The epoxides may be used as modifiers for epoxy resins and as polyether ingredients for polyurethanes.

3 CHLORINATED OLEFINS

Various alpha olefins from C_{10} through C_{24} are chlorinated to produce chlorinated olefins, which can be used as replacements for chlorinated paraffins. One major use for chlorinated paraffins or olefins is as additives to the hydraulic oil for high-pressure presses and other expensive equipment. These materials may also be used as secondary plasticizers in PVC formulations and in various flame-retardant applications.

4 WAXES

Some of the Chevron/Gulf higher olefins have high melting points and exhibit good wax properties. They have been used in specialty wax applications. The following data are taken from the Gulf product bulletin: The C_{24-28} melts at 145°F and the C_{30+} melts at 180°F. "C-30+ improves wax blends for candles, crayons, and coatings. These fractions may be further reacted to simulate more expensive carnauba or Montan waxes."

5 FUEL ADDITIVES

To improve the performance in smaller, more-fuel-efficient engines, there is a trend to using fuel injectors in automobile engines. This trend leads some automotive experts to forecast that 90% of U.S. automobiles will be converted to fuel injection systems by 1990. These injectors have very small openings that are easily fouled and subsequently are expensive to clean by mechanical methods. This factor is creating a much greater demand than expected for efficient fuel additives and fuel antifouling packages. Many types of fuel additives may be produced by the use of alpha olefin-derived oligomers, alkenylsuccinic anhydrides, and fatty amines.

6 DRAG FLOW REDUCERS

Flow through long pipelines is significantly impeded due to viscosity and friction. Addition of a flow improver at strategically located pumping stations can result in dramatic increases in observed pipeline capacity. Literature reports indicate that C_4, C_8, C_{10}, and C_{12} alpha olefin polymers have been utilized to reduce pipeline drag.

7 OTHER USES

Additional, less-defined areas felt to have potential for increased alpha olefin usage include: (1) agricultural chemical intermediates/synergists, (2) antioxidents, (3) miscellaneous automotive uses, and (4) water treatment chemicals. While current volumes for alpha olefin consumption are very small in these markets, projected growth for these developmental categories in the United States might reasonably require an additional 15 to 20 million pounds per year of alpha olefins by 1990–1995.

CHAPTER 17

Toxicology

MICHAEL N. PINKERTON Pinkerton International, Inc., Baton Rouge, Louisiana
ANN M. PETTIGREW Ethyl Corporation, Baton Rouge, Louisiana

1 INTRODUCTION

There is a limited amount of information in the open literature on the toxicology of alpha olefins. The majority of the data cited in this review was obtained through the Freedom of Information Act. Manufacturers of alpha olefins submitted these data in 1984 in response to a U.S. Environmental Protection Agency Data Call In of health effects data on hexene, octene, decene, and dodecene.

Evaluation of the acute toxicity of alpha olefins is based on data from animal studies using hexene, decene, and dodecene and structure-activity relationships analysis. Mutagenicity evaluations are based on data developed from laboratory studies with hexene and decene. Limited information from metabolic detoxification studies is reported for butene and octene. Hexene is the only alpha olefin that has been evaluated in subchronic studies. No data were available for evaluation of the carcinogenicity, reproductive, developmental, or chronic toxicity of the alpha olefins reviewed in this chapter.

2 GENERAL TOXICOLOGY

2.1 Acute Toxicity

Alpha olefins are members of the series of aliphatic hydrocarbons, comprising both the saturated alkane (paraffinic) compounds and the unsaturated alkene (olefinic) compounds. Toxicological properties of the alpha olefins are based largely on structure-activity relationship analysis of similar alkane compounds and limited data from product safety studies with laboratory animals.

Alpha olefins do not exhibit a high degree of acute toxicity. Alpha olefins are practically nontoxic by oral and dermal routes of exposure and are expected to be only minimally irritating to the skin and eyes. Acute inhalation studies with alpha olefins have not demonstrated adverse toxicological effects. Workplace exposures are not expected to exceed these concentrations.

The lower members of the alpha olefin series, which include ethylene, propylene, and l-butene, are characterized toxicologically as simple asphyxiants or weak anesthetics. Anesthetic potency increases to narcosis with increasing chain length (Sandmeyer, 1981).

2.2 Mutagenicity

Based on limited experimental data and structure-activity relationship analysis of similar compounds, alpha olefins are not expected to exhibit mutagenic activity.

2.3 Subchronic Toxicity

Repeated or prolonged exposure to vapors of alpha olefins at high concentrations are expected to result in narcosis. This conclusion is based on toxicological evaluation of hexene and structure-activity considerations.

2.4 Metabolism

Alpha olefins have been shown to be metabolized by an oxidative pathway. This metabolism is mediated by the cytochrome P450-dependent mixed-function oxidase system. The primary metabolite is the epoxide (Oesch, 1973; Lawley and Jarman 1972; Del-Monte et al., 1985).

2.5 Chronic Toxicity

There are no reports in the literature that indicate adverse carcinogenic, reproductive, developmental, or other chronic effects from occupational exposure to alpha olefins.

3 COMPOUND-SPECIFIC TOXICOLOGY

3.1 Ethylene and Propylene

Ethylene and propylene have been used to induce surgical anesthesia at concentrations of 60% and greater in oxygen. The industrial hazards from the flammability of these compounds exceed the toxicological concerns from occupational exposures. Asphyxiation is caused by ethylene and propylene lowering the oxygen concentration. ACGIH has suggested a maximum allowable exposure limit of 1000 ppm due to the asphyxiation hazard *(Threshold Limit Values for Chemical Substances and Physical Agents in the Workroom Environment,* 1978). An excellent and comprehensive review of the toxicological properties of ethylene and propylene can be found in *Patty's Industrial Hygiene and Toxicology* (Sandmeyer, 1981).

3.2 1-Butene

Based on a structure-activity analysis, 1-butene is expected to exhibit a low degree of acute toxicity. As with all alpha olefins the greatest acute health hazard arises from the possibility of aspiration of 1-butene into the lungs after ingestion of the liquid. The aspiration hazard of 1-butene is less than other olefins due to its physical state as a gas at room temperature. Direct contact of liquid 1-butene with the skin or eyes can cause frostbite-type burns if the contact is prolonged or repeated.

At concentrations above its limits of flammability, inhalation of 1-butene vapors causes an anesthetic effect (Deichmann and Gerade, 1969). Inhalation exposure of mice to 1-butene causes incoordination, confusion, and hyperexcitability at 15%. These effects were found to be reversible. After an exposure to 20% 1-butene, test animals exhibited anesthetic effects within 8 to 15 min and respiratory failure in 2 hr. Exposure to 30% 1-butene resulted in anesthetic effects in 2 to 4 min and respiratory failure within an hour. Complete anesthesia within 30 sec and death within 15 min resulted from inhalation exposure of 40% 1-butene (Riggs, 1925).

Metabolic pathways for biodegradation of 1-butene are similar to those of other alpha olefins. 1-Butene is metabolized through an epoxide to the diol derivative (Sandmeyer, 1981). This pathway is expected to form a metabolite that has greater water solubility promoting rapid elimination. In humans, symptoms of overexposure to high concentration of 1-butene vapors include headache, dizziness, drowsiness, asphyxia, narcosis, unconsciousness, and frostbite (*Hazardline Data Base,* 1986).

3.3 1-Hexene

Animal studies have demonstrated that 1-hexene has a low level of acute toxicity. 1-Hexene is practically nontoxic by the oral and dermal routes of exposure. The rat oral LD_{50} of 1-hexene is greater than 5.0 g/kg body weight. The rabbit dermal LD_{50} is greater than 2.0 g/kg body weight (Ethyl, 1984; Shell, 1984). Application of 1-hexene to the skin of rabbits causes minimal skin irritation. Eye irritation studies with 1-hexene show moderate to no irritation in albino rabbits (Shell, 1984; Ethyl, 1984; Sandmeyer, 1981). One percent 1-hexene in ethanol failed to cause skin sensitization in guinea pigs (Shell, 1984).

1-Hexene is an aspiration hazard. Vomiting subsequent to 1-hexene ingestion may lead to aspiration of 1-hexene liquid into the lungs and result in chemical pneumonia and possible death (Gerarde, 1963).

Inhalation of 1-hexene vapors at high concentration has been shown to cause narcotic effects in laboratory animals. The minimal narcotic concentration in mice is 29,100 ppm. The minimal fatal concentration in mice is 40,800 ppm (Lazarew, 1929). Rats exposed for 11 days to concentrations of 300, 1000, 3000, or 8000 ppm of 1-hexene exhibited no clinical signs of toxicity (Shell, 1984). Chronic inhalation exposure of rats to concentrations as high as 8000 ppm of 1-hexene did not cause treatment-related gross or histopathological lesions. Some dose-related clinical signs were reported (Shell, 1984).

Inhalation of high concentrations of 1-hexene vapors by humans causes narcosis, mucous membrane irritation, vertigo, vomiting, cyanosis, and central nervous system effects. Symptoms of the central nervous system effects reported from overexposure to 1-hexene include headache, dizziness, loss of appetite, weakness, and loss of coordination (Sandmeyer, 1981; Chevron, 1984).

1-Hexene is not mutagenic in in vitro short-term assays. No mutagenic activity was demonstrated in the Ames *Salmonella*/microsomal and the Chinese hamster ovary chromosome aberration assays (Shell, 1984).

3.4 1-Octene

There is no information in the toxicology literature on the acute toxicity of 1-octene. It is expected to exhibit the same low degree of acute toxicity as other alpha olefins. The greatest health hazard associated with 1-octene is the aspiration hazard (Gerarde, 1963).

1-Octene is reported to be more irritating to the eyes, skin, and mucous membranes than 1-butene and 1-hexene (Sandmeyer, 1981). This is probably due to the increased contact time on exposed tissues. Increased contact results from the lower vapor pressure and retarded evaporation rate of 1-octene compared to 1-hexene and 1-butene.

1-Octene is metabolized via the 1,2-epoxide to the 1,2-diol, the route common to other alpha olefins (Maynert et al., 1970). More recent studies have demonstrated an alternative pathway wherein 1-octene is metabolized first to oct-1-en-3-ol and subsequently to oct-1-en-3-one (White et al., 1986).

Symptoms of overexposure in humans to 1-octene include eye irritation, vomiting, unconsciousness, central nervous system depression, dizziness, and nausea (*Hazardline Data Base,* 1986).

3.5 1-Decene

Animal studies have demonstrated that 1-decene has a low level of acute toxicity. 1-Decene is practically nontoxic by the oral and dermal routes of exposure. The rat oral LD_{50} is greater than 10 g/kg body weight. The rabbit dermal LD_{50} is greater than 10 g/kg body weight (Shell, 1984). Not unlike other alpha olefins, l-decene is also an aspiration hazard (Gerarde, 1963). Sandmeyer has reported that 1-decene is irritating to the eyes and skin, and exhibits narcotic properties when inhaled at high concentrations (1981). 1-Decene is reported to be nonmutagenic in the Ames *Salmonella/* microsomal assay in the presence and absence of exogenous metabolic activation (Burghardtova et al., 1984).

In humans, overexposure of the eyes, skin, and mucous membrane to vapors of 1-decene causes mild irritation. Inhalation of vapors of 1-decene causes an anesthetic effect and narcotic properties at high concentrations (*Hazard Chemical Data Manual,* 1978).

3.6 1-Dodecene

Animal studies show 1-dodecene to be practically nontoxic. The rat oral LD_{50} is greater than 3200 mg/kg body weight (Shell, 1984; Eastman Kodak, 1984). 1-Dodecene presents the same aspiration hazard as the lower molecular-weight alpha olefins. The rabbit dermal LD_{50} is greater than 10 g/kg body weight (Shell, 1984). Application of 1-dodecene to the skin of rabbits causes minimal skin irritation (Ethyl, 1984) and eye irritation ranging from none to mild-transient (Ethyl, 1984; Eastman Kodak, 1984). Eastman Kodak (1984) also reported that dermal exposure to 1-dodecene elicited skin irritation and death in guinea pigs. The cause of death was not reported.

High vapor concentrations are reported to cause irritation and central nervous system effects in humans (Hawley, 1977). Symptoms of overexposure include eye and skin irritation, abdominal pain, coughing, and nausea (*Hazardline Data Base,* 1986).

3.7 1-Tetradecene and Higher Olefins

No human or animal data were found in the literature on the toxicity of tetradecene or higher molecular-weight alpha olefins, with the exception of a single Hazardline citation. This reference lists eye irritation as a symptom of overexposure to 1-tetradecene (*Hazardline Data Base,* 1986). Based on the physical properties and toxicological evaluation of structurally related chemicals, the higher molecular-weight alpha olefins are not expected to be acutely toxic. The eye and skin irritation potential is expected to be low.

REFERENCES

Burghardtova, K., Horvathova, B., and Valachova, M. (1984). *Biologia [Bratislava]*, *39:* 1121

Chevron (1984). Unpublished data on the production, use, occupational exposure, releases and toxicity of alpha olefins. Submitted to ITC.

Deichmann, W. B., and Gerarde, H. W. (1969). *Toxicology of Drugs and Chemicals,* Academic Press, New York.

Del-Monte, M., Citti, L., and Gervasi, P. G. (1985). *Xenobiotica, 15:* 591.

Eastman Kodak Company (1984). Unpublished toxicity data on 1-dodecene. Submitted to ITC.

Ethyl Corporation (1984). Unpublished data on the production, exposure, and toxicity of alpha olefins. Submitted to ITC.

Gerarde, H. (1963) *Arch. Environ. Health, 6:* 329.

Hawley, G .G. (1977). *The Condensed Chemical Dictionary,* 9th ed., Van Nostrand Reinhold, New York, 434.

Hazard Chemical Data Manual (1978).

Hazardline Data Base (1986).

Lawley, P. D., and Jarman, M. (1972). *Biochem. J., 126:* 895.

Lazarew, N. W. (1929). *Arch. Pathol. Pharm., 143:* 223.

Maynert, E. W., Foreman, R. L., and Watabe, T. (1970). *J. Biol. Chem., 245:* 5234.

Oesch, F. (1973). *Xenobiotica, 3:* 305.

Reynolds, C. (1927). *Anesth. Analg., 6:* 121.

Riggs, L. K. (1925). *J. Am. Pharm. Assoc., 14:* 380.

Sandmeyer, E. E. (1981). Aliphatic hydrocarbons, in *Patty's Industrial Hygiene and Toxicology, 3rd rev. ed.,* Vol. 2B (G. D. Clayton and F. E. Clayton, eds.) p. 3202.

Shell Chemical Company (1984). Unpublished data on the production, exposure, and toxicity of alpha olefins. Submitted to ITC.

Threshold Limit Values for Chemical Substances and Physical Agents in the Workroom Environment (1978). American Conference of Governmental Industrial Hygienists, Cincinnati, Ohio.

White, I. N. H., Green, M. L., Bailey, E., and Farmer, P. B. (1986). *Biochem. Pharmacol. 35:* 1569.

CHAPTER 18

Storage and Handling

GEORGE R. LAPPIN Ethyl Corporation, Baton Rouge, Louisiana

1 INTRODUCTION

Proper storage and handling of alpha olefins is important because improper practices account for the majority of quality problems with alpha olefins. Delivery of acceptable products requires a cooperative team effort involving producer, shipper, and customer. The alpha olefin producer must manufacture products of good quality and protect them from atmospheric contamination. The shipper and customer must handle them properly to prevent quality loss. Care must be taken to prevent contamination, spills, releases, and sources of ignition. Requirements of regulatory agencies must be met. Safety precautions must be heeded to avoid harm to workers and equipment.

The industry has an excellent record of handling alpha olefins safely. Effective procedures have been developed for delivering alpha olefins of good quality. For the purpose of prevention, emphasis is placed here on proper handlng of olefins. Although this discussion is based mainly on Ethyl Corporation experience, it is believed to be consistent with information supplied by the other U.S. suppliers, Chevron and Shell.

Fire is the most important hazard when handling any of the alpha olefins. The flammable limits are given in Tables 1 and 2. Smoking and other ignition sources should be prohibited near transfer operations in-

Table 1 Autoignition Temperatures

Alpha olefin	Fahrenheit	Celsius
Butene-1	830	443
Hexene-1	521	272
Octene-1	493	256
Decene-1	471	244
Dodecene-1	468	242
Tetradecene-1	463	239
Hexadecene-1	464	240

volving alpha olefins. Vessels and transportation equipment should be grounded to prevent buildup of static electricity. Safety goggles should be worn. Personal respiratory equipment should be used when the more volatile alpha olefins are involved in confined work areas. By definition, olefins exposed to air above their autoignition temperature burst into flame immediately.

The alpha olefins are thermally stable compounds and will not change form when heated as high as 400°C (750°F) in the absence of catalytic materials. However, depending on the catalyst to which it may be exposed, an alpha olefin may isomerize to internal olefins and/or oligomerize to dimers, trimers, tetramers, or higher oligomers.

When considering the proper handling of alpha olefins, it is helpful to think of them as having safety characteristics similar to those of more commonly handled petroleum products with similar boiling ranges. The alpha olefins are clear liquids with a noticeable olefinic odor. Their properties resemble the corresponding paraffins.

Butene is a liquefied petroleum gas.

Hexene and octene are flammable liquids having flammability characteristics similar to those of gasoline.

Decene is a combustible liquid by DOT (the U.S. Department of Transportation) standards but is a flammable liquid by IMCO (the international guidelines) classification.

Dodecene, tetradecene, and hexadecene are not regulated by DOT or IMCO and are similar to kerosene and diesel fuel in flammability characteristics.

Octadecene and higher olefins are similar to heavier fuel oils or waxes.

The higher-carbon-number linear alpha olefins are solids at ambient temperatures.

Some C_{18+} olefin mixtures are slushy liquids at room temperature because they contain branched alkyl chains.

[illegible] Properties of Alpha Olefins

Carbon number:	C_4 Butene	C_6 Hexene	C_8 Octene	C_{10} Decene	C_{12} Dodecene	C_{14} Tetra decene	C_{16} Hexa decene	C_{18} Octa decene	C_{20} Eicosene
Density									
g/mL at 20°C	0.595[a]	0.673	0.715	0.741	0.759	0.760	0.782	0.788	0.795
lb/gal at 68°F	4.96[a]	5.61	5.96	6.17	6.32	6.33	6.52	6.57	6.62
Viscosity									
cP at 20°C	—	0.281	0.492	0.728	1.18	est. 3	est. 5	est. 8	est. 10
cP at 50°C	—	0.212	0.313	0.491	0.741	est. 1.5	est. 2	est. 3	est. 5
Heat of vaporization									
cal/g at 25°C	4.87	7.32	9.7	12.1	14.4	est. 16	est. 18	est. 20	est. 22
cal/g at normal boiling point	5.24	6.76	8.07	9.24	10.3	est. 11	est. 12	est. 13	est. 14
Vapor pressure									
mmHg at 20°C	1910	140	15.2	1.7	0.2	0.1	0.1	0.1	0.1
Boiling range (°C)									
5%	—	63	121	170	213	245	276	298	
95%	—	64	123	171	216	250	283	316	
Pour point (°C)									
Ethyl	Liquefied gas	−140	−102	−66	−37	−18	−2	5	
Melting point (°C)									
Chevron							4	18	90
Flash point (°F)[b]	Flammable gas	−15	50	114	171	225	263	290	

[a]As liquid.
[b]Flash points: tag closed cup < 200°F, Pensky-Martens closed cup >200°F.

Based on the low level of toxicity (see Chapter 17) as well as negligible to mild skin and eye irritation, alpha olefins can be properly handled using good industrial hygiene practices. Eye protection and protective gloves are recommended. In case of contact with skin, thoroughly wash with soap and water. Should alpha olefins come in contact with the eyes, flush immediately and for 15 minutes with copious amounts of water. It is good practice to wash clothing contaminated with alpha olefins before wearing them again.

Butene, hexene, and octene present hazards due to fire, formation of explosive vapor mixtures with air, and suffocation because of their high vapor pressure. Butene is handled as a liquefied gas under pressure and it will cool when released into the atmosphere. It may cause frostbite on contact because of cooling as it volatilizes. Twenty years of commercial experience shows the lighter olefins to be the most hazardous because of the potential for fire and explosion. Still the higher olefins can be fuel for a fire and should be handled with care. If any of the alpha olefins must be handled inside, adequate forced ventilation should be provided.

2 CONTAMINATION BY AIR: PEROXIDES AND CARBONYLS

Oxygen and water in the air are absorbed by alpha olefins, with saturation limits in the vicinity of 100 to 300 ppm for water and under 100 ppm for oxygen. Oxygen reacts slowly with the olefins under storage conditions to form hydroperoxides, which can decompose to carbonyl-containing impurities. Hydroperoxide formation is autocatalytic, but it can also be catalyzed by other agents. Contaminated heels in storage tanks and other vessels appear to be key causes of the formation of peroxides and carbonyls in that they may seed peroxide formation in subsequent shipments. Once started, peroxide formation has been observed at a rate of 1 ppm per day but the rate is highly variable. Purging vessels with nitrogen may decrease or avoid buildup of peroxides. Ultraviolet light increases the rate of formation of peroxides. Samples for checking for peroxides should be stored in dark bottles and in the dark.

The peroxides and carbonyls may be removed from the alpha olefins through several techniques, including chemical reaction, adsorption, and distillation. (Caution: If using distillation, avoid buildup of peroxides). When adsorption is used, care must be taken to regenerate or replace the adsorbent periodically to prevent elution of contaminants through the bed. The olefin suppliers are able to assist users with removal as well as prevention of formation of the oxygen-containing contaminants.

The oxygen-containing contaminants interfere with reactions by killing some catalysts—thus reducing some reaction rates—while they may enhance other catalysts—thus increasing other reaction rates. They may result in higher ash content in polyolefins made by coordination catalysis, variable reaction rates in oligomerizations, higher catalyst usage in alkylation, and other difficulties, such as increased color. Thus their formation is to be avoided.

Besides interfering with reactions, theoretically there is a potential danger of explosion from unstable peroxides. However, to our knowledge there has never been an explosion caused by a buildup of peroxides in the higher alpha olefins. Distillation of the olefin will result in the peroxides concentrating in the bottoms. Peroxides may be destroyed by heat in the reboilers, but this may also be dangerous, as concentrated peroxides may explode.

Prevention of contamination from air, including the water and oxygen in air, is worthwhile and possible if an inert atmosphere is maintained. Alpha olefin producers routinely use nitrogen to avoid contamination with the atmosphere. This practice is difficult in some situations, such as in the many transfers that occur on international bulk shipments. Antioxidants are frequently added to the alpha olefins for some applications, such as alkylation. Some antioxidants may be useful in reducing side reactions, such as dimerizations during alkylation. Inhibition of autocatalytic oxidation of fuels has been translated to inhibiting alpha olefins. Ethyl antioxidant 733 has been successfully used at a concentration of 25 ppm to avoid peroxide problems with olefins stored in API tanks without nitrogen or inert blanketing.

3 STORAGE

Since fire is regarded as the primary hazard, storage vessels should be designed with the standard property and personnel safeguards commonly used in the petroleum and petrochemical industries. Vessels should have safety relief devices adequate to avoid vessel rupture if a fire occurs. The vessels should be placed inside concrete or earthen dikes large enough to hold the contents of the vessel. Drainage should be designed to allow for containment of the liquid if there is a release and yet allow drainage of rainwater. Drainage becomes a big problem if there is an explosion followed by fire, with subsequent dousing of the area by firefighters. Adequate firefighting equipment should be provided. Care should be taken in selecting the location of the vessels, with avoidance of too close spacing and of power lines or other strategic plant utilities.

All olefins except butene may be stored in either API or pressure tanks. Butene requires a pressure or refrigerated vessel. Hexene and octene usually require floating-roof tanks for environmental protection and loss-prevention reasons. Depending on the ambient conditions, heating coils may be required for the higher olefins, starting with C_{14}.

Dry nitrogen blanketing systems should be carefully designed to avoid exposure of the alpha olefins to water and oxygen in the atmosphere. Multistage nitrogen regulators are usually required to maintain adequate control. Nitrogen systems have been mistakenly undersized for several reasons, including underestimating peak loads on the nitrogen supply as more than one vessel is emptied, having pumps of higher actual capacity than the manufacturer's rating, having piping networks with less resistance to flow than estimated in design, and not considering that a full tank will empty faster than a partially full tank. For these reasons and others, storage tanks should be monitored for peroxide formation on a routine basis. If peroxides are found, a thorough engineering study should be made to determine the cause. It may be necessary to empty the vessel completely to avoid the aforementioned peroxide seed effect. These suggestions are especially appropriate when dealing with commercial terminals.

4 MATERIALS OF CONSTRUCTION

Carbon steel is inert to alpha olefins and thus is satisfactory as a material of construction for their storage. Since the olefins may dislodge any rust remaining on the inside of the vessel, it may be necessary to remove the rust or to prewash the vessel with alpha olefins if rust (and the contained oxygen) would constitute problems in the process.

Glass or any of the common metals (aluminum, brass, copper, iron, various stainless steels and alloys) may be used with alpha olefins and probably are preferred, as they do not rust, but they are more expensive. Being hydrocarbons, the alpha olefins will damage many rubbers, paints, and lining materials. Nitrile rubber is a satisfactory elastomer for alpha olefin service. Asbestos gaskets are satisfactory (although less frequently used because of health hazards). Flourocarbon resin seats and seals are successfully used in ball valves. When in doubt, compatibility tests are recommended. Again, alpha olefin suppliers have extensive experience and may offer additional suggestions.

5 SHIPPING CONTAINERS: Tank Cars, Tank Trucks, Drums

Alpha olefins are shipped in almost any commercial container from liter or quart sample bottles to 4000-metric ton parcels on ocean freighters. Containers include 55-gallon drums; iso-tanks; 6500-gallon tractor trailers; 10,000-, 20,000-, 30,000-, and 33,000-gallon tank cars; and barges. Care should be taken in selecting the container. Above all, it should be inspected for cleanliness prior to loading. Use of 17E 55-gallon drums for hexene and octene is permitted by DOT regulations. Because hexene and octene are flammable and very volatile, handling them in 55-gallon drums is regarded as dangerous and as being very similar to handling gasoline or other flammable liquids in 55-gallon drums.

Precautions must be taken to avoid buildup of static electricity in olefins. Grounding is a must for vessels, trucks, and tank cars, especially during loading and unloading operations (Fig. 1). Precise spotting of tank cars is desirable to avoid strains on unloading lines. The car brakes should be set and the wheels blocked. For DOT regulated products (C_4–C_{12}), caution signs should be placed in accordance with paragraph 174.67 of CFR 49.

Maintaining a dry blanket of nitrogen will not only assure better product quality with regard to oxygen and water contamination but will provide for improved safety. This is especially important with hexene, octene, and decene, as they are likely to fall within flammability concentrations as shown by the Reid vapor pressure plot of upper and lower flammability limits. Based on this chart, these olefins are most hazardous with respect to vapor flammability at the following temperatures:

Alpha olefin	Fahrenheit range	Celsius range
Hexene-1	−40 to 18	−40 to −8
Octene-1	55 to 105	10 to 42
Decene-1	118 to 180	48 to 82

From this view, octene is the most hazardous of the alpha olefins, as its range encompasses the normal ambient temperatures. Hexene becomes as severe a hazard only at low temperatures.

Butene tank cars tend to be very large, such as 33,000 gallons, because of butene's low density (5 lb/gal). LPG-type tank cars are required for butene and in the United States are typically used for hexene. Ethyl uses a 105A300W or 112J340W car for both butene and hexene (Fig. 2). Hexene may be shipped in open dome cars but the closed dome LPG cars provide better quality assurance, especially for comonomers. LPG cars are sometimes difficult to unload because they have flow-limiting check valves in the liquid dip pipes (Fig. 3). Attempts to unload at rates higher than allowed by these check valves will result in no flow as the check valve seats. Then flow cannot be obtained until the pressure on either side of the check valve is allowed to equalize. Equalization requires blocking valves between the tank car and the storage area and increasing the pressure on the downstream side of the check valve or decreasing the pressure on the car until the check valve is unseated. Operators should be taught about

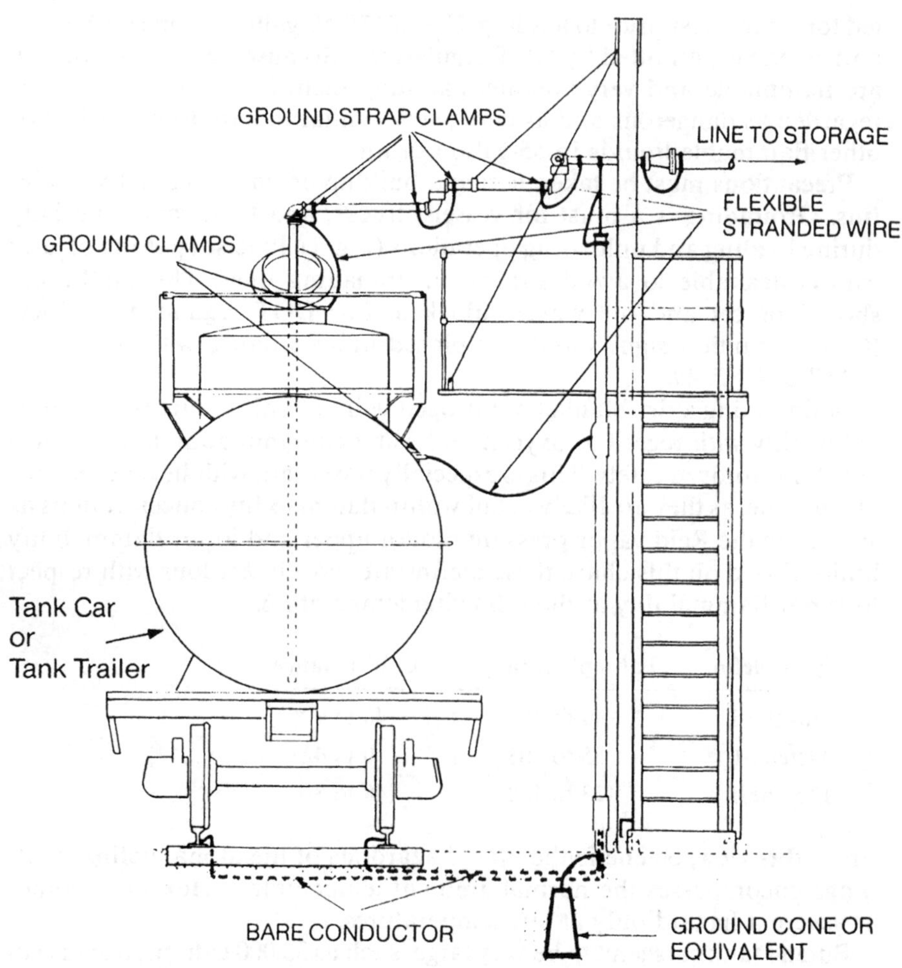

Figure 1 Grounding arrangement for unloading LPG.

Figure 2 Typical tank car (105A300W type).

these check valves. Operators should be cautioned not to hammer on unloading piping, as such will not correct the problem and the hammering may create sparks. These LPG-type cars are more expensive and somewhat more difficult to unload.

Butene unloading requires a motive force in the form of a compressor, vaporizer, or a pump. Based on economic and environmental considerations, the generally preferred methods involve pressurizing the car with butene from storage with either a compressor or a vaporizer. Less desirable are using dry nitrogen, as the nitrogen must be vented, or transferring liquid into the suction of a pump or directly into storage, as these will not completely empty the car. Unloading hexene from an LPG car may be accomplished by using dry nitrogen alone or in combination with a pump.

Hexene and octene may be shipped in open dome cars but are frequently shipped in closed dome cars to avoid contamination, especially in comonomer service. Decene and higher olefins are generally shipped in open domed cars. Coiled cars are necessary in winter for tetradecene and higher carbon numbers. These cars may be unloaded from either the top or the bottom. A dip pipe with an external valve (protected by a dome) provides top unloading of the car. A ball valve provides for unloading from the bottom. The bottom ball valves are capped and sealed in the closed position when shipped. Top unloading is recommended by Ethyl, and many major companies require top unloading, epecially for hexene and octene.

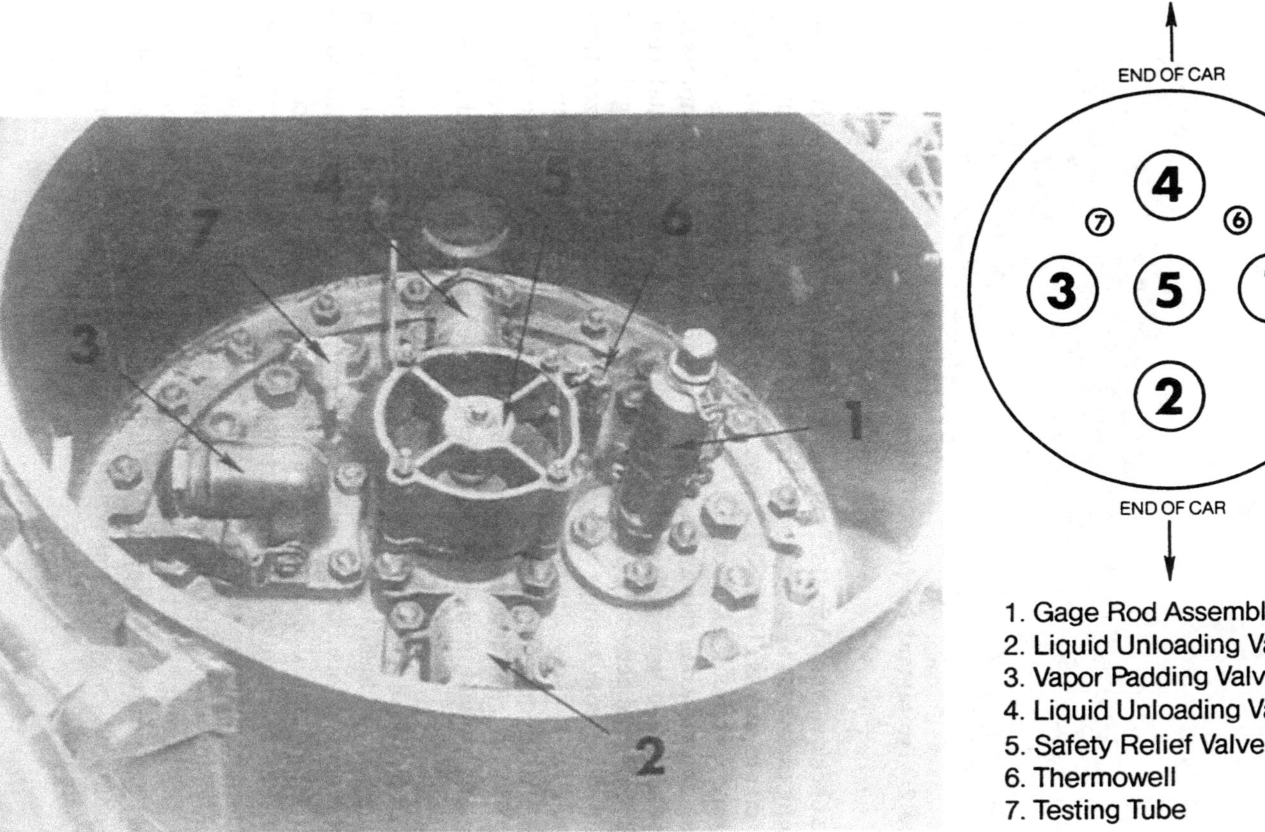

Figure 3 Typical tank car dome arrangement (105A300W type).

Appendix: Physical Properties

This appendix gives the following physical property charts for alpha olefins

All temperatures are in degrees Fahrenheit and the other units are typical of those used by American chemical engineers in the design of chemical processes.

These charts are presented courtesy of Ethyl Corporation, whose Engineering and Mathematical Science Section produced the originals.

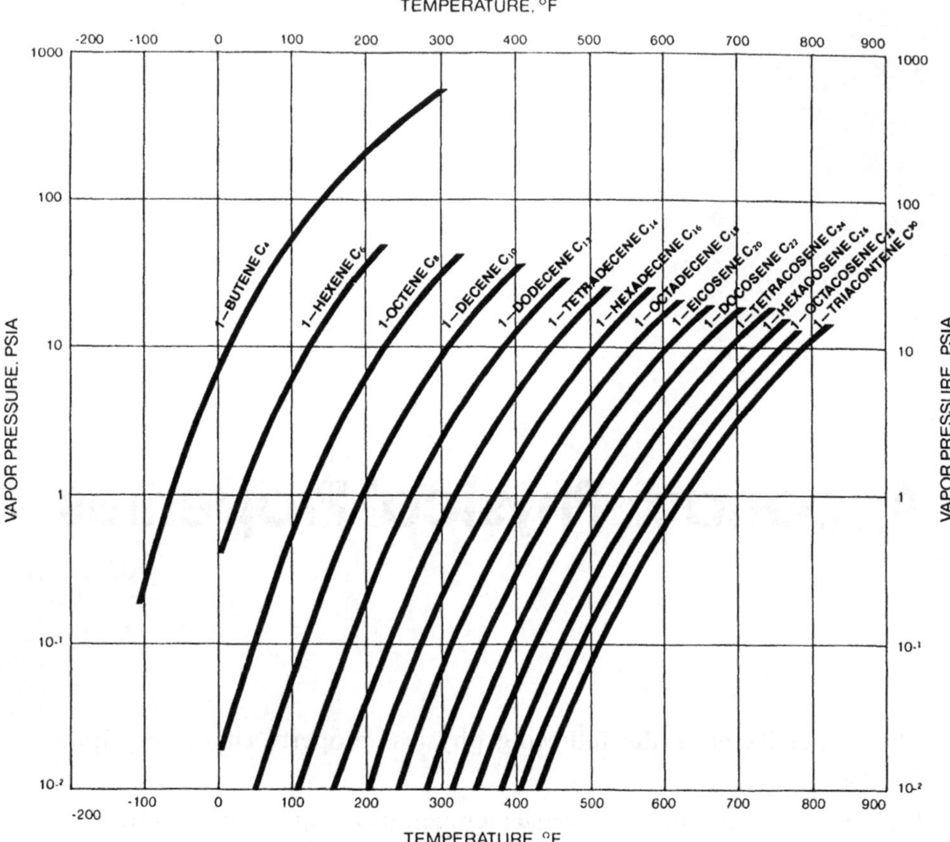

Figure A.1 Vapor pressure versus temperature: high pressure range.

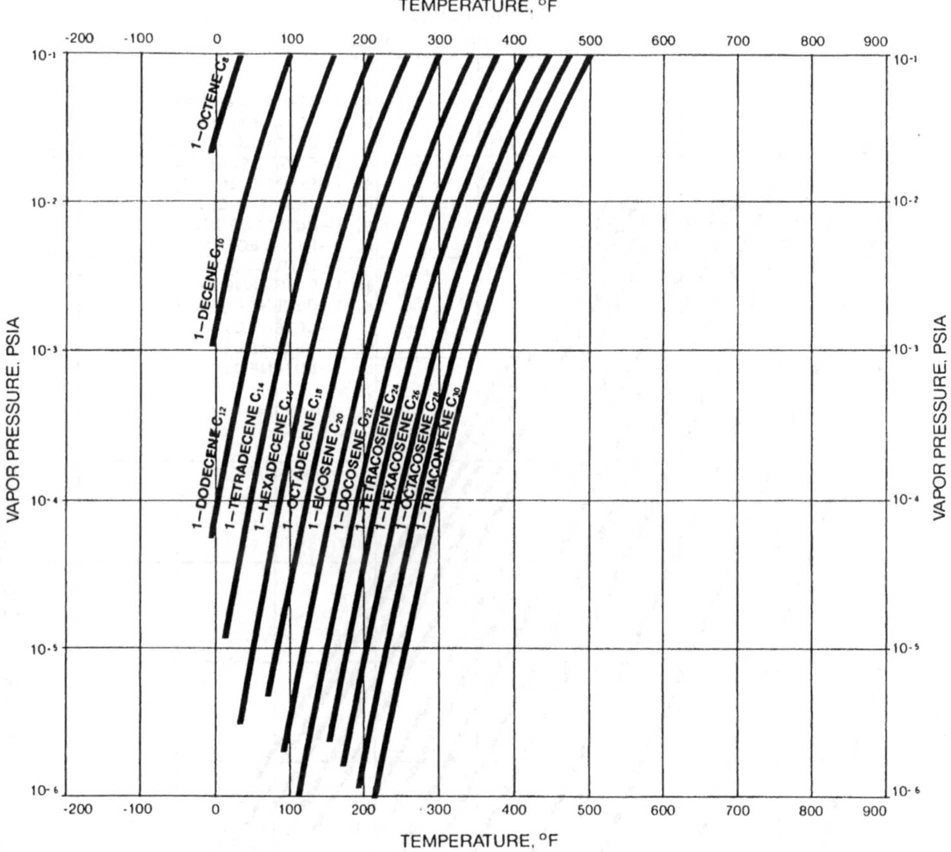

Figure A.2 Vapor pressure versus temperature: low pressure range.

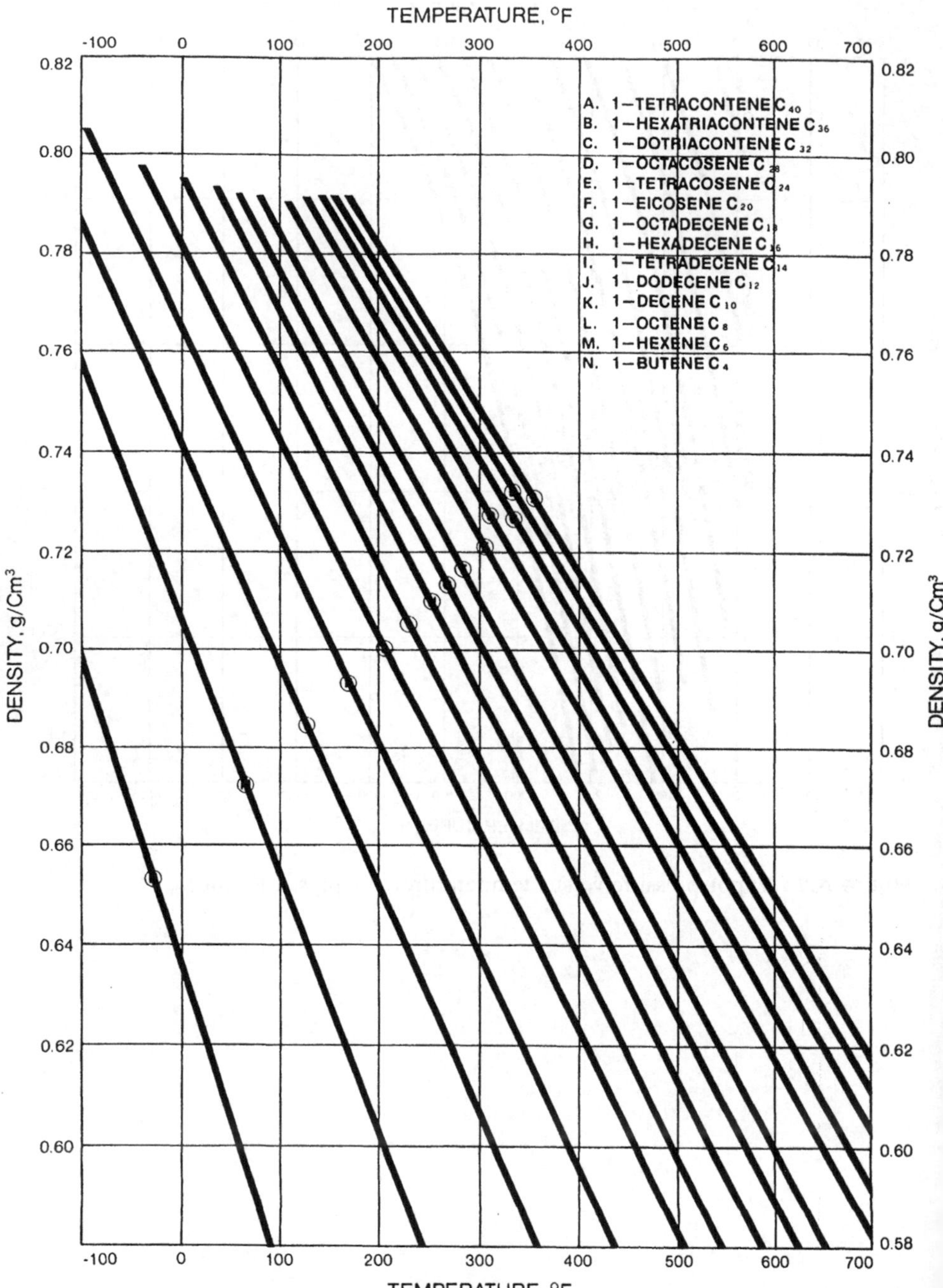

Figure A.3 Liquid density versus temperature.

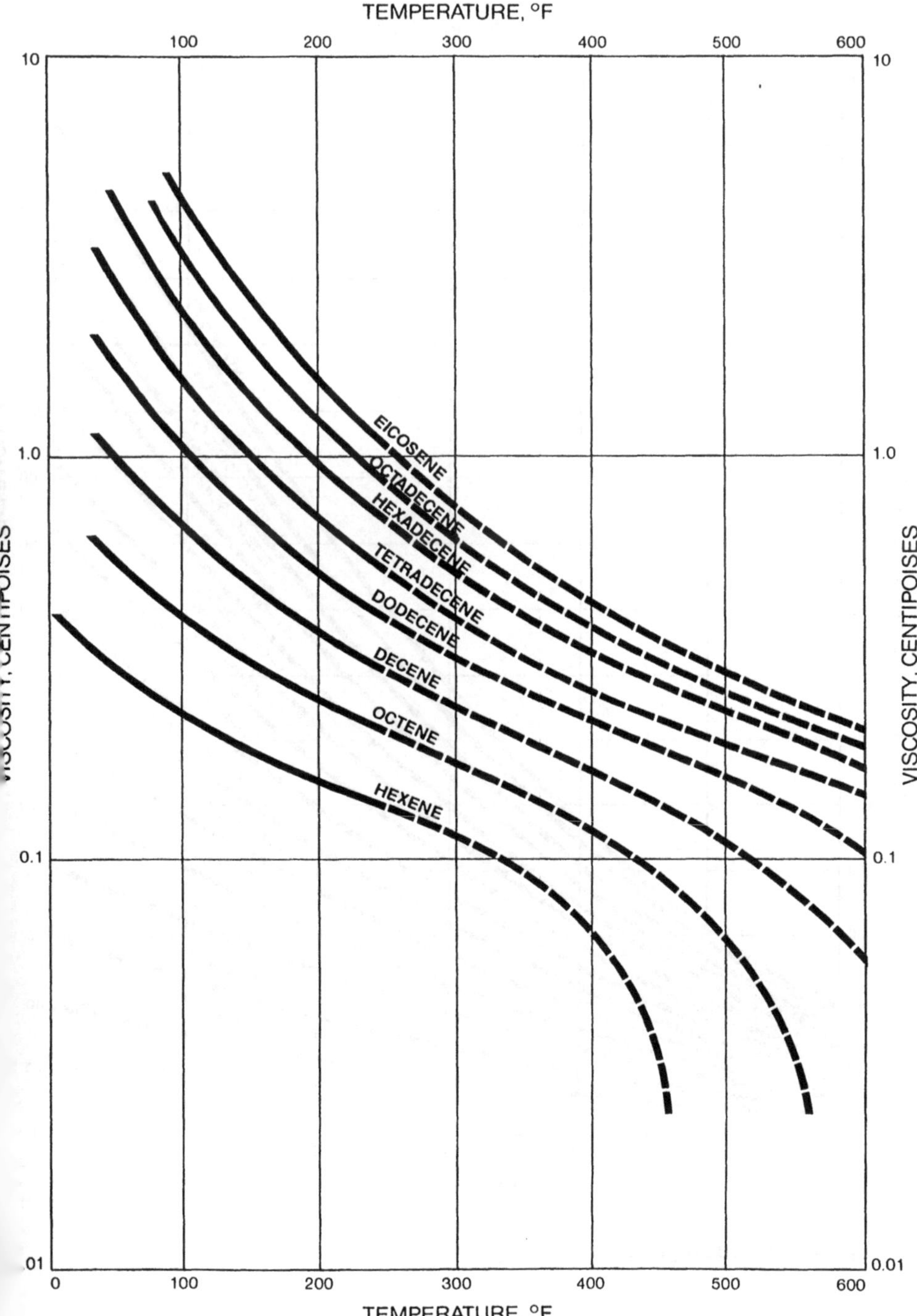

Figure A.4 Saturated liquid viscosity versus temperature.

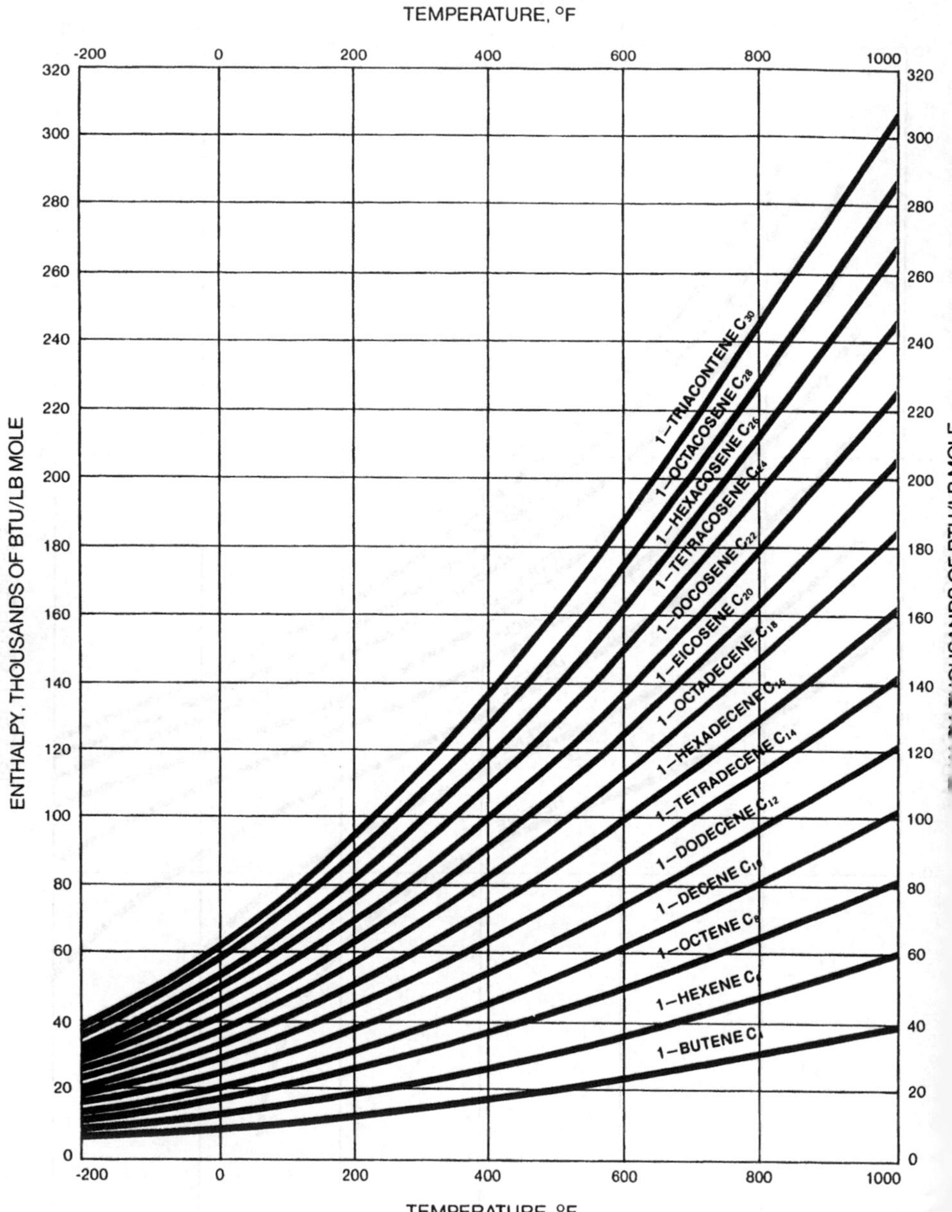

Figure A.5 Ideal gas state enthalpy versus temperature.

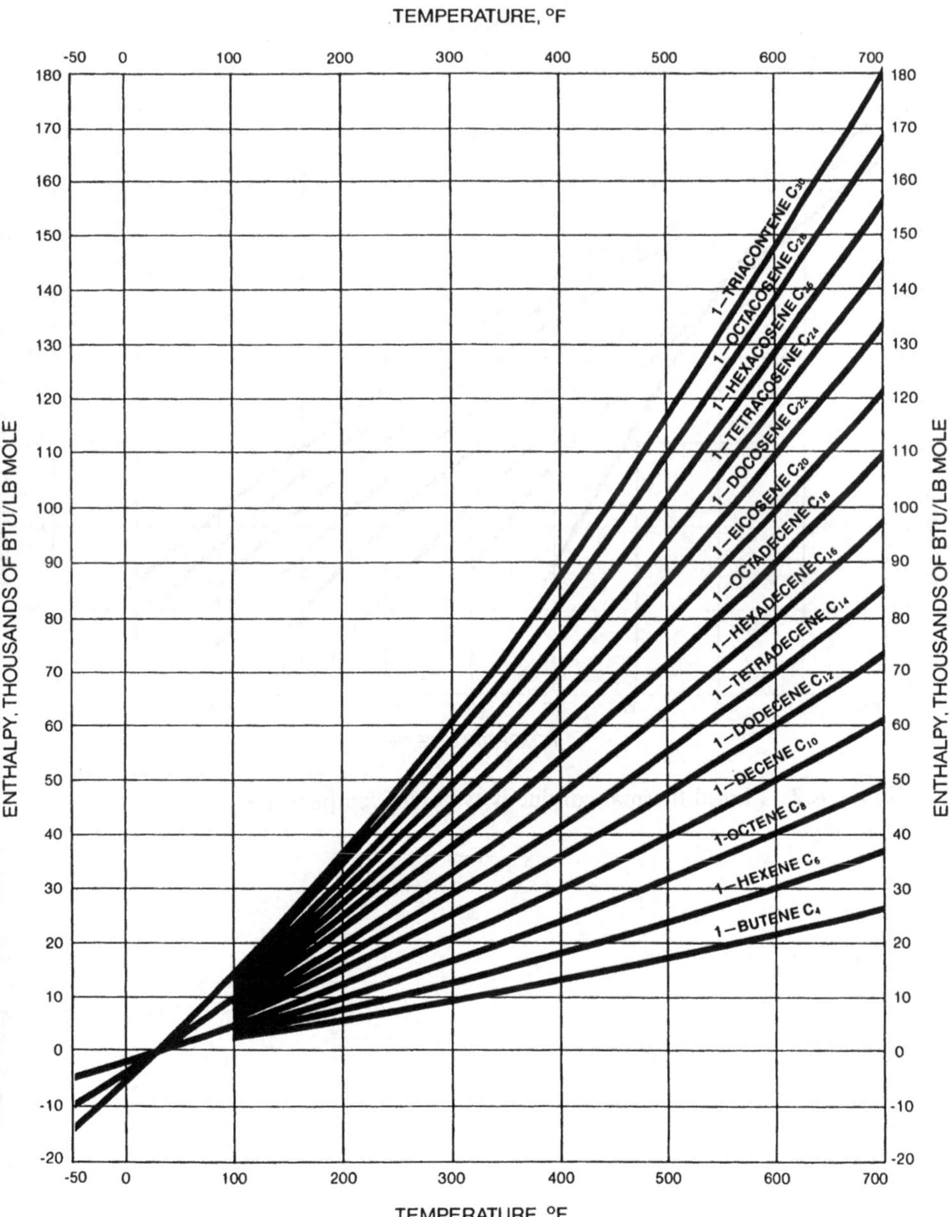

Figure A.6 Liquid enthalpy versus temperature.

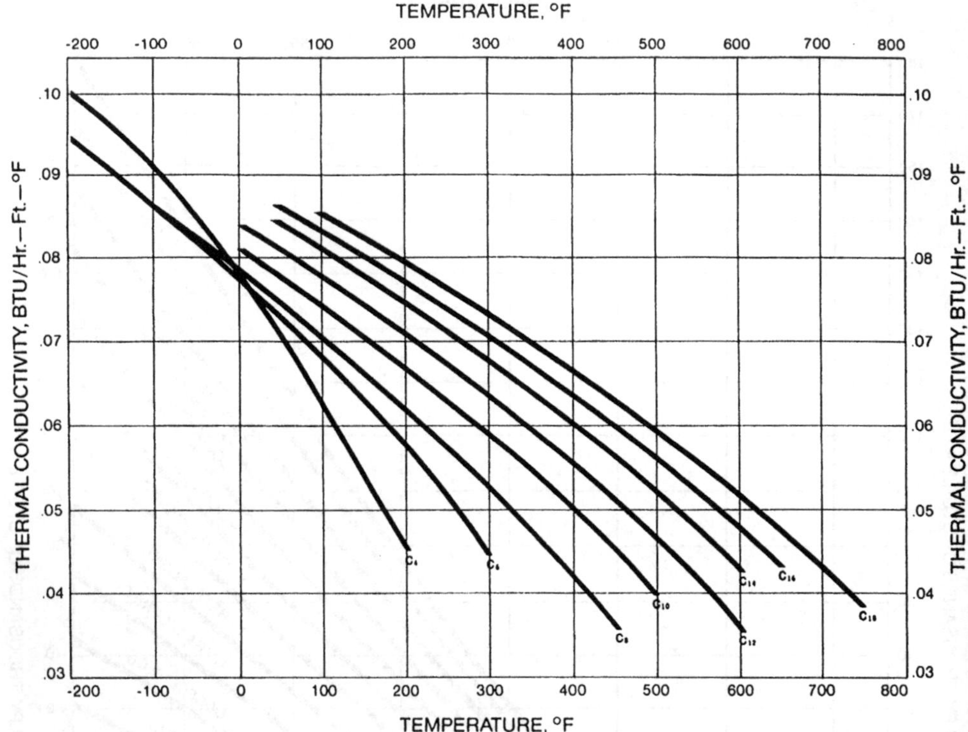

Figure A.7 Liquid thermal conductivity versus temperature.

INDEX

Milton Keynes UK
Ingram Content Group UK Ltd.
UKHW020009071024
449327UK00031B/2713